Design Space Exploration and Resource Management of Multi/Many-Core Systems

Design Space Exploration and Resource Management of Multi/Many-Core Systems

Editors

Amit Kumar Singh
Amlan Ganguly

MDPI • Basel • Beijing • Wuhan • Barcelona • Belgrade • Manchester • Tokyo • Cluj • Tianjin

Editors
Amit Kumar Singh
University of Essex
UK

Amlan Ganguly
Rochester Institute of
Technology
USA

Editorial Office
MDPI
St. Alban-Anlage 66
4052 Basel, Switzerland

This is a reprint of articles from the Special Issue published online in the open access journal *Journal of Low Power Electronics and Applications* (ISSN 2079-9268) (available at: hhttps://www.mdpi.com/journal/jlpea/special_issues/NOC).

For citation purposes, cite each article independently as indicated on the article page online and as indicated below:

LastName, A.A.; LastName, B.B.; LastName, C.C. Article Title. *Journal Name* **Year**, *Volume Number*, Page Range.

ISBN 978-3-0365-0876-4 (Hbk)
ISBN 978-3-0365-0877-1 (PDF)

Cover image courtesy of Amit Kumar Singh.

Contents

Preface to "Design Space Exploration and Resource Management of Multi/Many-Core Systems"

The reliance of computing systems of various scales, e.g., embedded to cloud computing, on multi-/many-core chips, mainly to satisfy the high-performance requirement of complex software applications, is increasing. At the same time, depending on the applications to be executed, these systems also demand energy efficiency, reliability and/or security. These demands can be fulfilled by exploring the design space by considering the software applications and the multi-/many-core chips to find the design points leading to efficiency in all the required metrics. Furthermore, considering varying workloads in the systems over time, efficient resource management methodologies need to be developed that can exploit the explored design points and/or perform online optimizations to meet the requirements of performance, energy efficiency, reliability and/or security. This Special Issue collects eight papers from leading researchers in this field and demonstrates the spectrum of design space exploration and resource management strategies for multi-/many-core systems. In addition, the collection provides visionary views of the computing landscape to be observed in future.

Amit Kumar Singh, Amlan Ganguly
Editors

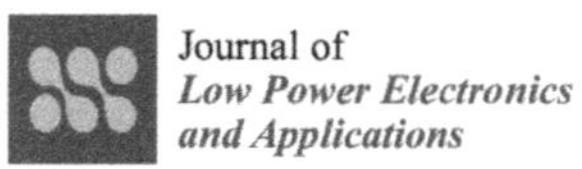

Journal of
*Low Power Electronics
and Applications*

Opinion

A Case for Security-Aware Design-Space Exploration of Embedded Systems

Andy D. Pimentel

Parallel Computing Systems Group, University of Amsterdam, Science Park 904, 1098 XH Amsterdam, The Netherlands; a.d.pimentel@uva.nl; Tel.: +31-20-5257578

Received: 22 June 2020; Accepted: 15 July 2020; Published: 17 July 2020

Abstract: As modern embedded systems are becoming more and more ubiquitous and interconnected, they attract a world-wide attention of attackers and the security aspect is more important than ever during the design of those systems. Moreover, given the ever-increasing complexity of the applications that run on these systems, it becomes increasingly difficult to meet all security criteria. While extra-functional design objectives such as performance and power/energy consumption are typically taken into account already during the very early stages of embedded systems design, system security is still mostly considered as an afterthought. That is, security is usually not regarded in the process of (early) design-space exploration of embedded systems, which is the critical process of multi-objective optimization that aims at optimizing the extra-functional behavior of a design. This position paper argues for the development of techniques for quantifying the 'degree of secureness' of embedded system design instances such that these can be incorporated in a multi-objective optimization process. Such technology would allow for the optimization of security aspects of embedded systems during the earliest design phases as well as for studying the trade-offs between security and the other design objectives such as performance, power consumption and cost.

Keywords: embedded computer systems; cyber security; system-level design and design-space exploration; multi-objective optimization; system trade-offs

---

1. Introduction

Embedded computer systems are ubiquitous and have a major impact on our society. Examples of such systems are close at hand—modern TVs contain one or multiple computer systems to handle functionality such as decoding the input signal, performing various image enhancement techniques as well as displaying and updating live information (e.g., program guide or weather forecast). Smart-phones rely on embedded computer systems to allow users to make phone calls, shoot photos and videos, perform GPS navigation, browse the Internet, execute apps, and so on. The use of embedded computer systems is, however, by no means restricted to consumer electronics: in industrial, medical, automotive, avionic, or defense applications they are equally pervasive.

The complexity of the underlying system architectures of modern embedded systems forces designers to start with modeling and simulating (possible) system components and their interactions in the very early design stages. This is often referred to as system-level design [1]. The system-level models typically represent application workload behavior, characteristics of the underlying computing platform architecture, and the relation (e.g., mapping, hardware-software partitioning) between application workload(s) and platform architecture. These models are applied at a high level of abstraction, thereby minimizing the modeling effort and optimizing the simulation speed. This is especially needed for targeting the early design stages since many design decisions are still open and, therefore, many design alternatives still need to be studied. High-level system modeling allows for the early verification of a design and can provide estimates on the extra-functional properties of a

design such as system performance and energy/power consumption. The system-level models are typically accompanied by a methodology for efficient design-space exploration (DSE) [2], which is the process of assessing alternative design instances with respect to (i) the platform architecture that will be deployed (e.g., the number and type of processing elements in the platform, the type of network to interconnect the processors, etc.) and(ii) the mapping of application tasks to the underlying platform components [3]. It is a multi-objective optimization problem that searches through the space of different implementation alternatives to find optimal design instances. Exploration of different design choices, especially during the early design stages where the design space is still at its largest, is of eminent importance. Wrong decisions early in the design can be extremely costly in terms of re-design effort, or even deadly to the product's success. Consequently, considerable research effort in the embedded systems domain has been spent in the last two decades on developing frameworks for system-level modeling and simulation that aim for early design-space exploration.

As embedded systems are becoming more and more ubiquitous and interconnected (illustrated by, e.g., the strong trend towards the Internet of Things), they also attract a world-wide attention of attackers. This makes the security aspect more important than ever during the design of these systems [4]. Moreover, given the ever-increasing complexity of the applications that run on modern embedded systems, it becomes increasingly difficult to meet all security criteria. While design objectives such as performance and power/energy consumption are usually taken into account during the early stages of design (as explained above), system security is still mostly considered as an afterthought. That is, security is typically not regarded in the process of (early) design-space exploration of embedded systems. However, any security measures that may eventually be taken much later in the design process do affect the already established trade-offs with respect to the other extra-functional properties of the system like performance, power/energy consumption, cost, and so forth [4]. Thus, covering the security aspect in the earliest phases of design is necessary to design systems that are, in the end, optimal with regard to all extra-functional objectives. However, this poses great difficulties because unlike the earlier mentioned conventional system objectives, like performance and power consumption, security is hard to quantify—there exists no single metric with which one can measure the degree of secureness of a design.

This position paper argues for the need for security-aware, system-level design-space exploration methods and techniques for embedded systems. To this end, we will discuss a multifaceted, scoring-based methodology for quantifying the degree of secureness of embedded system design instances. This methodology allows for incorporating the secureness quantifications in a multi-objective optimization process and would thus enable optimization of the security aspect during the earliest phases of design. However, we want to emphasize the fact that this is a position paper and therefore does not present an actual implementation of the proposed solution nor any experimental results.

The remainder of this paper is organized as follows. In the next section, we will provide a brief introduction to the concept of design-space exploration. In Section 3, we will describe our proposal for a security-aware DSE approach, focusing on a method to quantify the secureness of embedded system design instances. Section 4 discusses related work, after which Section 5 concludes the paper.

2. Design-Space Exploration

During the design-space exploration (DSE) of embedded systems, multiple optimization *objectives*—such as performance, power/energy consumption, and cost—should be considered simultaneously. This is called multi-objective DSE [2]. Since the objectives are often in conflict, there cannot be a single optimal solution that simultaneously optimizes all objectives. Therefore, optimal decisions need to be taken in the presence of trade-offs between design criteria.

2.1. Multi-Objective Optimization

Given a set of m decision variables, which are the degrees of freedom (e.g., parameters like the number and type of processors in the system, application mapping, etc.) that are explored during DSE, a so-called *fitness function* must optimize the n objective values [2]. The fitness function is defined as:

$$f_i : R^m \rightarrow R^1. \tag{1}$$

A potential solution $x \in R^m$ is an assignment of the m decision variables. The fitness function f_i translates a point in the solution space X into the i-th objective value (where $1 \leq i \leq n$). For example, a particular fitness function f_i could assess the performance or energy efficiency of a certain solution x (representing a specific design instance). The combined fitness function $f(x)$ subsequently translates a point in the solution space into the objective space Y. Formally, a multi-objective optimization problem (MOP) that tries to identify a solution x for the m decision variables that minimizes the n objective values using objective functions f_i with $1 \leq i \leq n$:

$$\text{Minimize } y = f(x) = (f_1(x), f_2(x), ..., f_n(x))$$
$$\text{Where } x = (x_1, x_2, ..., x_m) \in X$$
$$y = (y_1, y_2, ..., y_n) \in Y.$$

Here, the decision variables x_i (with $1 \leq i \leq m$) usually are constrained. These constraints make sure that the decision variables refer to valid system configurations (e.g., using not more than the available number of processors, using a valid mapping of application tasks to processing resources, etc.), that is, x_i are part of the so-called feasible set. In the remainder of this section, we assume a minimization procedure, but without loss of generality, this minimization procedure can be converted into a maximization problem by multiplying the fitness values y_i with -1.

With an optimization of a single objective, the comparison of solutions is trivial. A better fitness (i.e., objective value) means a better solution. With multiple objectives, however, the comparison becomes non-trivial. Take, for example, two different embedded system architecture designs: a high-performance system and a slower but much cheaper system. In case there is no preference defined with respect to the objectives and there are also no restrictions for the objectives, one cannot say if the high-performance system is better or the low-cost system. A typical MOP in the context of embedded systems design can have a variety of different objectives, like performance, energy consumption, cost and reliability. To compare different solutions in the case of multiple objectives, the Pareto dominance relation is generally used. Here, a solution $x_a \in X$ is said to dominate solution $x_b \in X$ if and only if $x_a < x_b$:

$$x_a < x_b \iff \forall i \in \{1, 2, ..., n\} : f_i(x_a) \leq f_i(x_b) \wedge$$
$$\exists i \in \{1, 2, ..., n\} : f_i(x_a) < f_i(x_b).$$

Hence, a solution x_a dominates x_b if its objective values are superior to the objective values of x_b. For all of the objectives, x_a must not have a worse objective value than solution x_b. Additionally, there must be at least one objective in which solution x_a is better (otherwise they are equal).

An example of the dominance relation is given in Figure 1, which illustrates a two dimensional MOP. For solution H the dominance relations are shown. Solution H is dominated by solutions B, C and D as all of them have a lower value for both f_1 and f_2. On the other hand, solution H is superior to solutions M, N and O. Finally, some of the solutions are not comparable to H. These solutions are better for one objective but worse for another.

The Pareto dominance relation only provides a partial ordering. For example, the solutions A to F of the example in Figure 1 cannot be ordered using the ordering relation. Since not all solutions $x \in X$

can be ordered, the result of a MOP is not a single solution, but a front of non-dominated solutions, called the *Pareto front*. A set X' is defined to be a Pareto front of the set of solutions X as follows:

$$\{x \in X' \mid \nexists x_a \in X : x_a < x\}.$$

The Pareto front of Figure 1 contains six solutions: $A - F$. Each of these solutions does not dominate the other. An improvement on objective f_1 is matched by a worse value for f_2. Generally, it is up to the designer to decide which of the solutions provides the best trade-off.

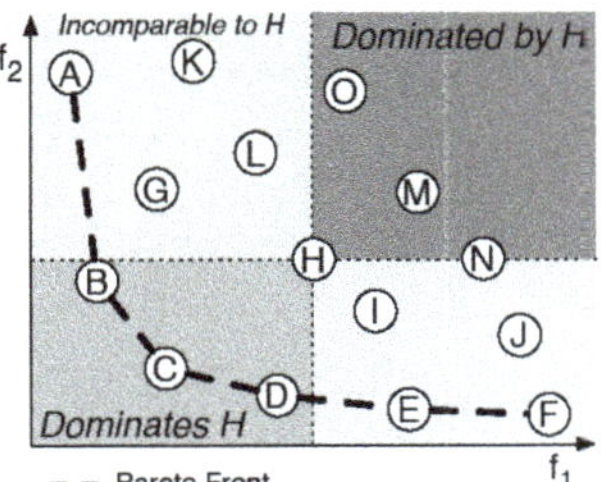

Figure 1. A Pareto front and an example of the dominance relation (taken from Reference [2]).

2.2. Search for Pareto Optimal Solutions

The search for Pareto optimal design points with respect to multiple design criteria entails two distinct elements [5]:

1. The evaluation of a single design point using the fitness function(s) $f(x)$ regarding all the objectives in question like system performance, power/energy consumption and so on. These evaluations are usually based on measurements using real systems or predictions from either analytical models or simulation models [2].
2. The search strategy for navigating through and covering the design space during the DSE process. Such search strategies can be based on exact, but typically unscalable, methods that guarantee finding the optimal solution(s). These exact methods can, for example, be implemented using integer linear programming (ILP) solutions (e.g., References [6,7]) or branch & bound algorithms (e.g., Reference [8]). Alternatively, so-called meta-heurisics, such as genetic algorithms (GA) or simulated annealing, can be used to search the design space for optimal solutions. They only perform a finite number of design point evaluations, and can thus handle larger design spaces. However, there is no guarantee that the global optimum will be found using meta-heuristics, and therefore the result can be a local optimum within the design space. GA-based DSE has been widely studied in the domain of system-level embedded design (e.g., References [9–12]) and has demonstrated to yield good results.

In this paper, we focus on the fitness evaluation aspect of DSE. More specifically, we argue that while there are well-established techniques and metrics for the fitness evaluation of traditional design objectives such as performance, power/energy consumption, cost, and reliability, this is not the case for evaluating the fitness of design instances in terms of how secure they are. This lack of security fitness evaluation methods and metrics inhibits the use of system security as a first-class citizen in the process of early design-space exploration of embedded systems. As was indicated before, such design practice leads to suboptimal products because any security measures that may be taken later in the design process do affect the already established trade-offs with respect to the other extra-functional properties of the system like performance, power/energy consumption, cost, and so forth.

In the next section, we will therefore argue for the development of a security-aware DSE approach, based on a multifaceted, scoring-based security quantification methodology. This methodology allows

for quantifying the degree of secureness of design instances such that these can be incorporated in the DSE's multi-objective optimization process. Eventually, once such a security-aware DSE would have been implemented, it would allow for optimization of security aspects of embedded systems in their earliest design phases as well as for studying the trade-offs between security and the other design objectives like performance, power consumption and cost. Evidently, such technology would provide a substantial competitive advantage in the embedded systems industry.

3. Towards Security-Aware, System-Level DSE

The envisioned approach for security-aware system-level design-space exploration, adopting a multifaceted, scoring-based security quantification methodology, is illustrated in Figure 2. Below, we will explain the different components of this approach. The blue parts of Figure 2 refer to the methodology components that only need to be specified or performed once, whereas the red parts refer to components that are dependent on the design-space exploration process and thus must be revisited every time a new design instance is evaluated in terms of extra-functional properties such as performance, power consumption, and of course, in the scope of this paper, also secureness. Before describing our approach in detail, however, we will first discuss several assumptions that delimit our proposed approach.

3.1. Assumptions

We focus on security threats in which the underlying embedded system architecture plays a central role, and do not consider any security flaws that can be exploited purely at the application level. This implies that we restrict ourselves to the following set of attack types:

1. *Side-channel attacks* like power analysis attacks, timing attacks such as the recent Spectre and Meltdown attacks, scan attacks, differential fault analysis attacks and electromagnetic analysis attacks (see References [13,14] for an overview of these side-channel attacks);
2. *Denial of service attacks* [15,16];
3. *Software-based attacks* such as buffer overflows for which protection mechanisms may be available at the system (architecture) level (e.g., Reference [13]);
4. *Attacks directed towards breaking encryption algorithms* [17].

For each of the above attacks, we subsequently consider a range of protection mechanisms—derived from literature—that can be applied to protect specific system components or the entire system against these attacks.

Moreover, we consider system-level DSE in which both the platform architecture (e.g., selection of platform components such processing elements, memories, and networking components) as well as the mapping of application tasks and communications to the selected platform components are optimized for traditional objectives such as performance, power consumption, and cost, but now also for secureness. Such system-level design-space exploration is depicted in the top-middle part of Figure 2 and could, for example, be performed with system-level DSE frameworks such as Sesame [18,19] or a similar environment (e.g, References [20–22]). Important to note here is that the performance, power and cost models used in the DSE also need to account for the effects of deploying specifically selected security protection mechanisms (as discussed above) inside a platform architecture.

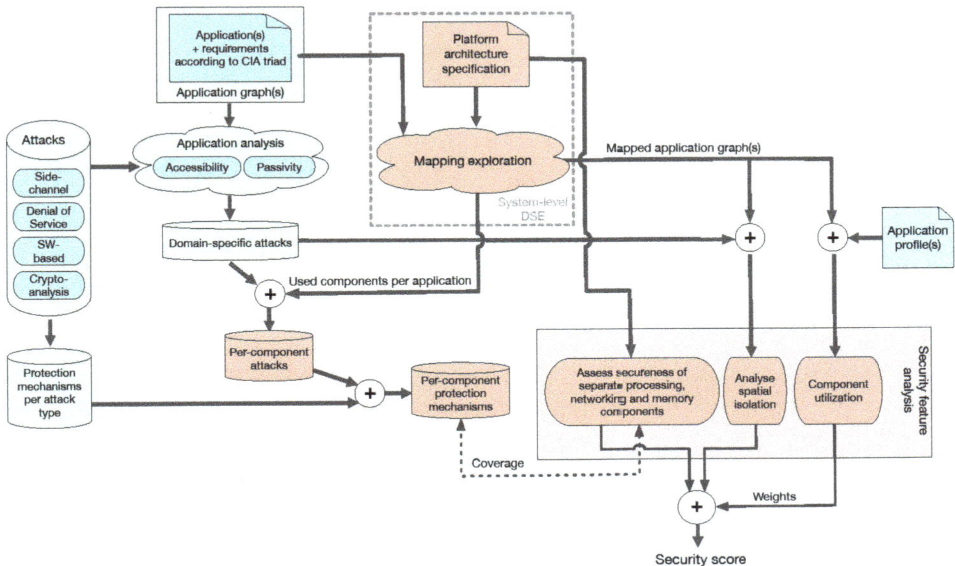

Figure 2. Proposed approach for security-aware system-level design-space exploration (DSE) using a multifaceted, scoring-based security quantification methodology.

3.2. A Multifaceted, Scoring-Based Methodology for Secureness Quantification

As a first step in our methodology, shown in the top left of Figure 2, the applications that need to execute on the target embedded system together with their extra-functional requirements are identified and specified. The specified extra-functional requirements include the traditional ones such as performance and real-time requirements, power/energy consumption budgets, and so forth, but also requirements in terms of secureness. Regarding the latter, one can use the well-known CIA triad to indicate the needs with respect to the security aspects **C**onfidentiality (preventing sensitive information from reaching the wrong people), **I**ntegrity (maintaining consistency, accuracy, and trustworthiness of data) and **A**vailability (ensuring timely and reliable access to, and use of, information) [23]. Here, a domain-specific language (DSL) could be developed to specify these extra-functional requirements. The application workloads themselves can be specified using task or process graphs, explicitly describing application tasks and their interactions (communications).

Given the attack types we consider in our proposed approach, as discussed in Section 3.1 and shown at the left in Figure 2, only those attacks that are relevant for the embedded system under design need to be identified, which we refer to as the so-called domain-specific attacks. To this end, we need to consider the security requirements of the target embedded system as specified using the CIA triad as well as the characteristics of the specific attack types in terms of, for example, passivity and accessibility. Here, passivity refers to what extent an attack manipulates the target system, either as a means or a goal of the attack. For example, a denial of service attack clearly is an active attack as its sole aim is to manipulate the system, whereas a side-channel attack based on power analysis is a passive attack. Accessibility refers to the access level that is required for an attack to be performed. Revisiting the example of a side-channel attack via power analysis, such an attack obviously requires physical access to the embedded system, whereas for example, a software-based attack does not require this. If we now consider, for example, an anti-lock braking system, then confidentiality is not a major concern as such a system does not process sensitive information. This makes passive attacks such as side-channel attacks not relevant and can therefore be excluded from the set of domain-specific attacks. However, the braking system may not be disrupted or manipulated (i.e., integrity and availability are

crucial CIA elements) thereby making active attacks highly relevant. Table 1 provides an overview of the required access level and passivity of the attack types we consider in this paper. Here, virtual access level attacks require access to one process that runs within an application on the embedded system in question. Cryptanalysis attacks often do not require access to the system. For example, in the case of public key cryptography, the public key is distributed to other systems and therefore freely available.

Table 1. Required access level and passivity of various types of attacks on embedded systems.

Attack	Sub-Type	Access Level	Passivity
Side-channel	Power analysis	Physical	Passive
	Timing attack	Virtual	Passive
	Scan attack	Physical	Active
	Fault Analysis	Physical & Virtual	Passive & Active
	Electromagnetic Analysis	Physical	Passive
Denial of Service		Virtual	Active
Software	Buffer overflow	Virtual	Active
Cryptanalysis		None	Passive

Once the set of domain-specific attacks has been determined, those attacks that are relevant to the different components in the underlying platform architecture can be determined: for example, a networking component is not susceptible to a software-based buffer overflow attack, whereas a microprocessor component is. To do so, we also need the mapping information (i.e., which application tasks and communications are mapped onto what platform components) of the design instance(s) that are currently being explored by the system-level DSE process. Subsequently, for each component in the platform architecture, we can now determine the set of possible security protection mechanisms that can be deployed to effectively increase its secureness ('Per-component protection mechanisms' in Figure 2). These sets of possible per-component protection mechanisms are an important ingredient of our envisioned scoring-based security quantification methodology: they allow for determining the coverage with respect to the protection mechanisms that are actually deployed in the design instances being explored by the system-level DSE. To achieve this, a scoring technique would be needed that can capture binary coverage relationships (i.e., a certain protection mechanism is available or not) as well as numerical coverage relationships. The latter applies in cases where, for example, a certain amount of random noise is added to a system component to disguise real power behavior in order to complicate or even prevent side-channel attacks based on power analysis [24] (here, the amount of noise forms a power/security trade-off) or when the strength of a cryptographic processor is identified as a function of the key-size it uses.

Besides the coverage of protection mechanisms deployed in the platform architecture components, one can take two other facets into account to determine the security score of a particular design instance. First, the spatial isolation realized in design instances can be considered. That is, reducing the amount of resource sharing between applications or even between tasks from a single application will increase the secureness of the system, as this will complicate certain types of attacks such as side-channel attacks. Therefore, a proper technique for quantifying the spatial isolation (using the mapping information from the system-level DSE) would be required such that it can be used for security scoring purposes. As a final ingredient of our anticipated security score, the platform component utilization can be used. The rationale behind this is that higher utilized components typically are more prone to certain attacks. Moreover, higher utilized components possibly also play a more important role in achieving the CIA-triad system requirements. To include the platform component utilization, we need to profile the application(s) to measure the activity of application tasks and communications, that is, the degree to which they utilize the underlying resources. Hereafter, this information is related to the mapping of

these application tasks and communications onto the platform architecture. The resulting component utilization can then be used to weight the protection mechanism coverage and spatial isolation of design instances in the final security scoring (as shown at the bottom of Figure 2).

3.3. Scoring the Security of Design Instances

Above, we described the ingredients of our envisioned security scoring methodology. We do realize that we have not provided any details on how such security scoring could actually be implemented. Actually, this remains a topic for future research, which will hopefully also be picked up by the community. Nevertheless, in this section, we would like to provide a rough sketch of a fairly simple approach to do such scoring.

Given a mapping of a (set of) application(s) to the underlying resources of a possible platform architecture, which includes the mapping of application tasks to computational resources as well as the mapping of inter-task communications to network and memory resources. Then, for each utilized component in the platform, we could calculate a security score along the following lines. Let AT_x be the set of Attack Types (see e.g., the second column of Table 1) to which component x is susceptible:

$$AT_x = \{ \bigcup_{\forall t_i | c_i \text{ mapped on x}} \text{Attacks}_{t_i | c_i} \}.$$

For AT_x, we only consider the application tasks t_i or inter-task communications c_i (dependent on whether x is a processing or communication component) that are mapped to component x. $\text{Attacks}_{t_i | c_i}$ refers to the possible attacks for task t_i or communication c_i, taking into account its security requirements according to the CIA triad as well as taking into account the access level and passivity of the various attacks (see Table 1). To determine a security score S_x for a component x, one could then perform the following calculation:

$$S_x = \frac{\sum_{p \in P_{AT_x}} \text{Protection-level}(p, x)}{|AT_x|} \cdot \frac{1}{\text{Utilization}_x}.$$

Here, p is a particular protection mechanism and is part of the set P_{AT_x} that consists of the possible protection mechanisms for the attacks listed in set AT_x for component x. The function Protection-level(p, x) returns a value that indicates the extent to which component x implements protection mechanism p. This function could, for example, return a value between 0 and 1: The value 0 would mean that the component does not implement the protection mechanism, implying that component x would be fully susceptible to the associated attack type. The value 1, on the other hand, would refer to an available implementation of protection mechanism p such that component x is fully protected against attacks of the associated type. Evidently, the returned value may also be in between 0 and 1, indicating partial protection. For example, in the case protection mechanism p adds a certain amount of random noise to a component to disguise real power behavior in order to prevent side-channel attacks based on power analysis [24], the level of added noise (which is a trade-off between power consumption and security) would determine the returned value of the function Protection-level(p, x). To calculate S_x, we also consider the reciprocal of the utilization of component x. Here, the utilization refers to the fraction the computing/communication component x contributes to the overall computing or communication time of an application. The rationale behind this is that one could argue that those components (both processing and communication components) that are less active over time will be less susceptible to certain types of attacks, like side-channel attacks. To determine the overall security score of a specific design instance (i.e., a particular application mapping to a selected platform architecture), one could simply accumulate the S_x scores for all components x used in the design instance.

The above scoring example is by no means meant to be a complete and fully-fledged solution to the security scoring problem. It is merely meant to act as an illustration for the direction of thought as presented in this paper. Moreover, in the above scoring example, we also do not consider spatial isolation as part of the security score. One direction to accomplish this, would for example, to penalize the mapping of multiple application tasks or communications to a single platform component.

The multifaceted, scoring-based security quantification methodology as outlined in this section could provide a real innovation to system-level embedded system design as it would facilitate designers to study the trade-offs between the performance, power consumption, cost, and secureness of design instances during the early stages of design.

4. Related Work

The need for recognizing security as a first-class citizen, next to traditional design objectives such as performance, cost and power consumption, in the design of embedded systems is not new. For example, quoting from Reference [4], "However, security is often misconstrued by embedded system designers as the addition of features, such as specific cryptographic algorithms and security protocols, to the system. In reality, it is a new dimension that designers should consider throughout the design process, along with other metrics such as cost, performance, and power." Nevertheless, the integration of security aspects in the process of system-level design-space exploration of embedded systems has never really got off the ground and is still a largely uncharted research ground. Only a few efforts exist that address this problem but, at best, most of them provide partial solutions or solutions to very specific security problems. For example, in Reference [25], the evaluation of security protocols is integrated in the design process. For instance, it rates the security of a system based on the probability of a hash collision. However, it does not cover other types of attacks, such as timing attacks and power analysis. The authors of Reference [26] try to neutralize several types of side-channel attacks by means of spatial isolation in a DSE setting but, again, they do not consider any other types of attacks/protection mechanisms. In Reference [27], a small number of attacker capabilities and corresponding requirements that refer to the CIA triad are defined in the context of DSE. The problem of this approach is that it is not trivial to relate types of attacks to those capabilities and requirements. The work of Reference [28] incorporates security in system-level DSE by first generating potential architecture configurations, after which an automated security analysis is performed to check the generated configurations against designer-specified security constraints.

In all the above works of References [26–28] security is modelled as a requirement in the DSE process, which does not allow for studying actual trade-offs between performance, power consumption or cost in relationship to secureness of a design.

An alternative approach for quantifying security is by means of a security risk assessment using a specific attack model [29]. For example, Reference [30], proposes an attack tree model to evaluate the user's privacy risks associated with an Internet-of-Things eco system. They evaluate the potential risks based on varying attack attributes, the probable considerations or preferences of an adversary, and the varying computational resources available on a device. Research efforts like this are, however, typically not focused on the process of (early) DSE.

To the best of our knowledge, only the works of References [31,32] and Reference [33] are similar to what we propose in terms of aiming at incorporating security as an objective that can be traded off with other objectives in the process of early DSE. In References [31,32], the authors introduce an UML-based approach in which application security requirements can be described together with security 'capabilities'—in addition to other extra-functional aspects such as performance and power consumption—of system components stored in a library. This then allows for a DSE process during which the application requirements are matched with the component capabilities. The very recent work of Reference [33] introduces a novel DSE framework that allows for considering security constraints, in the form of attack scenarios, and attack mitigations, in the form of security tasks. Based on the

descriptions of the system's functionality and architecture, possible attacks, and known mitigation techniques, the framework tries to find the optimal design for an embedded system.

5. Conclusions

As embedded systems are becoming increasingly ubiquitous and interconnected, they attract a world-wide attention of attackers. This makes the security aspect during the design of these systems more important than ever. However, state-of-the-art design tools and methodologies for embedded systems do not consider system security as a primary design objective. This is especially true for the early design phases in which the process of design-space exploration is of eminent importance for performing trade-off analysis. Any security measures that may eventually be taken much later in the design process will then affect the already established design trade-offs with respect to the other, and more traditional, design objectives like system performance, power consumption and cost. It goes without saying that such a design practice leads to suboptimal products.

In this position paper, we therefore argued for security-aware design methods for embedded systems that will allow for the optimization of security aspects of embedded systems in their earliest design phases as well as for studying the trade-offs between security and the other design objectives such as performance, power consumption and cost. To this end, we proposed a multifaceted, scoring-based methodology for quantifying the degree of secureness of embedded system design instances, which would allow for incorporating these secureness quantifications in early design-space exploration of embedded systems. The proposed methodology has not yet been implemented, and would require further research to do so. We do hope, however, that this position paper will be a trigger for more wide-spread research on techniques that allow for incorporating security as a first-class citizen in the process of early design-space exploration of embedded systems.

Funding: This research received no external funding.

Conflicts of Interest: The authors declare no conflict of interest.

References

1. Keutzer, K.; Newton, A.; Rabaey, J.; Sangiovanni-Vincentelli, A. System-level design: Orthogonalization of concerns and platform-based design. *IEEE Trans. Comput.-Aided Des. Integr. Circuits Syst.* **2000**, *19*, 1523–1543. [CrossRef]
2. Pimentel, A.D. Exploring Exploration: A Tutorial Introduction to Embedded Systems Design Space Exploration. *IEEE Des. Test* **2017**, *34*, 77–90. [CrossRef]
3. Singh, A.K.; Shafique, M.; Kumar, A.; Henkel, J. Mapping on multi/many-core systems: survey of current and emerging trends. In Proceedings of the 2013 50th ACM/EDAC/IEEE Design Automation Conference (DAC), Austin, TX, USA, 29 May–7 June 2013; pp. 1–10.
4. Ravi, S.; Raghunathan, A.; Kocher, P.; Hattangady, S. Security in embedded systems: Design challenges. *ACM Trans. Embed. Comput. Syst. (TECS)* **2004**, *4*, 461–491. [CrossRef]
5. Gries, M. Methods for evaluating and covering the design space during early design development. *Integration* **2004**, *38*, 131–183. [CrossRef]
6. Niemann, R.; Marwedel, P. An Algorithm for Hardware/Software Partitioning Using Mixed Integer Linear Programming. *Des. Autom. Embed. Syst.* **1997**, *2*, 165–193. [CrossRef]
7. Lukasiewycz, M.; Glass, M.; Haubelt, C.; Teich, J. Efficient symbolic multi-objective design space exploration. In Proceedings of the 2008 Asia and South Pacific Design Automation Conference, Seoul, Korea, 21–24 March 2008; pp. 691–696.
8. Padmanabhan, S.; Chen, Y.; Chamberlain, R.D. Optimal design-space exploration of streaming applications. In Proceedings of the ASAP 2011-22nd IEEE International Conference on Application-specific Systems, Architectures and Processors, Santa Monica, CA, USA, 11–14 September 2011; pp. 227–230.
9. Palesi, M.; Givargis, T. Multi-objective Design Space Exploration Using Genetic Algorithms. In Proceedings of the CODES02: 10th International Symposium on Hardware/Software Codesign, Estes Park, CO, USA, 6–8 May 2002; pp. 67–72.

10. Erbas, C.; Cerav-Erbas, S.; Pimentel, A.D. Multiobjective Optimization and Evolutionary Algorithms for the Application Mapping Problem in Multiprocessor System-on-Chip Design. *IEEE Trans. Evol. Comput.* **2006**, *10*, 358–374. [CrossRef]

11. Thompson, M.; Pimentel, A.D. Exploiting Domain Knowledge in System-level MPSoC Design Space Exploration. *J. Syst. Archit.* **2013**, *59*, 351–360. [CrossRef]

12. Quan, W.; Pimentel, A.D. Towards Exploring Vast MPSoC Mapping Design Spaces using a Bias-Elitist Evolutionary Approach. In Proceedings of the 2014 17th Euromicro Conference on Digital System Design, Verona, Italy, 27–29 August 2014; pp. 655–658.

13. Ravi, S.; Raghunathan, A.; Chakradhar, S. Tamper resistance mechanisms for secure embedded systems. In Proceedings of the 17th International Conference on VLSI Design. Proceedings, Mumbai, India, 9 January 2004; pp. 605–611. [CrossRef]

14. Yang, B.; Wu, K.; Karri, R. Scan based side channel attack on dedicated hardware implementations of Data Encryption Standard. In Proceedings of the 2004 International Conferce on Test, Charlotte, NC, USA, 26–28 October 2004; pp. 339–344. [CrossRef]

15. Papp, D.; Ma, Z.; Buttyan, L. Embedded systems security: Threats, vulnerabilities, and attack taxonomy. In Proceedings of the 2015 13th Annual Conference on Privacy, Security and Trust (PST), Izmir, Turkey, 21–23 July 2015; pp. 145–152. [CrossRef]

16. Koopman, P. Embedded systems security. *IEEE Comput.* **2004**, *37*, 145–152. [CrossRef]

17. Swenson, C. *Modern Cryptanalysis: Techniques for Advanced Code Breaking*; Wiley Publishing: Hoboken, NJ, USA, 2008.

18. Pimentel, A.D.; Erbas, C.; Polstra, S. A Systematic Approach to Exploring Embedded System Architectures at Multiple Abstraction Levels. *IEEE Trans. Comput.* **2006**, *55*, 99–112. [CrossRef]

19. Erbas, C.; Pimentel, A.D.; Thompson, M.; Polstra, S. A Framework for System-level Modeling and Simulation of Embedded Systems Architectures. *EURASIP J. Embed. Syst.* **2007**, *2007*, 082123. [CrossRef]

20. Mohanty, S.; Prasanna, V.K.; Neema, S.; Davis, J. Rapid Design Space Exploration of Heterogeneous Embedded Systems Using Symbolic Search and Multi-granular Simulation. *ACM SIGPLAN Not.* **2002**, *37*, 18–27. [CrossRef]

21. Mariani, G.; Brankovic, A.; Palermo, G.; Jovic, J.; Zaccaria, V.; Silvano, C. A correlation-based design space exploration methodology for multi-processor systems-on-chip. In Proceedings of the Proc. of the Design Automation Conference (DAC), Anaheim, CA, USA, 13–18 June 2010; pp. 120–125.

22. Jia, Z.J.; Bautista, T.; Núñez, A.; Thompson, M.; Pimentel, A.D. A System-level Infrastructure for Multi-dimensional MP-SoC Design Space Co-exploration. *ACM Trans. Embed. Comput. Syst.* **2013**, *13*, 1–26. [CrossRef]

23. Whitman, M.E.; Mattord, H.J. *Principles of Information Security*, 4th ed.; Course Technology Press: Boston, MA, USA, 2011.

24. Benini, L.; Macii, A.; Macii, E.; Omerbegovic, E.; Pro, F.; Poncino, M. Energy-Aware Design Techniques for Differential Power Analysis Protection. In Proceedings of the 40th Annual Design Automation Conference (DAC), Anaheim, CA, USA, 2–6 June 2003; pp. 36–41.

25. Lin, C.W.; Zheng, B.; Zhu, Q.; Sangiovanni-Vincentelli, A. Security-Aware Design Methodology and Optimization for Automotive Systems. *ACM Trans. Des. Autom. Electron. Syst.* **2015**, *21*, 1–26. [CrossRef]

26. Weichslgartner, A.; Wildermann, S.; Götzfried, J.; Freiling, F.; Glaundefined, M.; Teich, J. Design-Time/ Run-Time Mapping of Security-Critical Applications in Heterogeneous MPSoCs. In Proceedings of the 19th International Workshop on Software and Compilers for Embedded Systems (SCOPES), Sankt Goar, Germany, 23–25 May 2016; pp. 153–162.

27. Stierand, I.; Malipatlolla, S.; Fröschle, S.; Stühring, A.; Henkler, S. Integrating the Security Aspect into Design Space Exploration of Embedded Systems. In Proceedings of the IEEE International Symposium on Software Reliability Engineering Workshops, Naples, Italy, 3–6 November 2014; pp. 371–376.

28. Tan, B.; Biglari-Abhari, M.; Salcic, Z. An Automated Security-Aware Approach for Design of Embedded Systems on MPSoC. *ACM Trans. Embed. Comput. Syst.* **2017**, *16*, 1–20. [CrossRef]

29. Rocchetto, M.; Ferrari, A.; Senni, V. Challenges and Opportunities for Model-Based Security Risk Assessment of Cyber-Physical Systems. In *Resilience of Cyber-Physical Systems. Advanced Sciences and Technologies for Security Applications*; Flammini, F., Ed.; Springer: Berlin, Germany, 2019.

30. Asif, W.; Ray, I.G.; Rajarajan, M. An Attack Tree Based Risk Evaluation Approach for the Internet of Things. In Proceedings of the 8th International Conference on the Internet of Things, Santa Barbara, CA, USA, 15–18 October 2018.

31. Ferrante, A.; Milosevic, J.; Janjšević, M. A security-enhanced design methodology for embedded systems. In Proceedings of the 2013 International Conference on Security and Cryptography (SECRYPT), Reykjavik, Iceland, 29–31 July 2013; pp. 1–12.

32. Ferrante, A.; Kaitovic, I.; Milosevic, J. Modelling Requirements for Security-Enhanced Design of Embedded Systems. In Proceedings of the International Conference on Security and Cryptography (SECRYPT), Vienna, Austria, 28–30 August 2014; pp. 315–320.

33. Gressl, L.; Steger, C.; Neffe, U. Security Driven Design Space Exploration for Embedded Systems. In Proceedings of the 2019 Forum for Specification and Design Languages (FDL), Southampton, UK, 2–4 September 2019; pp. 1–8.

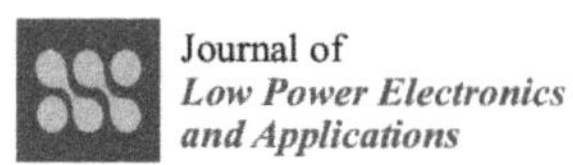

Journal of
*Low Power Electronics
and Applications*

Article

Hybrid Application Mapping for Composable Many-Core Systems: Overview and Future Perspective

Behnaz Pourmohseni [1,*], Michael Glaß [2], Jörg Henkel [3], Heba Khdr [3], Martin Rapp [3], Valentina Richthammer [2], Tobias Schwarzer [1], Fedor Smirnov [4], Jan Spieck [1], Jürgen Teich [1], Andreas Weichslgartner [5] and Stefan Wildermann [1]

[1] Chair for Hardware/Software Co-Design, Department of Computer Science, Friedrich-Alexander-Universität Erlangen-Nürnberg (FAU), 91058 Erlangen, Germany; tobias.schwarzer@fau.de (T.S.); jan.spieck@fau.de (J.S.); juergen.teich@fau.de (J.T.); stefan.wildermann@fau.de (S.W.)

[2] Institute of Embedded Systems/Real-Time Systems, Faculty of Engineering, Computer Science and Psychology, Ulm University, 89081 Ulm, Germany; michael.glass@uni-ulm.de (M.G.); valentina.richthammer@uni-ulm.de (V.R.)

[3] Chair for Embedded Systems, Department of Computer Science, Karlsruhe Institute of Technology (KIT), 76131 Karlsruhe, Germany; henkel@kit.edu (J.H.); heba.khdr@kit.edu (H.K.); martin.rapp@kit.edu (M.R.)

[4] Distributed and Parallel Systems Group, Institute of Computer Science, University of Innsbruck, 6020 Innsbruck, Austria; fedor@dps.uibk.ac.at

[5] AUDI AG, 85045 Ingolstadt, Germany; andreas.weichslgartner@audi.de

* Correspondence: behnaz.pourmohseni@fau.de

Received: 25 September 2020; Accepted: 13 November 2020; Published: 17 November 2020

Abstract: Many-core platforms are rapidly expanding in various embedded areas as they provide the scalable computational power required to meet the ever-growing performance demands of embedded applications and systems. However, the huge design space of possible task mappings, the unpredictable workload dynamism, and the numerous non-functional requirements of applications in terms of timing, reliability, safety, and so forth. impose significant challenges when designing many-core systems. Hybrid Application Mapping (HAM) is an emerging class of design methodologies for many-core systems which address these challenges via an incremental (per-application) mapping scheme: The mapping process is divided into (i) a design-time Design Space Exploration (DSE) step per application to obtain a set of high-quality mapping options and (ii) a run-time system management step in which applications are launched dynamically (on demand) using the precomputed mappings. This paper provides an overview of HAM and the design methodologies developed in line with it. We introduce the basics of HAM and elaborate on the way it addresses the major challenges of application mapping in many-core systems. We provide an overview of the main challenges encountered when employing HAM and survey a collection of state-of-the-art techniques and methodologies proposed to address these challenges. We finally present an overview of open topics and challenges in HAM, provide a summary of emerging trends for addressing them particularly using machine learning, and outline possible future directions. While there exists a large body of HAM methodologies, the techniques studied in this paper are developed, to a large extent, within the scope of invasive computing. Invasive computing introduces resource awareness into applications and employs explicit resource reservation to enable incremental application mapping and dynamic system management.

Keywords: Hybrid Application Mapping (HAM); many-core systems; embedded systems; composability; design space exploration (DSE); resource management; Network-on-Chip (NoC); real-time guarantees; predictability; machine learning

1. Introduction

The ever-increasing computational power requirements of embedded applications have substantially changed the design process of embedded systems over the past decade. To address the performance demands of emerging applications, embedded domains have undergone a paradigm shift from single-core platforms to many-core platforms. Many-core platforms such as Tilera TILE-Gx [1], Kalray MPPA-256 [2], Intel SCC (Single-chip Cloud Computer) [3], the KiloCore [4], or the upcoming SiPearl Rhea processor family [5] integrate tens, hundreds, or even thousands of processing cores on a single chip with a highly scalable communication scheme. This enables them to deliver a scalable computational power which is required to meet the progressively growing performance demands of emerging embedded applications and systems. Along the same line, modern platforms also incorporate heterogeneous processing resources to cater to the specific functional and non-functional requirements of applications from different domains of computing, see, for example, Reference [6]. In addition to the current practice of integrating various types of general-purpose cores on a chip, many-core platforms are also on the verge of incorporating domain/application-specific processing resources, for example, Digital Signal Processor (DSP) cores for signal/image processing [7], Graphics Processing Units (GPUs) for graphics processing and AI acceleration in deep learning [8,9], and Coarse-Grained Reconfigurable Arrays (CGRA) for the acceleration of (nested-)loops [10,11]. Moreover, Field Programmable Gate Arrays (FPGAs) have also been incorporated to provide a reconfigurable fabric for hardware acceleration [9,12]. While a large number of (possibly heterogeneous) processing resources increases the available computational power dramatically, it also introduces a significant number of additional design decisions to be made during the phases of system design as well as application mapping. In the following, an overview of the major challenges of many-core application mapping is provided.

1.1. Many-Core Application Mapping Challenges

1.1.1. Task Mapping Complexity

To exploit the massive computational power of many-core platforms, parallel computation models have been increasingly adopted in the development of embedded applications [13]. In such models, each application is partitioned into several (processing) tasks that can be executed concurrently on different cores and exchange data with each other. In this context, the optimization of the application mapping, that is, finding a suitable assignment of an application's tasks to platform resources, becomes a particularly challenging effort. This is due to the number of possible mappings for an application growing exponentially with the application size (number of application's tasks) and the platform size (number of cores). In fact, the many-core application mapping problem is known to be an *NP-hard* [14,15] optimization problem which renders an enumeration of all possible mappings for realistic problem sizes impractical, if at all feasible. As a consequence, system designers resort to meta-heuristic optimization algorithms, for example, evolutionary algorithms [16–18] and particle swarm optimization [19], for the automated Design Space Exploration (DSE) of possible application mappings. Meta-heuristic optimization algorithms have proven effective in finding satisfactory mappings at an affordable computational effort.

1.1.2. Complexity of Evaluation and Verification Techniques

In addition to functional correctness, embedded applications typically also need to satisfy a set of non-functional requirements, often provided in the form of upper/lower bound constraints on timing,

reliability, security, safety, and other qualities [20,21]. For each mapping of an application, the satisfaction of the application's non-functional requirements must be verified, for example, by means of measurement, simulation, or formal analysis. Subject to the characteristics of the application and the platform resources, the choice of non-functional requirements, and the strictness of their constraints, the verification process may become fairly complex and/or demand a considerable amount of computational effort and time.

1.1.3. Workload Dynamism

Another factor contributing significantly to the design complexity of modern embedded systems is the growing dynamism of workload. In these systems, a mix of applications—each with its own set of non-functional requirements—must typically be executed concurrently. Recent years give evidence of a rapid increase in the number of concurrent applications in embedded systems with different requirements. In these systems, the system's *workload scenario*, that is, the mix of concurrently executed applications (also known as the system's *use-case* [22]), tends to change over time such that, at each point in time, only a fraction of all applications in the system are active. These workload variations, including the activation and the termination of applications, often happen in reaction to external events whose arrival pattern cannot be predicted, for example, user requests or changes in the environment with which the system is interacting. In general, the number of system workload scenarios, that is, possible mixes of concurrently active applications, grows exponentially with the number of applications in the system [23]. The increasing trends in (i) the number of applications in the system and in (ii) the dynamic workload of the applications each contribute exponentially to the complexity of the process of finding optimal mappings of the applications to system resources [24]. To alleviate the design complexity w.r.t. the increased number of applications, the *integrated design* approach, where the mappings of all applications are considered at the same time, has been gradually replaced by an *incremental (constructive) design* approach in which the mapping process is partitioned into a phase with a per-application mapping optimization step followed by a system integration step.

1.2. Hybrid Application Mapping

Application mapping methodologies for multi/many-core systems are generally classified into two categories, namely, design-time (static) and run-time (dynamic) approaches, see Reference [25]. In this paper, in the context of application mapping, the terms *static*, *offline*, and *design-time* are used interchangeably to denote that the operation in question is performed at design time. Likewise, the terms *dynamic*, *online*, and *run-time* are used interchangeably to denote that the operation in question takes place at run time.

In *design-time (static) mapping approaches*, all mapping decisions are conducted statically at design time (offline). These approaches employ compute-intensive optimization and verification techniques to find an optimal mapping for each application which is also verified to satisfy the application's requirements. Since each system design generated by static approaches is tailored to a single scenario, these approaches either cannot at all be used for the design of dynamic systems or have to resort to single solutions compromising between different expected run-time scenarios.

In the second class of mapping approaches, namely, *run-time (dynamic) mapping approaches*, all mapping decisions are made dynamically at run time (online) when an application must be launched. These approaches take into account the current system workload in their mapping decisions. This offers an adaptive solution for the design of dynamic systems and eliminates the need to statically compromise between different workload scenarios. This advantage, however, comes at the expense of increased time pressure, since the time overhead of the application mapping process has a direct impact on the system's performance. Due to this time pressure, run-time mapping methodologies cannot afford powerful

mapping optimization and non-functional verification procedures. Instead, they are limited to lightweight (incremental and/or iterative) mapping heuristics to find an acceptable mapping at a low computational effort, see, for example, Reference [26]. Consequently, they mostly yield sub-optimal mappings and will often not strictly provide non-functional guarantees which require compute-intensive verification processes, for example, reliability analysis or worst-case timing verification.

Hybrid Application Mapping (HAM) is a new class of mapping approaches which addresses the aforementioned many-core application mapping challenges (discussed in Section 1.1) by combining static and dynamic mapping approaches to exploit the individual strengths of each, see Reference [25]. In HAM, a set of Pareto-optimal mappings is computed for each application at design time where compute-intensive mapping optimization and non-functional verification techniques are affordable. These mappings are then used at run time to launch the application on demand by selecting one of the precomputed mappings which fits best to the current system workload state and resource availability. Combining (Pareto-)optimal mappings with guaranteed non-functional properties while coping with the workload dynamism, HAM is regarded as a promising paradigm for the design of future embedded systems. In this paper, we provide an overview of HAM and the design methodologies developed in line with it.

Since the introduction of many-core platforms in embedded domains, numerous proposals for application mapping on these platforms have been registered, addressing a broad range of application mapping challenges. In the same line, new programming paradigms have emerged to enable a systematic design approach for the incremental mapping of applications to embedded many-core systems. *Invasive computing* [6] is an emerging many-core programming paradigm in which *resource awareness* is introduced into application programming, and dynamic per-application *resource reservation* policies are employed to achieve not only a high utilization of resources but also providing isolation between resources and applications on demand in order to create predictability in terms of timing, safety, or security [27,28]. This setup is particularly promising for HAM and has served as the base for a large number of works in the scope of HAM.

1.3. Paper Overview and Organization

1.3.1. Paper Overview

In this paper, we introduce the fundamentals of HAM and elaborate on the way HAM addresses the major design challenges in mapping applications to many-core systems. The fusion of offline mapping optimization and online mapping selection in HAM, however, also gives rise to new challenges that must be addressed to boost its effectiveness. In this paper, we also provide an overview of the main challenges encountered when employing HAM and survey a collection of state-of-the-art techniques and methodologies proposed to address these challenges. We also discuss a collection of open topics and challenges in HAM, present a summary of emerging trends for addressing them particularly using machine learning, and outline some promising future directions. The majority of the techniques studied in this paper are developed within the scope of invasive computing which serves as an enabler for HAM and incremental design. An early overview of HAM techniques can, for example, be found in Reference [25].

1.3.2. Paper Organization

The remainder of this paper is organized as follows. In Section 2, a review of static and dynamic application mapping schemes—which can be considered HAM predecessors—is given and schemes for incremental design are presented. Section 3 provides an overview of the application and architecture models commonly used to describe the application mapping problem in embedded many-core systems. In Section 4, the basics of HAM are presented. Sections 5 and 6 present an overview of state-of-the-art

methodologies and techniques in HAM—Techniques discussed in Section 5 aim at reducing the complexity of HAM at different stages of design while the approaches presented in Section 6 focus on enabling HAM for real-time systems as a predominant class of embedded systems. Section 7 presents open topics and challenges in HAM, discusses a collection of emerging trends for addressing them particularly using machine learning, and outlines promising future directions. The paper is concluded in Section 8.

2. Related Work

Design methodologies for many-core systems generally deal with the problem of mapping multiple applications to the resources of a many-core platform. In addition to functional correctness, a high-quality mapping must also satisfy the non-functional requirements of the application, for example, timing, reliability, and security, while exhibiting a high performance w.r.t., for example, resource utilization and energy efficiency. Prior to the introduction of HAM, application mapping methodologies for embedded systems were typically classified into two categories: *design-time (static) approaches* and *run-time (dynamic) approaches*. An elaborate survey of these techniques is presented in Reference [25].

The majority of existing application mapping methodologies fall into the category of static approaches, see, for example, References [29–33]. In these approaches, all mapping decisions are made offline. These approaches rely on a global view of the whole system, and in particular, the complete set of applications in the system, and exploit this knowledge to find an optimal mapping of all applications in the system to platform resources [25]. Due to their offline scheme, they can afford compute-intensive mapping optimization and non-functional verification techniques which are often inevitable, for example, in the case of applications with hard real-time constraints. Given the NP-hard nature of the many-core application mapping optimization problem [14,15], static approaches employ DSE techniques based on *meta-heuristic optimization* approaches, for example, evolutionary algorithms [16–18], simulated annealing [34], or particle swarm optimization [19], to find high-quality mapping solutions with a reasonable computational effort. For instance, for the DSE in References [15,35], genetic algorithms are used. In Reference [36], simulated annealing is used, while Reference [37] adopts particle swarm optimization. In spite of its advantages, this static design scheme is practical only for systems with a relatively static workload profile and is, thus, impractical for systems with dynamic workload scenarios or systems in which the complete set of applications is not known statically [25,38].

Dynamic mapping approaches offer a flexible and adaptive application mapping scheme that can be used for the design of systems with dynamic workload scenarios (use cases). In these approaches, mapping decisions are conducted online at the time an application must be launched, see, for example, References [26,39,40]. In spite of their flexibility, the processing power available for the online decision processes is limited to the resources of the underlying platform, and the time for their decision making is restricted by application-specific deadlines. Therefore, dynamic approaches cannot afford compute-intensive techniques for their mapping decisions and, hence, typically yield application mappings of inferior quality, compared to static approaches.

Hybrid Application Mapping (HAM) is an emerging category of mapping methodologies which employs a combination of offline mapping optimization and online mapping selection to address the shortcomings of both static and dynamic mapping methodologies [25]. A large body of work exists on HAM, for example, References [23,41–47]. At design time, HAM approaches employ DSE to compute a set of high-quality mappings for each application. Similarly to static approaches, the offline DSE in HAM benefits from compute-intensive optimization and verification techniques, for example, for worst-case timing verification. By using statically verified mappings as candidates for online application mapping, HAM enables the dynamic mapping of a broad range of applications for which the verification of

non-functional requirements involves time-consuming analyses and, hence, cannot be done online. An overview of a collection of state-of-the-art HAM techniques is given in Sections 5 and 6.

The majority of many-core design methodologies, including those listed above, follow an *incremental design* approach consisting of a per-application mapping computation step and a subsequent system integration step. An incremental design, verification, and integration approach alleviates the design complexity significantly [24]. To enable this design scheme, *system composability* is essential. Composability is a system property ensuring that the non-functional behavior of each application in the system is not affected by other applications [24,48]. For instance, in a *timing-composable* system, for example, CoMPSoC [49] or T-CREST [50], concurrent applications are decoupled from each other w.r.t. their (worst-case) timing behavior. This allows timing verification of each application to be performed individually and irrespectively of the other applications.

Composability can be established using temporal/spatial isolation between applications [24]. In the case of *spatial isolation*, resources are exclusively reserved for applications. Spatial isolation has been used to eliminate inter-application timing interferences for real-time applications [51] and side-channel attacks for security-critical applications [52]. In the case of *temporal isolation*, resource sharing among applications is allowed under exclusive resource budget reservation per application and a timing-composable arbitration/scheduling policy. Examples of such policies include Time Division Multiple Access (TDMA) used in References [49,50] or Weighted Round Robin (WRR) used in References [41,51,53]. In the same line, new programming paradigms have emerged which promote the isolation of applications in favor of composability to enable an incremental design scheme. For instance, in the paradigm of invasive computing [6,28], application programs can exclusively allocate (invade) resources and later release them again (retreat). Invasion establishes spatial isolation between concurrently executed applications to achieve timing composability by means of explicit resource reservation per application, see, for example, Reference [27]. Such support is particularly crucial for HAM which relies on separate mapping optimization of individual applications at design time. The HAM techniques discussed in this paper (in Sections 5 and 6) are developed based on these principles.

3. System Model

Most approaches for design automation require a formal system model which serves as a basis for the optimization and verification processes performed throughout the DSE. The design problem of heterogeneous embedded systems, including the many-core application mapping problem, is typically represented using a graph-based system model, referred to as a *specification* which consists of (i) an *application graph* representing the application, (ii) an *architecture graph* representing the target many-core platform architecture and (iii) a set of *mapping edges* which connect these two graphs to reflect the task-to-core assignment options, see References [15,21]. This section provides an overview of the application and architecture models commonly used in the embedded many-core domain and demonstrates how these models can be converted into the graph-based specification.

3.1. Application Model

In the parallel-processing paradigm of multi/many-core systems, an application is typically partitioned into a set of (processing) tasks which communicate with each other via a set of messages [13]. To reflect this structure, each application is modeled by an acyclic, directed, and bipartite graph $G_P(T \cup M, E)$ called the *application graph* (also known as task graph or problem graph) [27]. Here, T denotes the set of tasks, M denotes the set of messages exchanged between the tasks, and $E \subseteq (T \times M) \cup (M \times T)$ is a set of directed edges which specify the data dependencies among tasks and messages. Figure 1a illustrates an exemplary application graph where $T = \{t_0, \ldots, t_5\}$ and $M = \{m_0, \ldots, m_5\}$.

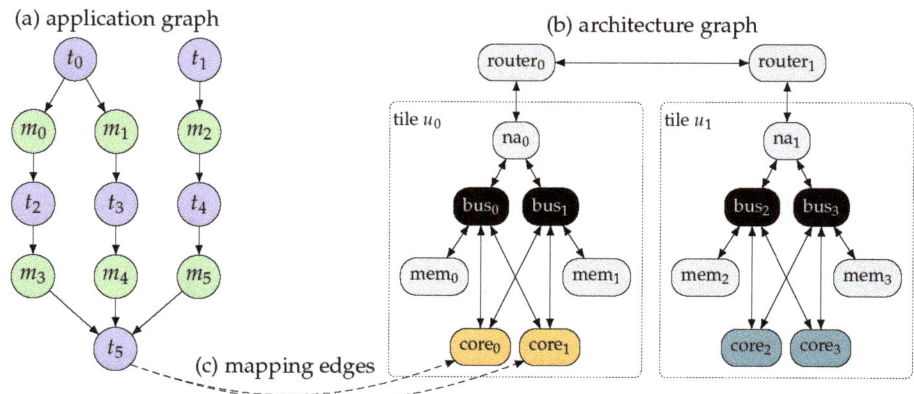

Figure 1. Example of a specification composed of (**a**) application graph, (**b**) architecture graph, and (**c**) mapping edges connecting them (depicted only for task t_5).

Tasks and messages in an application graph may also be annotated with additional information that might be required for the generation and/or the evaluation of the mappings. For instance, in the case of periodic applications, the execution period of each task and the production period of each message is provided to be used for resource scheduling. Similarly, in the case of real-time applications, the Worst-Case Execution Time (WCET) of each task and the maximum number of its memory operations may be provided to be used in the latency analysis and timing verification of the generated mappings. Also, the maximum payload size of a message and the maximum number of memory operations necessary to read/write the message from/to memory might be provided.

3.2. Architecture Model

Many-core platforms, for example, References [1–3], typically follow a regular and two-dimensional organization of resources in which resources are partitioned into a set of so-called *(compute) tiles*, interconnected by a two-dimensional Network-on-Chip (NoC) with a mesh topology, see, for example, Figure 2. Each tile in a many-core platform consists of a set of processing cores, one or multiple shared memories, and a Network Adapter (NA) which connects the on-tile resources to the routers of the NoC. The resources on each tile are interconnected via one or multiple (memory) buses. Each core may have a private L1 cache, and each tile may have an L2 cache shared among the resources located on that tile. In a heterogeneous many-core platform, the cores within each tile are typically from the same type (homogeneous) while different tiles comprise cores of different types. In addition to the compute tiles, a many-core platform may also contain one or multiple memory tiles to provide mass storage on the chip. Moreover, a set of I/O tiles may be available for connectivity with off-chip media, for example, cameras or sensors.

Many-core platforms employ a NoC interconnection infrastructure for inter-tile connectivity, chiefly due to its scalability [54]. A NoC consists of a set of routers and NAs which are interconnected via a set of links. Each router is associated with one tile of the platform and is connected to the NA located on that tile. In a mesh NoC, each router is also connected to its adjacent routers in the four cardinal directions. Each NoC connection enables the exchange of data in both directions using two uni-directional links.

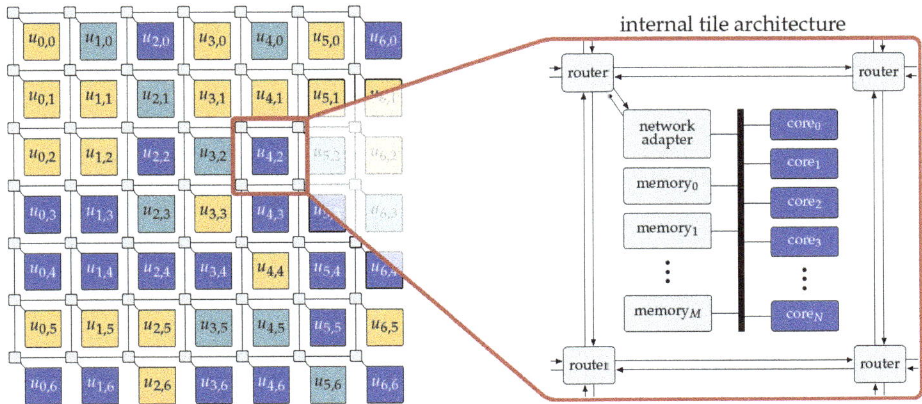

Figure 2. Example of a 7 × 7 heterogeneous tiled many-core architecture. Tiles are interconnected by a 2D mesh Network-on-Chip (NoC). Each tile consists of a set of processing cores, a set of memories, and a Network Adapter (NA), interconnected via one or more (memory) buses. Different tiles may contain different types of cores, here denoted by color.

The architecture of a many-core platform is typically modeled by a graph $G_A(R, L)$ called *architecture graph*, see, for example, References [27,41]. Here, R denotes the set of resources on the platform, that is, cores, memories, buses, NAs, and routers. The connections between these resources are reflected by the set of edges $L \subseteq R \times R$. Resources which access the NoC over the same NA are grouped into one tile $u \in U$. In heterogeneous systems, the processing cores C on different tiles can differ in architecture, instruction set, frequency, energy consumption, and so forth. To reflect this, each tile $u \in U$ can be of a certain resource type $r\_type \in P$ with $|P|$ different resource types in the system. Tiles that contain the same types of resources have the same resource type. Figure 1b illustrates the architecture graph of an exemplary heterogeneous many-core platform which is (for simplicity of illustration) composed of only two tiles: tile$_0$ and tile$_1$. Each tile consists of two cores (the type of each core is indicated by color), a NA, two shared memories, and two memory buses. Each bus connects the cores and the NA to one of the memories. The two tiles are each composed of different types of cores and are, thus, of different resource types.

3.2.1. Memory Model

Due to their scalability, distributed memory schemes are widely used in many-core systems. These schemes—also known as No Remote Memory Access (NORMA) memory architectures—restrict the accessibility of memories in each tile to resources located on that tile only [55]. The memory space in each tile may also be further partitioned into regions dedicated to individual resources or individual tasks on that tile.

3.2.2. Communication Model

The memory scheme of a many-core platform heavily impacts its viable choices for the communication of messages between tasks. Given the shared-memory scheme inside each tile, an exchange of data between resources located on one tile (*intra-tile communications*) is realized by the producer (sender) writing the data in a given space in the tile's shared memory and the consumer (receiver) reading the data from that memory location afterwards. For an exchange of data between different tiles (*inter-tile communications*), however, explicit message passing between the producer and the consumer is necessary due to the distributed

memory scheme between tiles. To that end, once the producer (sender) writes the data into the memory of the source tile, the local NA reads the data and injects it into the local router. Then, the data is forwarded through a chain of NoC routers on a hop-by-hop basis and in a pipeline fashion towards the destination router which provides the data to the NA on the destination tile. The destination NA stores the data in a dedicated memory space from where the consumer (receiver) can read the data thereafter.

3.3. Mapping

The application graph and architecture graph in the specification are connected by so-called *mapping edges V*. Each mapping edge $v = (t, c) \in V$ denotes that task $t \in T$ can be executed on core $c \in C$. Moreover, each mapping edge $v = (t, c)$ may be annotated with a set of attributes which reflect the execution characteristics of task t when executed on core c. For instance, in the context of heterogeneous many-core systems, mapping edges of a task t can also reflect the execution time of t on different types of cores in the system. Figure 1c, illustrates two exemplary mapping edges for task t_5, indicating that t_5 can be executed on $core_0$ or $core_1$.

4. Fundamentals of Hybrid Application Mapping

Hybrid Application Mapping (HAM) methodologies employ a pseudo-dynamic application mapping strategy, embodying a combination of offline mapping computation and online mapping selection. The standard flow of HAM is illustrated in Figure 3. This flow consists of (i) a design-time (offline) Design Space Exploration (DSE) step per application, followed by a (ii) run-time (online) system management step. These steps are detailed in the following.

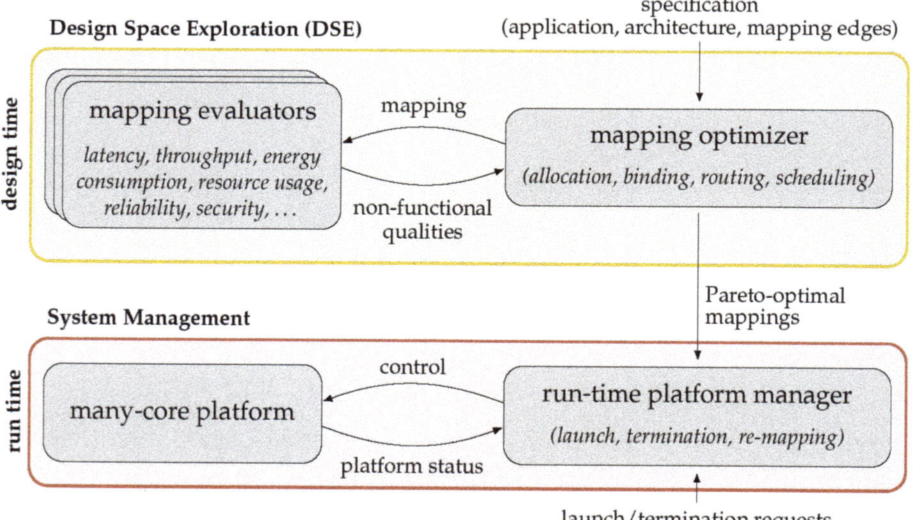

Figure 3. Flow of Hybrid Application Mapping (HAM). At design time, Design Space Exploration (DSE) is used to compute a set of Pareto-optimal mappings per application (**top**). At run time, a Run-time Platform Manager (RPM) launches applications using their precomputed mappings (**bottom**).

4.1. Offline Design Space Exploration (DSE)

In HAM, the computation of mappings for each application is performed in a DSE at design time (offline), see Figure 3 (top). The DSE takes as input the specification (detailed in Section 3) describing the space of design decisions of the currently considered application. During the DSE, various mappings of the application on the platform are generated by a *mapping optimizer* and are, subsequently, examined by a (set of) *mapping evaluator(s)* which assess the quality of each mapping w.r.t. a given (set of) non-functional design objective(s), for example, latency, throughput, and energy consumption. Subject to the application domain and the type of each non-functional objective, the respective evaluation can be performed using simulation, measurement, formal analysis, or a combination of them.

Each mapping candidate of the application on the given platform is generated in the course of four steps of design decisions, namely, resource *allocation*, task-to-core *binding*, message *routing*, and task/message *scheduling*, following the classical practice of system-level synthesis [15]. In the (i) *allocation* step, a set of platform resources, for example, cores, NAs, and routers, are specified and allocated for the execution of the application's tasks and/or for the communication of messages among them. In the (ii) *binding* step, the assignment of each task to the allocated cores is specified. In the (iii) *routing* step, a NoC route (sequence of connected routers) is specified for the communication of each message exchanged between data-dependent tasks which are bound to different tiles. Recall that messages communicated between tasks bound to the same tile are exchanged implicitly through the shared memories on that tile and, therefore, do not require a NoC route. Finally, in the (iv) *scheduling* step, the schedule of tasks $t \in T$ and messages $m \in M$ on their respective resources is specified, for example, a periodic (time) budget is computed for each task/message in case of a periodic application.

Given the NP-hard complexity of the many-core mapping optimization problem [14,15], the DSE usually employs a meta-heuristic optimization technique to find high-quality (Pareto-optimal) mappings at an acceptable computational effort. The majority of these meta-heuristic techniques operate based on an *iterative* optimization of a so-called *population* of solutions (mappings). Here, in the course of several optimization iterations, the population is used to generate new mappings (e.g., through genetic operators of mutation and crossover), and is updated with the new mappings. The set of non-dominated Pareto-optimal mappings iteratively generated thus far is always preserved. Generally, a large share of mappings generated this way could be infeasible (invalid) solutions, for example, mappings lacking the necessary NoC links for inter-tile communications. This harms the optimizer's performance as it would strive for finding feasible mappings rather than the optimization aspect [56]. As a remedy, hybrid optimization approaches combining exact (e.g., SAT or ILP) and meta-heuristic (e.g., evolutionary algorithm) techniques have emerged. These approaches, for example, SAT-decoding [57], implement powerful repair mechanisms which are capable of unambiguously mapping every point in the search space to a feasible solution, see also Reference [58].

4.2. Online System Management

In HAM, the statically computed set of mappings for each application is used at run time to launch that application on demand. For this purpose, the mappings are provided to a so-called *Run-time Platform Manager (RPM)*, see Figure 3 (bottom). Whenever an application shall be launched, the RPM selects one of the precomputed mappings of that application for which the required resources are currently available and uses that mapping to launch the application. In addition to launching applications, the RPM also terminates applications on demand and, if necessary, modifies the mapping of running applications (re-mapping) in reaction to unexpected events, for example, resource failures, or to enhance the system utilization, for example, through load/thermal balancing.

5. Tackling the Complexity

The task of application mapping is to find an allocation, binding, routing, and scheduling that is best with respect to the objectives of interest. However, with the huge amount of resources on many-core systems and more and more parallel tasks and messages of applications in modern use cases, the amount of possible mappings is immense. In particular, the large number of task-to-resource assignment options contributes significantly to the size of the search space. Finding the best or even only an optimized mapping out of this huge search space is, thus, a complex and time-consuming task. The HAM scheme described in Section 4 allows us to split the problem of application mapping between design time and run time. The general idea is to explore as much as possible of the search space already at design time. At the same time, there should be sufficient options left for the RPM to react to dynamic and unforeseeable system scenarios.

To achieve this, the DSE has to efficiently find the Pareto-optimal mapping options within this huge search space to be handed to the RPM. At the same time, the RPM must be able to efficiently find a feasible mapping candidate that can be realized on the available system resources. This section summarizes a selection of techniques that cope with the immense search space of application mapping by eliminating architectural symmetries (i.e., recurring resource-organization patterns) as well as applying architecture decomposition to decompose the complex problem into more tractable sub-problems. These techniques are likewise applicable as part of the design-time DSE and the run-time management.

The design-time DSE typically produces a huge number of Pareto-optimal mappings due to the large number of design objectives. Considering all Pareto-optimal mappings for RPM is not practical since a large set of candidates can quickly exhaust the available storage and computational capacity of the RPM. Therefore, only a fraction of the Pareto-optimal mappings must be retained, whose choice is particularly crucial for the system performance and requires new multi-objective truncation techniques tailored to many-core mapping selection. In this section, we also present a technique called *mapping distillation* that aims at reducing the number of mapping options determined by DSE so that the RPM can actually benefit from the DSE-based pre-optimization.

5.1. Constraint Graphs for Symmetry Elimination from the Search Space

Many-core architectures are heterogeneous and composed of different types of resources interconnected via a communication infrastructure. However, with the increasing parallelism, also an increasing amount of resources of equivalent resource types will be present on the chip, appearing in recurring patterns across the architecture. This means that the search space contains a large degree of redundancy in terms of symmetries, that is, mappings with equivalent resource requirements and non-functional properties. A major solution to deal with the scalability issue is, therefore, to choose a mapping representation that eliminates such symmetries. The classical application mapping as presented in Section 3 represents every possible mapping of tasks to resources. The representation introduced in Reference [41] and applied for DSE in Reference [45] instead uses a *task-cluster-to-resource-type* representation. A task cluster thereby describes a subset of application tasks which must be mapped on the same resource at run time. Each task cluster is also annotated with a resource type which specifies the type of the resource to which it must be mapped.

Figure 4 presents an example where an application consisting of tasks t_0, t_1, and t_2 should be mapped onto an architecture containing four tiles $u_{0,0}$, $u_{1,0}$, $u_{0,1}$, and $u_{1,1}$. A classical task-to-resource application mapping results in $4^3 = 64$ mapping combinations. Figure 4c illustrates four different concrete mappings. In each mapping, tasks t_0 and t_1 are mapped to a resource of type r_type_0 while t_2 is mapped to a resource of type r_type_1. The mappings differ from each other in their choice of resource instances. However, all four mappings are identical in terms of the number of allocated resources (one instance of resource type

r_type_0 and one of r_type_1), the assignment of tasks to the allocated resource types, and the hop distance, direction, and allocated bandwidth for the messages exchanged between tasks $t_0 \rightarrow t_2$ and $t_1 \rightarrow t_2$. This is indicated in the abstract representation in Figure 4d which is referred to as a *constraint graph*.

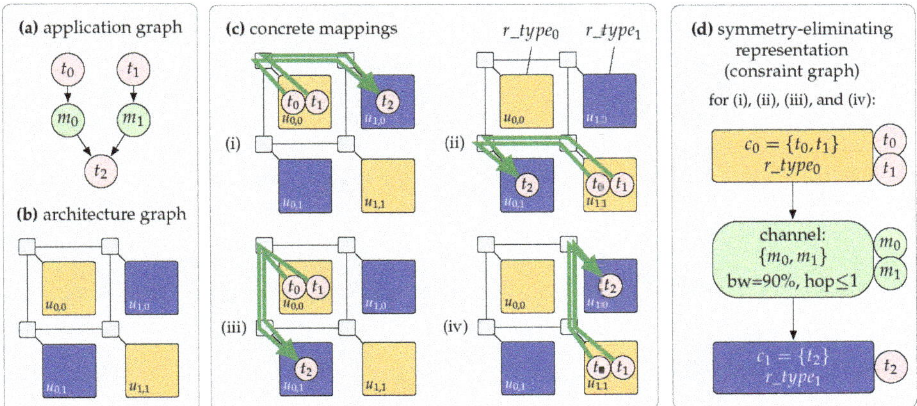

Figure 4. Illustration of application mapping and architectural symmetries, adopted from Reference [45]. The specification consists of (**a**) an application graph and (**b**) an architecture graph. (**c**) A design-time DSE explores mappings. Shown are four different concrete mappings for the three tasks t_0, t_1, and t_2 of the application onto the architecture containing four tiles $u_{0,0}$, $u_{1,0}$, $u_{0,1}$, and $u_{1,1}$ where tile types are distinguished by color. (**d**) Each mapping is transformed into an intermediate representation called *constraint graph* [41]. The constraint graph encodes rules for the RPM on how to feasibly embed the application. As depicted, the resulting constraint graph is identical for all four concrete mappings in (**c**).

The constraint graph is a representation that allows us to remove the symmetries from the search space when performing DSE based on this representation. However, it is also a representation that abstracts from concrete positional information. Determining a concrete application mapping based on a constraint graph is referred to as *constraint graph embedding*.

5.2. Symmetry-Eliminating DSE Using Constraint Graphs

Symmetry-eliminating DSE based on constraint graphs is introduced in Reference [45]. The main idea of symmetry-eliminating DSE is to explore symmetric task-cluster-to-resource-type mappings on a given target architecture based on constraint graphs instead of exploring concrete task-to-resource-instance mappings. As depicted in Figure 5, the steps to construct a mapping candidate in symmetry-eliminating DSE include (i) to explore on which resource type to bind each task (task assignment problem), (ii) to cluster subsets of tasks that are assigned to the same resource type together into task clusters (task clustering problem), and (iii) to combine the messages exchanged between the resulting task clusters to message clusters (to be routed over the NoC) and construct the constraint graph. The task assignment, task clustering, and message routing problems can be formulated as a 0-1 Integer Linear Program (ILP). The DSE can then work on this formulation by making use of SAT-decoding (see Section 4.1).

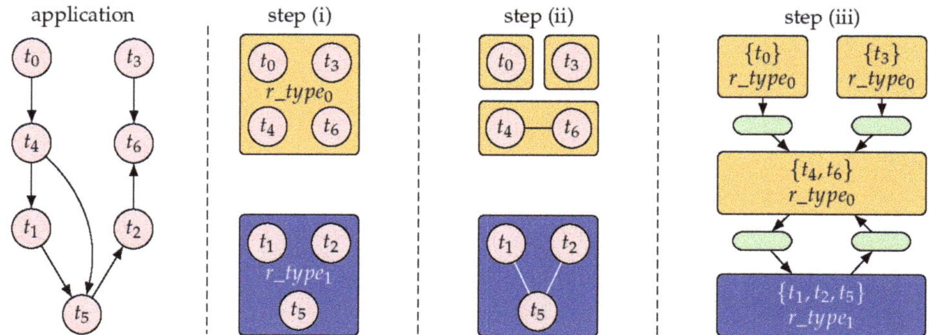

Figure 5. Symmetry-eliminating exploration consisting of three steps: (**i**) mapping tasks to resource types, (**ii**) clustering of tasks, and (**iii**) construction of constraint graph. For the sake of clarity, messages are not shown in the application. The example is adopted from Reference [45].

While the task-cluster-to-resource-type based mapping representation significantly reduces the search space, it is shown in Reference [45] that it still over-approximates the search space of feasible mappings: Due to the platform-independence of the representation (constraint graphs just encode a set of mapping solutions, but do not provide a concrete feasible one), the search space may still contain solutions, that is, constraint graphs, that cannot be feasibly mapped to a given target architecture instance due to topological constraints in a concrete architecture. Figure 6 depicts such an example of a constraint graph that cannot be feasibly mapped to the concrete architecture since resources of the required type are only available at a minimum hop distance of 4 in the target architecture, whereas the routing constraint in the constraint graph restricts the allowed distance to a maximum hop distance of 3. As a remedy, the DSE also has to perform a formal *feasibility check* to ensure that all considered solutions can be feasibly mapped to a concrete instance on the given target architecture. Techniques for determining feasible constraint graph embeddings on a given target architecture are discussed in Section 5.3. However, by means of Satisfiability Modulo Theories (SMT) techniques, it is possible to take the result of such a feasibility check as feedback for improving the DSE subsequently. For this purpose, the conditions that render a solution infeasible are extracted, and then this knowledge is added to the 0-1 ILP formulation so that not only this single but all other solutions that fulfill these conditions are removed from the search space. For the example in Figure 6, it can be deduced that all solutions with identical clustering of tasks and an identical mapping but a lower maximum hop distance (i.e., hops ≤ 1 and hops ≤ 2) will only be harder to embed and, thus, can also be excluded from the search space. Since also bandwidth requirements, hop distances, and the task clustering are included in the 0-1 ILP formulation, this can be learned by formulating respective constraints and adding them after each failed feasibility check. In Reference [45], it has been shown that this problem-specific learning technique has the potential of excluding large parts of the search space with much fewer feasibility checks.

5.3. Constraint Graph Embedding

Embedding a constraint graph in a given architecture requires (i) binding of task clusters to resources and (ii) routing of messages between them on the NoC. Predictable application execution is only possible when embedding follows the resource reservation configuration of the constraint graph. This basically means that sufficient computation resources have to be provided to bind all task clusters as well as sufficient bandwidth on communication resources to route all message clusters with the maximum allowed hop distance. Selection of resources could be done by counting the required number of resources of each

resource type and, then, selecting available resources on the architecture, that is, treating this problem as a knapsack problem as, for example, done by References [59–61]. However, these approaches neglect the routing of messages between the selected resources. Also, restricting the resource selection to resources which lie within a maximal hop distance (as, for example, done by Reference [46]) neglects constrained availability of shared resources as well as resource consumption.

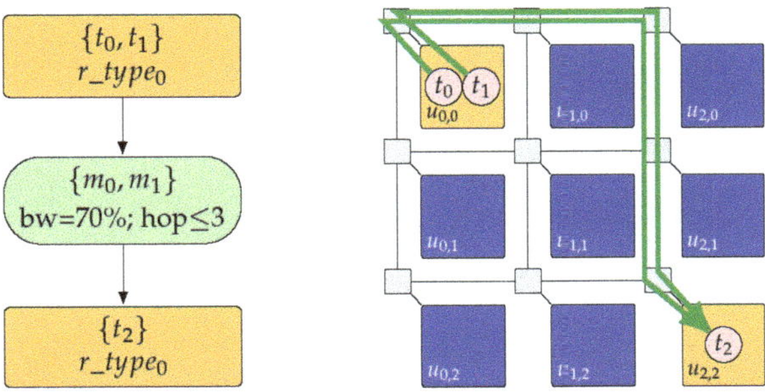

Figure 6. Illustration of a constraint graph which cannot be embedded on a given architecture: Since the constraint graph (**left**) formulates the requirement for a maximum hop distance of 3 for the transfer of m_0 and m_1 between the two task clusters, it is not embeddable on the given architecture (**right**), where the minimum hop distance between the corresponding resource types is 4.

5.3.1. Constraint Satisfaction Problem (CSP)

The above approaches may serve as a preliminary test for deciding whether there exists a feasible embedding of a constraint graph at all, as they have polynomial time complexity and form at least a *necessary condition* for feasibility. However, for determining the actual embedding, all constraints for a feasible binding have to be tested. Therefore, Reference [41] proposes to handle the embedding problem as a *Constraint Satisfaction Problem (CSP)* based on the constraint graph. Generally, a CSP is the problem of finding an assignment for a given set of variables which does not violate a given set of constraints.

The specific problem of *constraint graph embedding* consists of finding a binding for each task cluster of the constraint graph, as well as a routing between the sender and the receiver of each message. For each task cluster, a feasible binding fulfills the following *binding constraints*: (i) The resource type of the selected resource matches the required type annotated to the task cluster. (ii) The target resource provides sufficient capacity for scheduling the tasks in the cluster. For each message cluster, a feasible routing fulfills the following *routing constraints*: (i) The hop distance between the resource of the sending task cluster and the resource of the receiving task cluster is no greater than the hop distance annotated to the message cluster. (ii) Each link along the route provides sufficient capacity to meet the bandwidth requirement of the message cluster.

5.3.2. Constraint Solving Techniques

There exists a smorgasbord of techniques for solving CSPs in general. For the specific problem of constraint graph embedding, two major techniques have been evaluated which are briefly introduced in the following.

Constraint Graph Embedding Using SAT Solvers

The authors of Reference [45] formulate the constraint graph embedding problem as a satisfiability (SAT) problem. Here, the binding and routing constraints are described by a set of linear

Pseudo-Boolean equations over binary decision variables, thus, forming a 0-1 ILP. This formulation is passed to a SAT solver that returns the binding of task clusters and the routing of messages, if existent.

Constraint Graph Embedding Using Backtracking Solvers

The constraint graph embedding problem can also be solved by a backtracking algorithm as initially proposed in Reference [41]. In contrast to SAT solving techniques which work on binary decision variables, backtracking techniques work directly based on an application-specific representation of the problem. They recursively try to find a binding of each task cluster of the constraint graph to a target resource of suitable type in the architecture, while ensuring that a feasible routing between assigned variables remains possible. In case no feasible binding can be determined, a backtracking step from the most recent assignment is performed, and then, it is recursively proceeded until either all task clusters are bound or it has been verified that no embedding exists at all.

The major advantage of backtracking approaches over SAT solving is that application-specific optimizations can be applied as suggested in Reference [62], for example, restricting the resource candidates for binding a task cluster to the hop distance of already mapped connected task clusters as well as executing parallel solvers which start their search in different partitions of the architecture. With such measures, backtracking solvers exhibit better scalability for RPM as they have less memory demands compared to SAT solving techniques and are even able to determine feasible embeddings at run time within a few milliseconds also for systems with more than 100 cores. A run-time management technique to manage the mapping of multiple applications in a dynamic many-core system by applying these backtracking solvers has been proposed in Reference [42].

5.4. Architecture Decomposition for Complexity Reduction

Another natural way to reduce the problem complexity in both the design-time DSE and the run-time management in HAM is a *decomposition* of the input specification. In particular, a decomposition of the target architecture (cf. Figure 7) is well-suited for large-scale many-core architectures since they oftentimes contain multiple instances of the same resource types in a (semi-)regular topology. A careful elimination of available resources from a specification via architecture decomposition significantly reduces the number of mapping possibilities so that speed-ups and quality improvements can be achieved for both the design-time DSE and the run-time embedding in HAM.

5.4.1. Design-Time Decomposition

In the design-time DSE, architecture decomposition can be applied to reduce the size of the search space by eliminating allocatable resources and, consequently, mapping possibilities from the input specification (see Section 3). This allows for a more efficient exploration of the reduced search space and, consequently, results in a better optimization of mapping candidates. A first approach to decompose the architecture is *static decomposition* as proposed in Reference [63]. This variant of architecture decomposition removes a predetermined number of computational resources from the input specification before performing DSE, so that a sub-architecture of predetermined topology and size remains, see, for example, the three statically determined sub-architectures in Figure 7. This approach works especially well for regular many-core architectures, since it can easily be ensured that at least one resource of each required resource type remains in each sub-architecture. By performing the DSE on a number of different sub-architectures—whilst aggregating the results—it can be ensured that a variety of optimized mapping candidates is derived.

The authors of Reference [64] propose a second possibility of architecture decomposition that is better suited for irregular architectural topologies or for cases where an a-priori decision about the number

and type of resource instances to be removed cannot be made. There, a *dynamic decomposition* approach is presented which utilizes information from a short preliminary DSE, that is, a pre-exploration based on the complete architecture, to determine resources to be pruned dynamically for the actual extensive DSE. During the pre-exploration, a *heat map* of the architecture is generated which stores information about resources allocated in high-quality mappings. Low-temperature areas of the heat map, that is, resources *not* part of high-quality mappings, are subsequently pruned from the architecture before the actual DSE is performed. State-of-the-art *data-mining* techniques are demonstrated to be able to extract suitable sub-architectures as well [65]. Similarly to dynamic architecture decomposition using heat maps, data mining is applied during a pre-exploration of the complete architecture. In particular, frequent-itemset mining and emergent-pattern mining are used to determine differences in resources allocated in high- vs. low-quality mappings during the DSE. Based on the obtained results, un-promising areas of the search space can thus be pruned while a reduced sub-architecture is used as input for the main DSE. All approaches discussed above are demonstrated to result in a higher quality of solutions derived by the DSE and reduce the exploration time of DSE significantly for many-core application mapping in the general case but also for constraint graphs in symmetry-eliminating DSE (cf. Section 5.1).

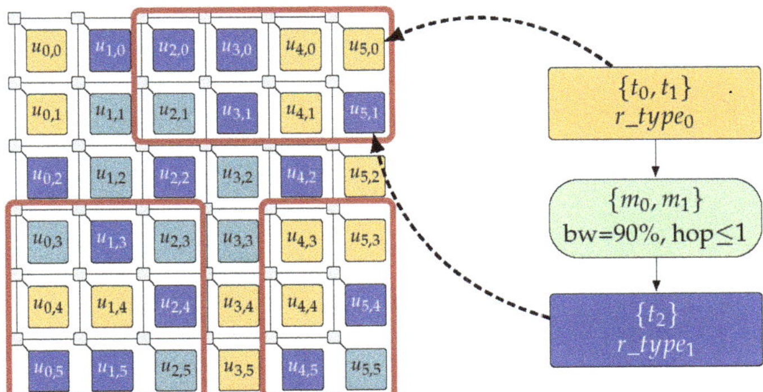

Figure 7. Constrain graph embedding on a 6 × 6 tiled many-core architecture with 3 possible sub-architectures (red boxes) created by *architecture decomposition*. The complexity of the constraint graph embedding problem for both the design-time DSE and the run-time embedding in HAM is significantly reduced by limiting the set of allocatable resources to such decomposed sub-architectures.

As mentioned in Section 5.2, a symmetry-eliminating DSE requires an additional *feasibility check* to guarantee that there exists at least one feasible concrete mapping on the given target architecture [45]. Since the complexity of the NP-complete constraint graph embedding problem grows exponentially with the number of resources in the architecture (see Section 5.3), architecture decomposition is a suitable method to reduce the complexity of such feasibility checks as well. For example, it is shown in Reference [66] how to apply architecture decomposition during feasibility checks by creating a large set of increasingly complex sub-architectures and searching for a feasible embedding on each of them. This achieves noteworthy speed-ups on average, despite the fact that the constraint graph embedding problem must eventually be solved for the complete architecture if no embedding on any generated sub-architecture exists. However, if embedding on a sub-architecture is possible, the embedding time is crucially reduced. Since each and every mapping out of the hundreds of thousands of mapping candidates generated during DSE must undergo this feasibility check, the speed-ups achieved for individual mappings accumulate to a tremendous speed-up of the overall DSE.

5.4.2. Run-Time Decomposition

At run time, the system synthesis problem must be solved to find a feasible mapping of an application on the target architecture which may already be partially occupied by concurrently running applications. The same holds true when using the constraint graph representation for run-time embedding. Architecture decomposition can decrease the embedding time in this scenario as well by limiting the search for a feasible embedding to selected parts of the architecture [62,66]. For both SAT- and backtracking-based formulations of the constraint graph embedding problem (cf. Section 5.3.2) with architecture decomposition, it is furthermore possible to parallelize the solving process by using separate solvers for different decompositions of the architecture and collating the results for even greater embedding speed-ups [62].

5.5. Mapping Distillation

The relatively high run-time overhead of constraint-graph embedding often restricts the number of mapping candidates that can be considered by the RPM. Yet, the offline DSE in HAM often delivers a huge set of Pareto-optimal mappings as it considers many design objectives: On the one hand, mappings are optimized w.r.t. several *quality objectives*, for example, latency, energy, and reliability, subject to the application domain. On the other hand, several *resource-related objectives* are often incorporated to diversify the resource demand of mappings for a better fit in various resource-availability scenarios [41,43,44]. The resulting high-dimensional objective space results in an immense number of Pareto-optimal mappings. Due to timing (and storage) restrictions at run time, only a fraction of these mappings can be provided to the RPM, necessitating the distillation (truncation) of the mappings set [67].

In the domain of multi-objective optimization, the truncation problem is well studied [68], and numerous techniques have been proposed for retaining a representative subset of Pareto-optimal points by maximizing the *diversity* of retained points in the space of design objectives, see, for example, References [69–71]. However, when adopted for mapping distillation, these well established yet generic truncation techniques typically retain mappings which exhibit a prohibitively low embeddability. The main problem here lies in the fact that these techniques regard all design objectives similarly, whereas quality objectives and resource-related objectives are of very different natures: Quality objectives denote *independent* qualities of a mapping where a high *diversity* of retained mappings is desired to offer a representative blend of quality trade-offs. In contrast, resource-related objectives *jointly* affect the embeddability of a mapping and, hence, must be considered collectively during the truncation process where both resource *diversity* and *efficiency* are desired.

In line with these observations, an automatic mapping distillation technique is presented in Reference [67] which operates as follows: The original set of Pareto-optimal mappings is first projected into the space of resource-related objectives where Pareto ranking [72] is used to sort the mappings. Then, the mappings are projected into the space of quality objectives where a grid-based selection scheme is employed to retain mappings from different regions of the quality space (ensuring diverse quality trade-offs) based on the previously computed Pareto ranks (ensuring resource efficiency and diversity). Experimental results in Reference [67] demonstrate that while retaining only a fraction (as few as only 3%) of the original set, this distillation technique highly preserves the embeddability and quality diversity of the original set and outperforms generic truncation techniques substantially.

6. Support for Hard Real-Time Applications

Embedded applications often have a set of non-functional requirements in terms of timing, safety, security, reliability, and so forth. A mapping of such applications is considered useful only if it is verified to satisfy the application's requirement(s). *Real-time applications* which have timing constraints, for example, w.r.t. their latency and/or throughput, are particularly prevalent in embedded systems.

While *soft real-time* applications can tolerate occasional violation of their timing constraints, for a *hard real-time* application, any timing violation can lead to a system failure which is not tolerable. In recent years, the rapid spread of many-core systems in various embedded domains, for example, safety-critical areas of automotive electronics, avionics, telecommunications, medical imaging, consumer electronics, and industrial automation, has led to a significant increase in the number and diversity of embedded applications with hard real-time requirements, see, for example, Reference [73] as an overview.

The hybrid (design-time/run-time) mapping scheme in HAM offers a unique opportunity for supporting hard real-time applications in dynamic embedded systems. As a result, several techniques have been proposed in recent years which enable a predictable and adaptive execution of hard-real time applications in dynamic embedded systems using HAM. In this section, we review a collection of these works after introducing the key system properties necessary for the adoption of HAM for real-time applications.

6.1. Predictability and Timing Composability

Hard real-time applications require *worst-case timing guarantees* to ensure a strict satisfaction of their timing constraints. Deriving temporal guarantees in many-core systems is particularly challenging due to their typically unpredictable execution context: On the one hand, uncertain resource behaviors, for example, (pseudo-)random cache replacement policies, branch prediction, or speculative execution, often lead to intractable variations in the timing behavior of applications such that *useful* worst-case timing guarantees cannot be derived. On the other hand, contention between concurrent applications for accessing shared resources renders the timing behavior of each application dependent not only on the arbitration policy of shared resources but also on the behavior of the concurrent applications. In a dynamic system, this dependence often results in an extensive number of possible execution scenarios which complicates the timing analysis of applications such that even if timing guarantees can be derived, they are typically too loose to be of any practical interest.

In this context, to enable deriving practical (useful) worst-case timing guarantees, two complexity-reducing system properties have been introduced: *predictability* and *timing composability*. Here, *predictability* ensures that each and every resource in the system has a predictable behavior which enables deriving useful bounds on the worst-case timing behavior of applications by means of *formal timing analysis and verification* [48]. *Timing composability*, on the other hand, ensures that concurrent applications are separated and, therefore, cannot affect the (worst-case) timing behavior of one another [48]. In a timing-composable system, resources (or resource budgets) are exclusively assigned per running application so that concurrent applications are temporally and/or spatially isolated from each other. This enables analyzing the worst-case timing behavior of each application based on its reserved resources, regardless of the presence or behavior of other applications in the system. Together, predictability and timing composability serve as the key system properties necessary for enabling an incremental design of systems with real-time applications.

Application mapping in such systems, on the one hand, involves compute-intensive timing analyses to examine the satisfaction of hard real-time constraints of each application. On the other hand, the typically dynamic nature of the application workloads in these systems necessitates workload-adaptive deployment and management of applications. These requirements render HAM methodologies particularly effective as they enable mapping optimization and timing verification for hard real-time applications at design time while empowering adaptive deployment and management of applications at run time. In this context, several HAM techniques have been proposed lately, enabling a predictable and adaptive execution of hard-real time applications in dynamic embedded systems using HAM. A collection of these works is presented in the following.

6.2. Adaptive Inter-Application Isolation and Timing Verification

In a many-core system, timing composability can be established by means of *spatial isolation* and/or *temporal isolation* among concurrent applications where resources (and/or resource budgets) are exclusively reserved for each running application at launch time. The resource reservation policy followed by the applications in a system is specified by the so-called *(inter-application) isolation scheme* selected for that system. Existing many-core systems typically employ one of the three following isolation schemes and a timing analysis tailored to their choice of isolation scheme to derive worst-case timing guarantees for real-time applications: (i) *tile reservation* in which each tile is exclusively reserved to one application, for example, References [41,46,47], (ii) *core reservation* in which each core is exclusively reserved to one application, for example [35], and (iii) *core sharing* in which core budgets are exclusively reserved per application such that a core may be shared among multiple applications. Noteworthy, sharing the NoC can hardly be avoided [73]. The choice of a system's isolation scheme regulates the amount of resources reserved for each application. This not only affects the timing behavior of that application, necessitating a timing analysis *tailored* to the system's isolation scheme to derive worst-case timing guarantees but also has a significant impact on other non-functional qualities, for example, resource utilization and energy efficiency.

A *fixed* isolation scheme imposes a single resource reservation policy on *each and every* application in the system where the amount of resources reserved per application cannot be fine-tuned according to its specific resource demands. Consequently, the majority of hereby obtained mappings either fail to satisfy the timing constraints of the application (common under a core-sharing scheme), or they exhibit an over-provisioning of resources which results in their poor performance w.r.t. other properties, for example, resource utilization and energy efficiency (common under core/tile-reservation schemes).

This issue can be lifted by *exploring the choices of isolation schemes* for each application during its mapping optimization process to find mapping solutions in which the amount of reserved resources is adjusted to the application's demands [51]. The advantage of this practice is exemplified in Figure 8 for an illustrative mapping of an application deployed on two adjacent tiles with a hard deadline of 600 ms. Figure 8c–e correspond to the three cases of fixed isolation schemes introduced above while, in Figure 8f,g, a combination of multiple isolation schemes is used. The resulting latency and resource cost reported below Figure 8c–g denote that the fixed-scheme solution in Figure 8e fails to meet the application's deadline, and those in Figure 8c,d are respectively outperformed by the ones in Figure 8f,g where isolation schemes are used in combination.

Applying isolation schemes in combination requires a timing analysis that is applicable to mappings with a mix of different isolation schemes. To address this, an *isolation-aware timing analysis* is presented in Reference [51] which is applicable to mappings with arbitrary combinations of isolation schemes on different used resources. This analysis captures the interplay between the applied mix of isolation schemes and automatically excludes inter-application timing-interference scenarios that are impossible under the given mix of isolation schemes. Reference [51] then extends the offline DSE of HAM to also perform *isolation-scheme exploration* during mapping optimization. During the DSE, the choice of isolation scheme for each resource (core/tile) is explored, and the worst-case timing behavior of each thereby obtained mappings is analyzed using the aforementioned timing analysis. This approach has been shown to improve the quality of the obtained mappings significantly (up to 67%) compared to classical fixed-scheme approaches [51].

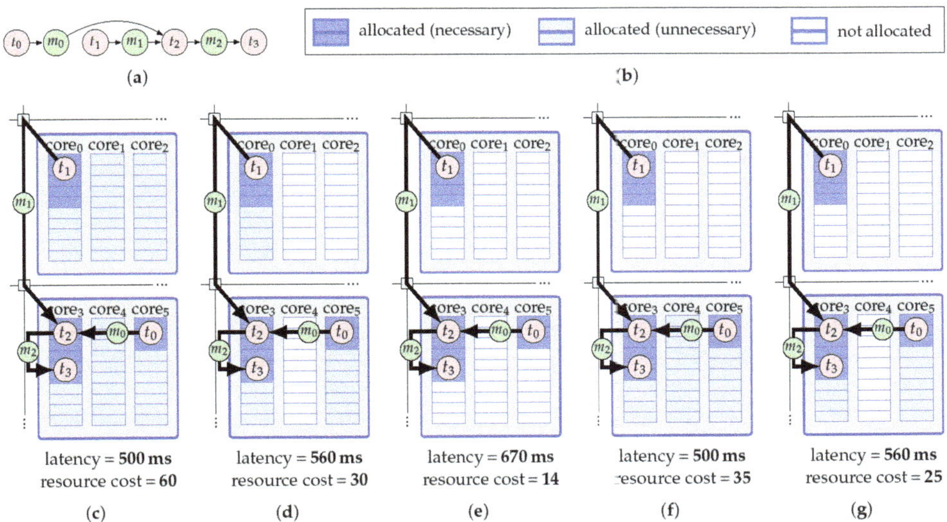

Figure 8. Example of inter-application isolation schemes for a mapping of an application (**a**). The mapping is visualized under five isolation-scheme scenarios: (**c**) fixed tile-reservation isolation scheme; (**d**) fixed core-reservation isolation scheme; (**e**) fixed core-sharing isolation scheme; (**f**) a combination of core sharing (core_0) and tile reservation (bottom tile) isolation schemes; (**g**) a combination of core sharing (core_0) and core reservation (core_3 and core_5) isolation schemes. The resulting latency and resource cost of the mapping under each scheme is given below the respective sub-figure. The description of the color code for the core (compute) budgets in (**c**)–(**g**) is given in (**b**).

6.3. Thermal Safety and Thermal Composability

The dense integration of resources in a many-core chip results in a high density of power consumption on the chip which, in turn, leads to an increased on-chip temperature. Due to their technological limitations, chip packaging and cooling systems often fail to dissipate the generated heat fast enough which may result in overheated regions (so-called hot spots) and even lead to a chip burn-down [74]. To preserve a thermally safe operation, many-core systems employ Dynamic Thermal Management (DTM) schemes which monitor the thermal state of the chip and use mechanisms such as power gating or Dynamic Voltage and Frequency Scaling (DVFS) to prevent or counteract hot spots [75]. Since DTM countermeasures interfere with the execution of applications running in the hot spots, they may lead to the violation of hard real-time constraints which is not acceptable.

In order to preserve both temperature and timing guarantees, it is necessary to ensure the thermal safety of real-time applications *proactively*, for example, using worst-case thermal analysis of their mappings during DSE. Moreover, due to heat transfer between adjacent regions of a chip, the thermal interactions between concurrent applications must also be accounted for to ensure that the thermal behavior of one application will never lead to DTM countermeasures which affect other (possibly real-time) applications. Such an indirect inter-application interference can arise in cases where the scope of DTM countermeasures extends beyond a single core, for example, tile-level or even chip-level DVFS. Moreover, in systems where a core can be shared between multiple applications (see the core-sharing isolation scheme in Section 6.2), such indirect interferences can happen even under core-level DTM countermeasures. To eliminate such

temperature-related inter-application interferences, *thermal composability* must be established and preserved in the system.

Figure 9 illustrates the significance of thermal composability in an example where a new application (gray) is to be launched in a system which is partially occupied by other running applications and is initially in a safe thermal state, see Figure 9a.

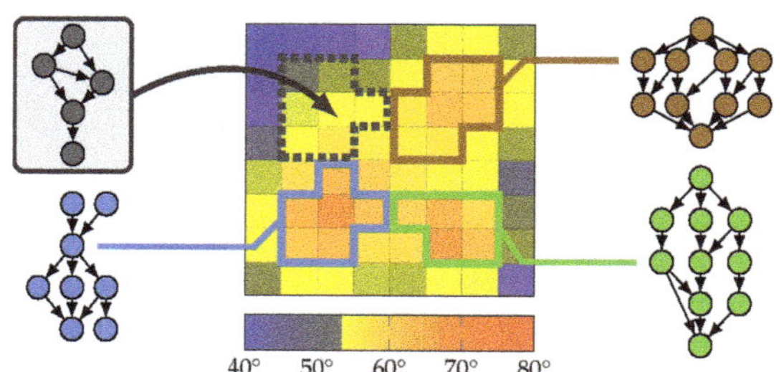

(**a**) thermal state of the chip before the new application (gray) is launched on the dashed region

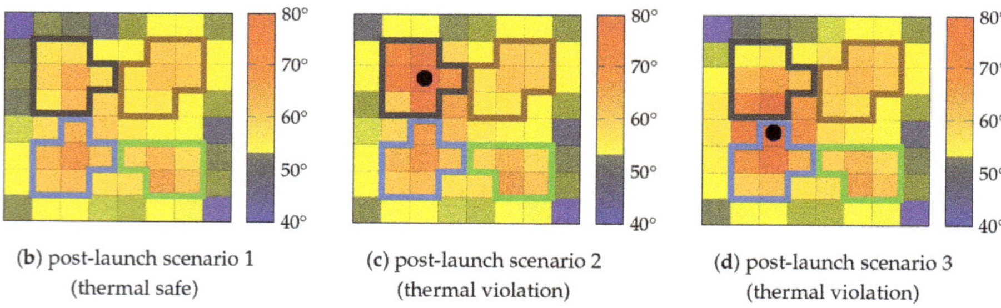

(**b**) post-launch scenario 1 (thermal safe)   (**c**) post-launch scenario 2 (thermal violation)   (**d**) post-launch scenario 3 (thermal violation)

Figure 9. Example of possible thermal scenarios subsequent to launching a new (gray) application. Dots denote overheated cores. Initially, the platform is partly occupied and in a safe thermal state (**a**). Three possible post-launch thermal scenarios are shown: no thermal violations occur (**b**), a core in use by the new application is overheated (**c**), or a core in use by another application is overheated (**d**).

The mappings of all applications are individually verified at design time to be thermally safe. Although in some scenarios, the thermal safety of the system remains unaffected by the launch (e.g., see Figure 9b), subject to the initial thermal state of the system and the thermal behavior of the new application, thermal scenarios may arise in which the heat transfer between cores in use by different applications leads to thermal violations. This can result in two types of dangerous situations: (i) An application can be launched using a mapping that causes thermal violations on one or more cores it uses, see Figure 9c. This triggers DTM countermeasures, for example DVFS, that may affect the execution of this application and may violate its real-time constraints. (ii) Due to heat transfer between adjacent cores, the mapping used to launch an application can affect the temperature profile of the neighboring cores used by other applications and cause a thermal violation there, see Figure 9d. This exposes the applications running on the affected core(s) to DTM countermeasures, though they have not induced the thermal violation in the first place.

Establishing *thermal composability* is a challenging task. Whereas timing composability can be achieved by exclusive resource reservation and/or proper choice of arbitration policies to regulate the timing impact

of concurrent applications on each other, achieving thermal composability is more difficult since heat transfer between neighboring cores used by different applications cannot be anticipated or controlled. To address this issue, Reference [76] presents a HAM approach which establishes thermal composability by introducing a (i) *thermal-safety analysis* to be used offline during/after the DSE and (ii) a set of *thermal-safety admission checks* to be used online by the RPM. There, the offline thermal-safety analysis computes a so-called *Thermally Safe Utilization (TSU)* for each mapping generated by the DSE. The TSU of a mapping denotes the maximum number n of active cores in the system for which that mapping is guaranteed not to lead to any thermal violations. By using the Thermal Safe Power (TSP) analysis from References [77,78], the TSU of each mapping is derived based on its power-density profile and for the worst-possible selection of n active cores (resulting in the highest temperature) so that the thermal-safety guarantee holds for *any* selection of n active cores. At run time, when launching a new application, the RPM uses a set of lightweight thermal-safety admission checks to examine the thermal safety of each mapping candidate for the current system state based on the TSU of that mapping, the TSU of other running applications, and the number of active cores in the system. By avoiding mappings that do not pass these checks, the RPM preserves thermal safety proactively and establishes thermal composability.

While TSU can be calculated for the mappings *after* the offline DSE before they are provided to the RPM, the authors of Reference [76] show that by incorporating TSU as an additional design objective to be maximized *during* the offline mapping optimization process, the DSE will deliver mappings with a higher TSU, meaning that these mappings are thermally safe for a higher number of active cores in the system. Therefore, they can be used in a larger number of system utilization levels, each corresponding to a given number of active cores in the system. This not only enables launching the application in a higher occupation of the system, but it also enhances the flexibility of the RPM as it enlarges the number of admissible mapping options available to it at different system states. This flexibility can be exploited towards secondary goals, for example, load balancing.

6.4. Online Mapping Adaptation with Hard Timing Guarantees

Run-time resource management approaches generally benefit from adapting the mapping of running applications *during* their execution, for example, for load balancing (see, for example, References [14,79]), temperature balancing (see, for example, References [80,81]), or to release the resources that are required for launching a new application. Besides such beneficial but often optional adaptions, in some situations, changing the mapping of a running application becomes inevitable. For instance, due to technology downsizing, many-core systems are subject to an increased rate of temporary/permanent resource failures as a consequence of, for example, overheating or hardware faults. A resource failure necessitates a mapping adaptation for the application(s) that depend on the affected resource in order to preserve their execution. Moreover, the performance requirements of an application may also change dynamically, for example, upon user request, such that in some cases the newly imposed requirements cannot be satisfied by the mapping already in use, thus, necessitating an online adaptation of the application's mapping.

Mapping adaptation involves changing the distribution of an application's task on the platform and is mainly realized by means of *task migration*. The adaptation process typically interferes with the timing behavior of the application and may lead to the violation of its hard real-time constraints which is not acceptable. Therefore, worst-case timing verification of the adaptation process and the post-adaptation mapping becomes necessary to ensure a seamless satisfaction of the real-time constraints. Such a timing verification, however, often relies on compute-intensive timing analyses that are not suitable for online use. In this context, the design-time/run-time scheme of HAM provides a unique opportunity to enable dynamic mapping adaptations with hard real-time guarantees at a negligible run-time compute overhead. In this line, References [82,83] present a methodology for hard real-time mapping adaptation in the form

of a *reconfiguration* between the statically computed mappings of an application. Since in HAM, the set of mappings to be used at run time for each application are computed offline, the timing verification of possible reconfigurations between the mappings can also be performed offline to obtain worst-case timing guarantees for each reconfiguration option. These guarantees can then be provided to the RPM to be used for conducting reconfiguration decisions, hence, eliminating the need for online timing verification.

Mapping reconfiguration between two mappings of an application is illustrated in Figure 10. In each mapping, the dashed red arrows denote the destination tile to which the respective task must be migrated if a reconfiguration to the other mapping is performed. The authors of References [82,83] present a (i) *deterministic reconfiguration mechanism* which enables the RPM to perform each reconfiguration (involving possibly several migrations) predictably so that worst-case reconfiguration latency guarantees can be derived using formal timing analysis. They also present an (ii) *offline reconfiguration analysis* developed based on the proposed reconfiguration mechanism. During the offline analysis, first, efficient migration routes with minimized allocation overhead and migration latency are identified for the migrating tasks of each reconfiguration. Then, the worst-case latency of the whole reconfiguration process is bounded base on the worst-case timing properties of the source and target mappings and the identified migration route for each migrating task. The computed migration routes and timing guarantee of each reconfiguration are then provided to the RPM. At run time, the RPM verifies the real-time conformity of each reconfiguration candidate based on this information, the current timing requirements of the application, and the actual resource availability.

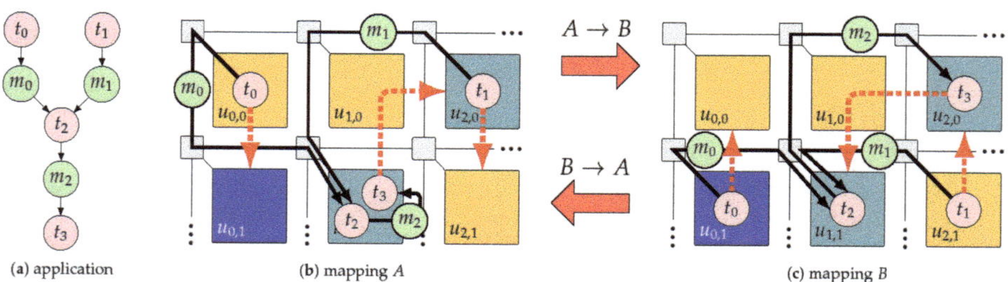

(a) application   (b) mapping *A*   (c) mapping *B*

Figure 10. Example of reconfiguration between two mappings of an application (**a**) on a 2×3 section of a many-core platform. In each mapping, namely *A* (**b**) and *B* (**c**), the red dashed arrows denote migrating tasks and their destination tile for a reconfiguration to the other mapping.

This mapping reconfiguration approach is improved upon in Reference [83]: Generally, a large part of the reconfiguration latency is imposed due to the migration of tasks between tiles over the NoC. Given that the latency analysis of migration routes in a composable system is a lightweight process, in Reference [83], this part of the reconfiguration analysis is postponed to run time where the *actual* NoC load is known. Therefore, instead of relying on pessimistic assumptions about the online NoC load, the actual available bandwidth of the NoC is considered to alleviate the pessimism in the reconfiguration latency guarantee. The resulting reduction in the derived latency bounds renders many reconfiguration options admissible which would have been rejected based on their statically derived latency guarantees.

Recently, it has been demonstrated that in a composable many-core system, task migrations can be performed in such a way that a *lightweight* analysis of worst-case migration latency becomes possible. In this line, the authors of Reference [84] present a (i) *deterministic task migration mechanism* supported by a (ii) *lightweight worst-case timing analysis* which enables on-the-fly timing verification for the migration of any arbitrary subset of an application's tasks. Using this approach, the RPM is able to conduct migration decisions dynamically at run time. Thus, instead of being restricted to a limited set of

reconfiguration options which were pre-explored at design time, the RPM can fine-tune its choice of migrating tasks according to the given situation at run time and verify the admissibility of the migration timing overhead on-the-fly.

7. Upcoming Trends and Future Directions: A Machine Learning-Based Perspective

Machine Learning (ML) techniques have recently gained tremendous attention from both academia and industry and are considered as promising solutions in many application domains. In this section, we discuss how ML techniques can be used to further enhance HAM methodologies. For the sake of brevity, we refrain from discussing approaches which focus on individual components of HAM, for example, the use of ML techniques for guiding the mapping optimizer during the offline DSE [85–88]. Instead, our discussion will be focused on some promising recent approaches which are more specifically tailored to a combination of mapping optimization at design time and dynamic system management at run time. In our following discussion, we categorize the approaches into two groups, see Figure 11: (i) Approaches which focus on learning the *properties* of individual mappings, the platform, or its environment and (ii) approaches which focus on learning the *actions* suited for different run-time conditions.

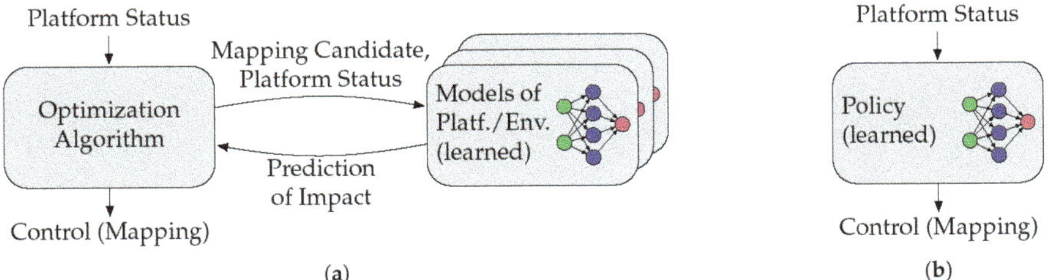

Figure 11. HAM can be supported by Machine Learning (ML)-based techniques in two ways. (**a**) Learning properties of mappings, the platform, or its environment (visualized here for learning platform and environment properties to predict the impact of mapping candidates). (**b**) Directly learning actions, i.e., learning the policy that decides the suitable action for each given situation, e.g., selecting mappings based on current platform status.

7.1. Learning Properties

The majority of challenges in HAM root in the increasing complexity of the *design space* of DSE and the *decision space* of the RPM. The exploration of the design/decision space in quest of near-optimal mappings involves the consideration of a large number of concrete mappings, each of which must be evaluated w.r.t. multiple design objectives. The extent of both spaces depends exponentially on the number of applications and tasks in the system, the number of cores on the platform, the number of possible core configurations (e.g., voltage/frequency-levels), and so forth, leading to a *combinatorial explosion* of the aforementioned spaces. Consequently, exploring the whole design/decision space in its entirety becomes impractical. Instead, a trade-off between the search overhead and the quality of the obtained solutions must be made which depends on factors such as the number of considered design points and the accuracy of their quality evaluation. This trade-off must be tackled differently by the DSE (at design time) and the RPM (at run time).

Traditionally, DSE relies on *accurate methods*, for example, simulation, to evaluate the quality of a mapping. The time required for the evaluation of each mapping can be considerably high and can even become the main timing bottleneck of the entire DSE, dictating whether the DSE is efficient, if at all

feasible. Reducing the complexity of evaluations while maintaining a high accuracy is quite challenging. In contrast to the DSE, the RPM always has a strong requirement for low overhead. Hence, *heuristic* policies with negligible overhead have emerged for online use, for example, policies for maximizing the power budget in a greedy manner [89]. However, such policies may result in a low quality of run-time decisions because heuristic metrics cannot accurately capture complex platform and environment behaviors and interdependencies. At the same time, these heuristics impose only a very low overhead. Hence, a slight increase in their overhead is affordable if this improves the quality of RPM's decisions. This, however, is fairly challenging to achieve. In summary, both DSE and RPM require a flexible trade-off between accuracy and overhead.

ML models built based on *Supervised Learning (SL)* are known for their capability in approximating black-box functions. Thereby, both the achievable prediction accuracy and the prediction overhead depend on the complexity of the chosen prediction model. Importantly, SL models facilitate the exploration of different overhead-accuracy trade-offs, for example, by varying the topological parameters of a Neural Network (NN). This is a valuable property for both the design-time DSE and the run-time management in HAM.

Figure 12 illustrates how SL can enhance the overhead-accuracy trade-off in different steps of HAM. The offline DSE inherently relies on the evaluation of mappings w.r.t. several design objectives. The traditional exact analyses (e.g., using simulation) offer high accuracy, yet suffer from high overhead. Ultimately, this overhead becomes the main timing bottleneck of the DSE. The so-called *surrogate approaches* employ a NN to substitute a time-consuming simulation with a fast quality assessment of mappings at a decreased accuracy, see Figure 12 (top). On the other hand, RPMs traditionally rely on heuristics, which commonly have very low overhead but also abstract from many aspects, resulting in sub-optimal run-time decisions and, hence, limiting the achievable overall system performance. A higher performance can be achieved if the impact of each decision on the system can be assessed with a higher accuracy. ML-based models promise increased accuracy, yet at the cost of an inflated overhead, see Figure 12 (bottom). In the following, both research directions are discussed in more detail.

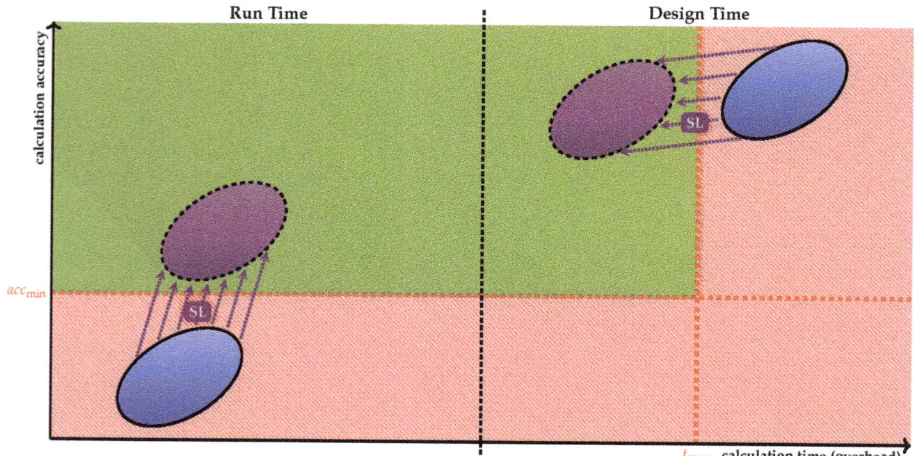

Figure 12. Learning properties using Supervised Learning (SL) can be used to enhance the overhead-accuracy trade-off for both the DSE at design time (**right**) and RPM at run time (**left**) in HAM. The ability of SL in approximating black-box functions enables a higher accuracy of run-time decisions or a faster evaluation of mappings during DSE.

7.1.1. Learning Properties for DSE

As outlined previously, reducing the evaluation time of mappings is an important research objective in the DSE community. One group of approaches which have recently been shown to be particularly effective for this purpose are *surrogate approaches* [90]. These approaches exploit the fact that most optimization techniques used in the context of DSE rely solely on the *relative* quality of each mapping in comparison with other mappings rather than the *exact* (absolute) quality of each mapping. Therefore, they rely on the *fidelity* of the evaluation function rather than its *accuracy*. In this context, surrogate approaches achieve significant evaluation-time reduction by (partially) replacing the computationally intensive exact evaluations with lightweight approximations with acceptable evaluation errors (to establish a high fidelity). The applicability of surrogates depends on the presence of patterns within the evaluation function, which must be detectable with an overhead justified by the speedup achieved through the incorporation of the surrogate method within the DSE. Naturally, their ability to predict/approximate the values of a black-box function based on previous observations makes ML-based approaches—in particular, from the domain of SL—such as linear/polynomial regressors, NNs, or Bayesian approaches, ideal candidates for the creation of surrogates [91–94].

7.1.2. Learning Properties of the Platform and its Environment

As discussed before, the limited quality of RPM decisions is a major restrictive factor of the achieved system performance at run time. In this scope, compared to traditional techniques, ML-based techniques may enhance the trade-off between the quality and the overhead of RPM decisions. One way to achieve this is to use ML-based techniques to learn models that predict the properties of the platform and its environment. These models may be used to predict the impact of a mapping candidate on metrics like power, performance, temperature, and so forth. The input to such a model is the current platform status and some features of the mapping candidate. The platform status also includes relevant features about the characteristics of the current workload. Using the prediction models, the optimization algorithm used by the RPM in its decision processes can consider the impact of many mapping candidates on various system quality metrics.

Models of the properties of the platform and its environment can be built with SL algorithms, where training data is extracted with the help of run-time or design-time profiling. Such techniques have been successfully employed, for example, for deciding task migrations [95,96]. In this scope, Reference [95] uses a lightweight NN to predict the steady-state temperature after a task migration. Reference [96] employs a NN-based model to predict the performance of a task after migrating it to another core. This model takes into account the complex workload-specific dependencies of the performance on average cache latency and power budget. Many migration candidates are assessed based on the performance prediction and the best one is selected for execution.

One advantage of learning properties is its good interpretability compared to learning actions directly. By learning properties, resource-management decisions can easily be understood by designers because the model outputs are physical properties like temperature or performance that are familiar to designers. Furthermore, since the models learn properties of the platform, they generalize to different management strategies. For instance, if a platform has several operation modes (e.g., high performance, low temperature, etc.), the models are valid in all modes and do not need to be retrained. Here, only the optimization algorithm used by the RPM needs to be adapted upon a mode change. A key drawback is that each mapping candidate needs to be assessed individually. This results in a high overhead if the number of potential actions is high [97]. To reduce this overhead, only a limited number of mapping candidates can be assessed. This, however, may result in sub-optimal mappings if mapping candidates are created at run time. HAM offers a potential to mitigate this problem through its offline pre-filtering of the possible

mappings during DSE such that only Pareto-optimal mappings will be provided to the RPM. However, future work is required in this direction. We highlight some future perspectives later in Section 7.3.

7.2. Learning Actions

Existing HAM approaches offer numerous advantages over purely static or purely dynamic application mapping approaches. However, in most existing HAM-based approaches, the offline design step and the online management (decision making) step are only weakly interlinked and still strongly resemble static and dynamic design approaches, respectively. In particular, in HAM, the RPM is provided with a set of Pareto-optimal mappings generated within the offline DSE, however, without receiving any information as to which mappings to use in which run-time situations. The amount of Pareto-optimal mappings can still have a considerable size, especially in cases where abstract design goals have to be transformed into (a large number of) quantifiable objectives. The ensuing necessity to search the decision space consisting of these mappings at run time compromises the responsiveness of the RPM and/or the quality of its decisions. In what follows, we discuss a few directions to address this issue by means of ML-based techniques, namely, *Imitation Learning (IL)* and *Reinforcement Learning (RL)*, which can be applied either at design time or at run time to refine the decision strategy of the RPM.

7.2.1. Imitation Learning (IL)

Imitation Learning (IL) uses Supervised Learning (SL) to construct an oracle for sequential mapping-decision processes. The prerequisite for IL is the availability of labeled training data that resembles the platform status occurring at run time. The training data is created at design time with the help of training benchmarks. Each training sample hereby represents a certain platform status and is labeled with a mapping which is considered optimal for this platform status. This typically involves brute-forcing a large number of mappings to find the optimal one (The optimal mapping for a given platform status can be found, e.g, by an enumeration of the available mappings). Since it is not possible to evaluate every existing mapping combination, only a reduced set of pre-optimized mappings found by the DSE are considered as labels. The RPM learns the actions of choosing mappings at design time and then imitates the actions at run time by adapting them to the given platform status.

The authors of Reference [98] propose a HAM approach that uses IL and incorporates—in addition to the platform status—the influence of input data onto the execution characteristics of the applications into the mapping-inference process. Here, no functional properties of the applications have to be known. To reduce the computation complexity, input data with similar execution properties are clustered into *data scenarios*. This allows for a finer granularity of mapping decisions since the workload dynamism induced by the varying input data can be captured by tailored mappings for each data scenario. The clustering of data into scenarios and the optimization of corresponding mappings are performed at design time using the data-driven scenario-aware DSE approach from Reference [99] based on training data.

This approach entails identifying the best-suited scenario for incoming data at run time before inferring the mapping. However, for complex input data like images, it may not be possible to determine the best scenario prior to processing the data and identifying/observing its execution properties. As a consequence, the scenarios are identified after processing the data based on the evoked execution properties. For that, SL is used where a NN is trained at design time to classify the execution data vectors of the training data depending on the current platform status into the best-suited scenarios. This NN is then used to identify the scenario of incoming input data at run time. The scenario for the next data is afterwards derived from the identified scenarios of the previously processed input data. Here, another model is utilized whose selection algorithm is optimized offline by genetic programming based on the sequence of

training data. If no correlation between subsequent input data is found in the training sequence, a scenario optimized for the average-case data is chosen.

Finally, a mapping is selected from the set of mappings tailored to the chosen scenario depending on the given objectives. It has been shown for a soft real-time setting in Reference [100], that IL with a NN outperforms a statistic-based approach in terms of both the number of deadline misses and the energy consumption. Here, the NN infers the mapping based on a fixed deadline and the history of execution properties of the previously processed input data. The combination of all three models of (i) scenario identification, (ii) scenario selection, and (iii) mapping selection forms the entity of a scenario-based RPM responsible for the online system management of the scenario-aware HAM. The structure of this HAM approach for a single application is shown in Figure 13 differentiating between the offline (design time) phase and the online (run time) phase.

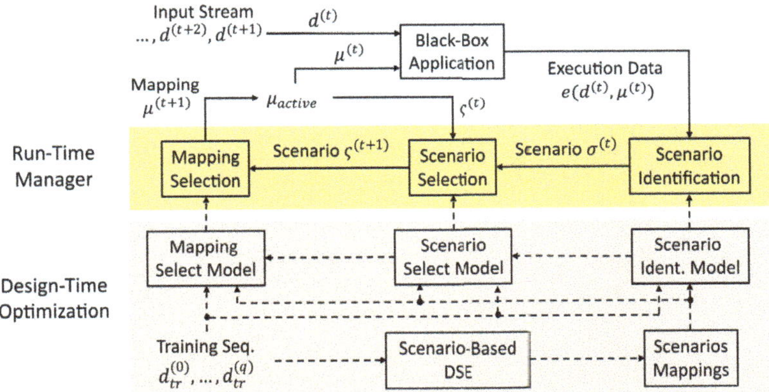

Figure 13. Structure of the data-driven scenario-based HAM. At design time (**bottom**), a DSE optimizes data scenarios and mappings, and ML models are optimized for the mapping inference using training data. At run time (**top**), the RPM uses the optimized models, scenarios, and mappings to select tailored mappings for the application depending on the incoming input data.

In summary, IL can help to tackle the uncertainty of workload distribution at run time by learning patterns in the interplay between input and application characteristics from training data at design time. In the HAM approach above, a mapping is not directly inferred by a single IL model, but instead, by a succession of three separate models specialized on different aspects of mapping selection. This facilitates the training and convergence of the models. Additionally, the whole mapping-decision process becomes more comprehensible which, for example, facilitates the detection of outliers.

7.2.2. Reinforcement Learning (RL)

The majority of existing HAM approaches, as outlined in the previous sections, are established based on a relatively straightforward combination of existing static (design-time) and dynamic (run-time) approaches. A major challenge encountered by these HAM approaches, during both the offline DSE and the online system management steps, is the evaluation of mappings, that is, the estimation of their impact on the overall system performance. In a *static system*, where exactly one mapping is used throughout the entire lifetime of the system, mappings can be evaluated purely based on their non-functional characteristics, for example, energy consumption, which are easy to quantify. The performance of a *dynamic system*,

on the other hand, is not determined based on the quality of a single mapping, but instead, based on the entire decision *strategy* of the RPM, that is, the set of available mappings and the rules dictating which mapping to use in which run-time situations. The usefulness of a single mapping can, therefore, (i) only be evaluated when taking into account the other mappings and the decision rules of the RPM's strategy and (ii) in some cases, may only become apparent after a prolonged time interval. The design and decision approaches used in most existing HAM solutions are not capable of (efficiently) considering these complex interrelations between individual mappings and the overall system performance, which significantly impairs their ability to generate RPM strategies.

In the domain of ML, *Reinforcement Learning (RL)* approaches have been specifically designed to generate sophisticated behavior strategies to address various conditions in complex dynamic environments. In particular, these approaches are designed to consider the non-trivial long-term effects of the chosen actions. In the following, we discuss opportunities to adopt RL techniques in HAM.

Reinforcement Learning (RL) at Run Time

When used at run time, RL offers the opportunity to incorporate learning capabilities into the decision process of the RPM in a dynamic system. By observing the effects of the mappings it selects in particular run-time situations, the RPM can, over time, estimate the *utility* of each available policy, denoting the performance impact of a particular choice of mapping in a particular situation. This ability enables the RPM to (i) generate a system management strategy which is precisely tailored to the observed online conditions and (ii) adapt the strategy in the case of (unforeseen) changes in the conditions. For an example of an adaptable system applying RL at run time, see Reference [101].

While RL approaches are capable of dynamically generating a suitable strategy for (previously unknown) dynamic conditions, the time (in terms of the number of mapping selection actions) until a high-quality strategy is found scales with the number of possible conditions and the number of possible actions and, hence, can become prohibitively long. Furthermore, the learning process typically involves a trial-and-error phase, during which the RPM is likely to take undesirable—or, in the case of safety-critical systems, even dangerous—actions. The long adaptation times and the necessity of an unstable exploration phase have, for a long time, been the main impediments to the application of RL in the area of (safety-critical) real-time embedded systems. With its capability to significantly reduce the decision space of the RPM—and, thereby, also the size of the state space of any RL algorithm used therein—HAM offers a unique opportunity to overcome these weaknesses and unleash the potential of run-time RL techniques for dynamically adapted systems.

Reinforcement Learning (RL) at Design Time

Most existing design approaches utilizing RL techniques use them exclusively for the run-time adaptation of the system. However, an integration of RL techniques into the offline design phase of HAM offers several advantages. Extending static optimizers with the concepts of long-term utility and a cooperative usage of the mappings can address the previously outlined weaknesses of HAM w.r.t. the evaluation of individual mappings. Furthermore, considering (parts of) the interactions between the system and its environment (which imposes run-time conditions onto the system) at design time enables the offline optimization of not only the mappings, but also their activation conditions. In such a scheme, the RPM is provided with a decision strategy which is pre-optimized for the (statically known portion of the) environment characteristics, so that a run-time learning phase can be significantly shortened, carried out within safe action bounds, or even completely avoided.

Based on these ideas, the authors of Reference [102] present a novel optimization framework, LOCAL, which is specifically tailored for HAM in adaptable systems. In its structure, LOCAL is inspired by Learning

Classifier Systems (LCSs) [103], optimization frameworks developed specifically for complex and dynamic problems. Instead of individual mappings, LOCAL co-optimizes a set of rule-action tuples, so-called *policies*, where the action corresponds to a concrete mapping and the rule describes the run-time conditions for the usage of this mapping. By emulating the iterative interactions between the adaptive system and its environment, LOCAL considers (i) the available information about the dynamic external events expected at run time, (ii) the way in which the adaptive system influences its environment, and (iii) the long-term utility of individual policies (rule-action tuples). Within each interaction with the environment (corresponding to new conditions imposed by the environment), LOCAL generates/optimizes policies in a *constrained DSE*, which is restricted to explore only the search space of mappings that are applicable under the given conditions. With this search-space reduction, LOCAL is capable of generating mappings which are exactly tailored to realistic run-time situations.

In Reference [102], LOCAL is used to optimize the embeddability of applications in a many-core system featuring a dynamic launch and termination of applications. It has been shown that, compared to the mapping set generated by a static optimizer, the strategy generated by LOCAL offers a higher embeddability, while containing a significantly smaller number of mappings.

7.3. Future Perspectives

While the approaches presented above already demonstrate the great potential of using ML techniques in HAM, there are still several open questions demanding further research and investigation. In the following, we discuss two particularly interesting research directions, both centered on the way that the design approach addresses the availability of information at design time.

7.3.1. Integration of Expert Knowledge

The first research direction focuses on the exploitation of problem-specific knowledge available at design time. Similarly to any other optimization process, the effectiveness of both offline and online steps of HAM is directly influenced by the amount of the injected problem-specific knowledge. Particularly, in embedded domains, where a large amount of experience and expertise has been accumulated over decades, the development of efficient ways to incorporate this information into the design process is of paramount importance. Techniques gathering information about different application classes or their launch patterns (see Sections 7.2.1 and 7.2.2) constitute the first steps in this direction. However, the integration of more complex knowledge such as known property dependencies (e.g., the fact that power increases monotonically with the voltage/frequency level) or functional application knowledge remains challenging. This applies in particular to optimization approaches comprising multiple interdependent phases (e.g., the scenario-based RPM presented in Section 7.2.1) and complex ML approaches such as NNs which, while offering multiple different points for the injection of expert knowledge [97], are especially difficult to customize and to interpret.

7.3.2. Balancing Offline and Online Learning

The second research direction which we would like to highlight focuses on finding the best way to address uncertainties in the run-time conditions. Thanks to the great versatility of ML approaches, uncertainties in the run-time conditions of the designed system can be addressed during either the offline or the online step of HAM. This introduces an interesting trade-off: On the one hand, the time and compute power available at design time can be used to train a complex model, enabling the system to react to any possible outcome of the uncertainty at hand. On the other hand, transferring the learning process to run time may enable the creation of a lightweight model which fits the actually observed situations. While a more lightweight model is likely to result in a faster and more energy-efficient

inference, unexpected changes in run-time conditions may render it useless or necessitate a retraining phase. The development of HAM approaches which are capable of automatically finding the optimal balance between offline and online learning—which is highly problem-dependent and has a strong impact on the efficiency and reliability of the designed system—constitutes one of the most challenging and promising directions for future research.

8. Conclusions

This paper provides an overview of Hybrid Application Mapping (HAM) as a promising approach for mapping emerging applications to embedded many-core systems. We introduced the fundamentals of HAM and the major design challenges addressed by HAM. An elaborate discussion of the new challenges encountered by HAM was presented together with an overview of a collection of state-of-the-art techniques proposed to address these challenges. The paper also outlined a series of open topics and challenges in HAM, presented a summary of some emerging machine-learning-based directions for addressing these challenges, and highlighted possible future directions.

Author Contributions: Conceptualization, methodology, investigation, visualization, and writing–review and editing: B.P., M.G., J.H., H.K., M.R., V.R., T.S., F.S., J.S., J.T., A.W., and S.W.; writing–original draft preparation: B.P., M.G., H.K., M.R., V.R., T.S., F.S., J.S., A.W., and S.W.; supervision, project administration, and funding acquisition: M.G., J.H., J.T., and S.W. All authors have read and agreed to the published version of the manuscript.

Funding: This work is funded by the Deutsche Forschungsgemeinschaft (DFG, German Research Foundation)—Project Number 146371743-TRR 89 Invasive Computing.

Conflicts of Interest: The authors declare no conflict of interest. The funders had no role in the design of the study; in the collection, analyses, or interpretation of data; in the writing of the manuscript, or in the decision to publish the results.

References

1. Tilera Corporation. *Tile Processor Architecture Overview for the TILE-Gx Series*; Tilera Corporation: San Jose, CA, USA, 2012.
2. De Dinechin, B.D.; Ayrignac, R.; Beaucamps, P.E.; Couvert, P.; Ganne, B.; de Massas, P.G.; Jacquet, F.; Jones, S.; Chaisemartin, N.M.; Riss, F.; et al. A clustered manycore processor architecture for embedded and accelerated applications. In Proceedings of the IEEE Conference on High Performance Extreme Computing (HPEC), Waltham, MA, USA, 10–12 September 2013; pp. 1–6.
3. Howard, J.; Dighe, S.; Hoskote, Y.; Vangal, S.; Finan, D.; Ruhl, G.; Jenkins, D.; Wilson, H.; Borkar, N.; Schrom, G.; et al. A 48-core IA-32 message-passing processor with DVFS in 45 nm CMOS. In Proceedings of the International Solid-State Circuits Conference (ISSCC), San Francisco, CA, USA, 7–11 February 2010; pp. 108–109.
4. Bohnenstiehl, B.; Stillmaker, A.; Pimentel, J.; Andreas, T.; Liu, B.; Tran, A.; Adeagbo, E.; Baas, B. A 5.8 pJ/Op 115 billion ops/sec, to 1.78 trillion ops/sec 32nm 1000-processor array. In Proceedings of the IEEE Symposium on VLSI Circuits (VLSI-Circuits), Honolulu, HI, USA, 15–17 June 2016; pp. 1–2.
5. Kovač, M.; Notton, P.; Hofman, D.; Knezović, J. How Europe is preparing its core solution for exascale machines and a global, sovereign, advanced computing platform. *Math. Comput. Appl.* **2020**, *25*, 46. [CrossRef]
6. Teich, J.; Henkel, J.; Herkersdorf, A.; Schmitt-Landsiedel, D.; Schröder-Preikschat, W.; Snelting, G. Invasive computing: An overview. In *Multiprocessor System-on-Chip*; Springer: New York, NY, USA, 2011; pp. 241–268.
7. Irza, J.; Doerr, M.; Solka, M. A third generation many-core processor for secure embedded computing systems. In Proceedings of the IEEE Conference on High Performance Extreme Computing (HPEC), Waltham, MA, USA, 12–14 September 2012; pp. 1–3.

8. Exynos 5 Octa (5422). 2019. Available online: http://www.samsung.com/exynos/ (accessed on 12 November 2020).

9. Hansen, L. Unleash the Unparalleled Power and Flexibility of Zynq UltraScale+ MPSoCs. *Cell* **2015**, *2*, 3–4.

10. Kissler, D.; Hannig, F.; Kupriyanov, A.; Teich, J. A highly parameterizable parallel processor array architecture. In Proceedings of the International Conference on Field Programmable Technology (FPT), Bangkok, Thailand, 13–15 December 2006; pp. 105–112.

11. Hannig, F.; Lari, V.; Boppu, S.; Tanase, A.; Reiche, O. Invasive tightly-coupled processor arrays: A domain-specific architecture/compiler co-design approach. *ACM Trans. Embed. Comput. Syst.* **2014**, *13*, 1–29. [CrossRef]

12. Oliver, N.; Sharma, R.R.; Chang, S.; Chitlur, B.; Garcia, E.; Grecco, J.; Grier, A.; Ijih, N.; Liu, Y.; Marolia, P.; et al. A reconfigurable computing system based on a cache-coherent fabric. In Proceedings of the International Conference on Reconfigurable Computing and FPGAs (ReConFig), Cancun, Mexico, 30 November–2 December 2011; pp. 80–85.

13. Martin, G. Overview of the MPSoC design challenge. In Proceedings of the Design Automation Conference (DAC), San Francisco, CA, USA, 24–28 July 2006; pp. 274–279.

14. Kobbe, S.; Bauer, L.; Lohmann, D.; Schröder-Preikschat, W.; Henkel, J. DistRM: Distributed resource management for on-chip many-core systems. In Proceedings of the International Conference on Hardware/Software Codesign and System Synthesis (CODES + ISSS), Taipei, Taiwan, 9–14 October 2011; pp. 119–128.

15. Blickle, T.; Teich, J.; Thiele, L. System-level synthesis using evolutionary algorithms. *Des. Autom. Embed. Syst.* **1998**, *3*, 23–58. [CrossRef]

16. Davis, L. *Handbook of Genetic Algorithms*; VNR Computer Library, Van Nostrand Reinhold: New York, NY, USA, 1991.

17. Fonseca, C.M.; Fleming, P.J. An overview of evolutionary algorithms in multiobjective optimization. *Evol. Comput.* **1995**, *3*, 1–16. [CrossRef]

18. Fonseca, C.M.; Fleming, P.J. Genetic algorithms for multiobjective optimization: Formulation, discussion and generalization. In Proceedings of the International Conference on Genetic Algorithms, Urbana-Champaign, IL, USA, 17–21 July 1993; Morgan Kaufmann Publishers Inc.: San Francisco, CA, USA, 1993; pp. 416–423.

19. Kennedy, J. Particle swarm optimization. In *Encyclopedia of Machine Learning*; Springer: Boston, MA, USA, 2010; pp. 760–766.

20. Marwedel, P. *Embedded System Design*; Springer: Dortrecht, The Netherlands, 2006; Volume 1.

21. Teich, J. Hardware/software codesign: The past, the present, and predicting the future. *Proc. IEEE* **2012**, *100*, 1411–1430. [CrossRef]

22. Hansson, A.; Coenen, M.; Goossens, K. Undisrupted quality-of-service during reconfiguration of multiple applications in networks on chip. In Proceedings of the Design, Automation and Test in Europe Conference and Exhibition (DATE), Nice, France, 16–20 April 2007; pp. 1–6.

23. Quan, W.; Pimentel, A.D. A hybrid task mapping algorithm for heterogeneous MPSoCs. *ACM Trans. Embed. Comput. Syst.* **2015**, *14*, 1–25. [CrossRef]

24. Akesson, B.; Molnos, A.; Hansson, A.; Angelo, J.A.; Goossens, K. Composability and predictability for independent application development, verification, and execution. In *Multiprocessor System-on-Chip*; Springer: New York, NY, USA, 2011; pp. 25–56.

25. Singh, A.K.; Shafique, M.; Kumar, A.; Henkel, J. Mapping on multi/many-core systems: Survey of current and emerging trends. In Proceedings of the Design Automation Conference (DAC), Austin, TX, USA, 2–6 June 2013; pp. 1–10.

26. Chou, C.L.; Ogras, U.Y.; Marculescu, R. Energy-and performance-aware incremental mapping for networks on chip with multiple voltage levels. *IEEE Trans. Comput.-Aided Des. Integr. Circuits Syst.* **2008**, *27*, 1866–1879. [CrossRef]

27. Teich, J.; Glaß, M.; Roloff, S.; Schröder-Preikschat, W.; Snelting, G.; Weichslgartner, A.; Wildermann, S. Language and compilation of parallel programs for *-Predictable MPSoC execution using Invasive Computing. In Proceedings of the International Symposium on Embedded Multicore/Many-core Systems-on-Chip (MCSOC), Lyon, France, 21–23 September 2016; pp. 313–320.

28. Wildermann, S.; Bader, M.; Bauer, L.; Damschen, M.; Gabriel, D.; Gerndt, M.; Glaß, M.; Henkel, J.; Paul, J.; Pöppl, A.; et al. Invasive computing for timing-predictable stream processing on MPSoCs. *IT-Inf. Technol.* **2016**, *58*, 267–280. [CrossRef]

29. Bonfietti, A.; Benini, L.; Lombardi, M.; Milano, M. An efficient and complete approach for throughput-maximal SDF allocation and scheduling on multi-core platforms. In Proceedings of the Design, Automation and Test in Europe Conference and Exhibition (DATE), Dresden, Germany, 8–12 March 2010; pp. 897–902.

30. Hu, J.; Marculescu, R. Energy-and performance-aware mapping for regular NoC architectures. *IEEE Trans. Comput.-Aided Des. Integr. Circuits Syst.* **2005**, *24*, 551–562.

31. Marcon, C.; Borin, A.; Susin, A.; Carro, L.; Wagner, F. Time and energy efficient mapping of embedded applications onto NoCs. In Proceedings of the Asia and South Pacific Design Automation Conference (ASP-DAC), Shanghai, China, 18–21 January 2005; pp. 33–38.

32. Lin, L.Y.; Wang, C.Y.; Huang, P.J.; Chou, C.C.; Jou, J.Y. Communication-driven task binding for multiprocessor with latency insensitive network-on-chip. In Proceedings of the Asia and South Pacific Design Automation Conference (ASP-DAC), Shanghai, China, 18–21 January 2005; pp. 39–44.

33. Hu, J.; Marculescu, R. Energy-aware mapping for tile-based NoC architectures under performance constraints. In Proceedings of the Asia and South Pacific Design Automation Conference (ASP-DAC), Kitakyushu, Japan, 21–24 January 2003; pp. 233–239.

34. Kirkpatrick, S.; Gelatt, C.D.; Vecchi, M.P. Optimization by simulated annealing. *Science* **1983**, *220*, 671–680. [CrossRef] [PubMed]

35. Mariani, G.; Avasare, P.; Vanmeerbeeck, G.; Ykman-Couvreur, C.; Palermo, G.; Silvano, C.; Zaccaria, V. An industrial design space exploration framework for supporting run-time resource management on multi-core systems. In Proceedings of the Design, Automation and Test in Europe Conference and Exhibition (DATE), Dresden, Germany, 8–12 March 2010; pp. 196–201.

36. Giannopoulou, G.; Stoimenov, N.; Huang, P.; Thiele, L.; de Dinechin, B.D. Mixed-criticality scheduling on cluster-based manycores with shared communication and storage resources. *Real-Time Syst.* **2016**, *52*, 399–449. [CrossRef]

37. Sahu, P.K.; Shah, T.; Manna, K.; Chattopadhyay, S. Application mapping onto mesh-based network-on-chip using discrete particle swarm optimization. *IEEE Trans. Large Scale Integr. Syst.* **2014**, *22*, 300–312. [CrossRef]

38. Singh, A.K.; Dziurzanski, P.; Mendis, H.R.; Indrusiak, L.S. A survey and comparative study of hard and soft real-time dynamic resource allocation strategies for multi-/many-core systems. *ACM Comput. Surv.* **2017**, *50*, 1–40. [CrossRef]

39. Al Faruque, M.A.; Krist, R.; Henkel, J. ADAM: Run-time agent-based distributed application mapping for on-chip communication. In Proceedings of the Design Automation Conference (DAC), Anaheim, CA, USA, 8–13 June 2008; pp. 760–765.

40. Schranzhofer, A.; Chen, J.J.; Thiele, L. Power-aware mapping of probabilistic applications onto heterogeneous MPSoC platforms. In Proceedings of the IEEE Real-Time and Embedded Technology and Applications Symposium, San Francisco, CA, USA, 13–16 April 2009; pp. 151–160.

41. Weichslgartner, A.; Gangadharan, D.; Wildermann, S.; Glaß, M.; Teich, J. DAARM: Design-time application analysis and run-time mapping for predictable execution in many-core systems. In Proceedings of the International Conference on Hardware/Software Codesign and System Synthesis (CODES+ISSS), Uttar Pradesh, India, 12–17 October 2014; pp. 1–10.

42. Wildermann, S.; Weichslgartner, A.; Teich, J. Design methodology and run-time management for predictable many-core systems. In Proceedings of the IEEE International Symposium on Object/Component/Service-Oriented Real-Time Distributed Computing Workshops, Auckland, New Zealand, 13–17 April 2015; pp. 103–110.

43. Weichslgartner, A.; Wildermann, S.; Gangadharan, D.; Glaß, M.; Teich, J. A design-time/run-time application mapping methodology for predictable execution time in MPSoCs. *ACM Trans. Embed. Comput. Syst.* **2018**, *17*, 1–25. [CrossRef]

44. Weichslgartner, A.; Wildermann, S.; Glaß, M.; Teich, J. *Invasive Computing for Mapping Parallel Programs to Many-Core Architectures*; Springer: Singapore, 2018.
45. Schwarzer, T.; Weichslgartner, A.; Glaß, M.; Wildermann, S.; Brand, P.; Teich, J. Symmetry-eliminating design space exploration for hybrid application mapping on many-core architectures. *IEEE Trans. Comput.-Aided Des. Integr. Circuits Syst.* **2017**, *37*, 297–310. [CrossRef]
46. Singh, A.K.; Kumar, A.; Srikanthan, T. Accelerating throughput-aware runtime mapping for heterogeneous MPSoCs. *ACM Trans. Des. Autom. Electron. Syst.* **2013**, *18*, 1–29. [CrossRef]
47. Khanh, P.N.; Singh, A.K.; Kumar, A.; Aung, K.M.M. Incorporating energy and throughput awareness in design space exploration and run-time mapping for heterogeneous MPSoCs. In Proceedings of the Euromicro Conference on Digital System Design (DSD), Los Alamitos, CA, USA, 4–6 September 2013; pp. 513–521.
48. Kopetz, H. *Real-Time Systems: Design Principles for Distributed Embedded Applications*, 2nd ed.; Springer: New York, NY, USA, 2011.
49. Hansson, A.; Goossens, K.; Bekooij, M.; Huisken, J. CoMPSoC: A template for composable and predictable multi-processor system on chips. *ACM Trans. Des. Autom. Electron. Syst.* **2009**, *14*, 1–24. [CrossRef]
50. Schoeberl, M.; Abbaspour, S.; Akesson, B.; Audsley, N.; Capasso, R.; Garside, J.; Goossens, K.; Goossens, S.; Hansen, S.; Heckmann, R.; et al. T-CREST: Time-predictable multi-core architecture for embedded systems. *J. Syst. Archit.* **2015**, *61*, 449–471. [CrossRef]
51. Pourmohseni, B.; Smirnov, F.; Wildermann, S.; Teich, J. Isolation-aware timing analysis and design space exploration for predictable and composable many-core systems. In Proceedings of the Euromicro Conference on Real-Time Systems (ECRTS), Stuttgart, Germany, 9–12 July 2019; Volume 12, pp. 1–24.
52. Weichslgartner, A.; Wildermann, S.; Götzfried, J.; Freiling, F.; Glaß, M.; Teich, J. Design-time/run-time mapping of security-critical applications in heterogeneous MPSoCs. In Proceedings of the International Workshop on Software and Compilers for Embedded Systems (SCOPES), Sankt Goar, Germany, 23–25 May 2016; pp. 153–162.
53. Heisswolf, J.; König, R.; Kupper, M.; Becker, J. Providing multiple hard latency and throughput guarantees for packet switching networks on chip. *Comput. Electr. Eng.* **2013**, *39*, 2603–2622. [CrossRef]
54. Borkar, S. Thousand core chips: A technology perspective. In Proceedings of the Design Automation Conference (DAC), San Diego, CA, USA, 4–8 June 2007; pp. 746–749.
55. Madalozzo, G.; Duenha, L.; Azevedo, R.; Moraes, F.G. Scalability evaluation in many-core systems due to the memory organization. In Proceedings of the IEEE International Conference on Electronics, Circuits and Systems (ICECS), Monte Carlo, Monaco, 11–14 December 2016; pp. 396–399.
56. Smirnov, F.; Pourmohseni, B.; Glaß, M.; Teich, J. Variety-aware routing encoding for efficient design space exploration of automotive communication networks. In Proceedings of the 5th International Conference on Vehicle Technology and Intelligent Transport Systems, Crete, Greece, 3–5 May 2019; Volume 1, pp. 242–253.
57. Lukasiewycz, M.; Glaß, M.; Haubelt, C.; Teich, J. SAT-decoding in evolutionary algorithms for discrete constrained optimization problems. In Proceedings of the IEEE Congress on Evolutionary Computation (CEC), Singapore, 25–28 September 2007; pp. 935–942.
58. Glaß, M.; Teich, J.; Lukasiewycz, M.; Reimann, F. Hybrid optimization techniques for system-level design space exploration. In *Handbook of Hardware/Software Codesign*; Ha, S., Teich, J., Eds.; Springer: Dortrecht, The Netherlands, 2017; pp. 217–246.
59. Wildermann, S.; Glaß, M.; Teich, J. Multi-objective distributed run-time resource management for many-cores. In Proceedings of the Design, Automation and Test in Europe Conference and Exhibition (DATE), Dresden, Germany, 24–28 March 2014; pp. 1–6.
60. Bini, E.; Buttazzo, G.; Eker, J.; Schorr, S.; Guerra, R.; Fohler, G.; Arzen, K.E.; Segovia, V.R.; Scordino, C. Resource management on multicore systems: The ACTORS approach. *IEEE Micro* **2011**, *31*, 72–81. [CrossRef]
61. Ykman-Couvreur, C.; Nollet, V.; Catthoor, F.; Corporaal, H. Fast multi-dimension multi-choice knapsack heuristic for MP-SoC run-time management. In Proceedings of the International Symposium on System-on-Chip, Tampere, Finland, 13–16 November 2006; pp. 1–4.

62. Schwarzer, T.; Roloff, S.; Richthammer, V.; Khaldi, R.; Wildermann, S.; Glaß, M.; Teich, J. On the complexity of mapping feasibility in many-core architectures. In Proceedings of the International Symposium on Embedded Multicore/Many-core Systems-on-Chip (MCSoC), Hanoi, Vietnam, 12–14 September 2018; pp. 176–183.

63. Richthammer, V.; Glaß, M. On search-space restriction for design space exploration of multi-/many-core systems. In *Workshop "Methoden und Beschreigungssprachen zur Modellierung und Verifikation von Schaltungen und Systemen" (MBMV)*; University of Tübingen: Tübingen, Germany, 13–14 May 2018.

64. Richthammer, V.; Fassnacht, F.; Glaß, M. Search-Space decomposition for system-level design space exploration of embedded systems. *ACM Trans. Des. Autom. Electron. Syst.* **2020**, *25*, 1–32. [CrossRef]

65. Richthammer, V.; Scheinert, T.; Glaß, M. Data mining in system-level design space exploration of embedded systems. In Proceedings of the International Conference on Embedded Computer Systems (SAMOS), Samos, Greece, 5–9 July 2020.

66. Richthammer, V.; Schwarzer, T.; Wildermann, S.; Teich, J.; Glaß, M. Architecture decomposition in system synthesis of heterogeneous many-core systems. In Proceedings of the Design Automation Conference (DAC), San Francisco, CA, USA, 24–28 June 2018; pp. 1–6.

67. Pourmohseni, B.; Glaß, M.; Teich, J. Automatic operating point distillation for hybrid mapping methodologies. In Proceedings of the Design, Automation and Test in Europe Conference and Exhibition (DATE), Lausanne, Switzerland, 27–31 March 2017; pp. 1135–1140.

68. Vachhani, V.L.; Dabhi, V.K.; Prajapati, H.B. Survey of multi objective evolutionary algorithms. In Proceedings of the International Conference on Circuit, Power and Computing Technologies (ICCPCT), Nagercoil, India, 19–20 March 2015; pp. 1–9.

69. Knowles, J.; Corne, D. Properties of an adaptive archiving algorithm for storing nondominated vectors. *IEEE Trans. Evol. Comput.* **2003**, *7*, 100–116. [CrossRef]

70. Zitzler, E.; Laumanns, M.; Thiele, L. SPEA2: Improving the strength Pareto evolutionary algorithm. *Eurogen* **2001**, *3242*, 95–100.

71. Deb, K.; Pratap, A.; Agarwal, S.; Meyarivan, T. A fast and elitist multiobjective genetic algorithm: NSGA-II. *IEEE Trans. Evol. Comput.* **2002**, *6*, 182–197. [CrossRef]

72. Abdou, W.; Bloch, C.; Charlet, D.; Spies, F. Multi-Pareto-ranking evolutionary algorithm. In Proceedings of the European Conference Evolutionary Computation in Combinatorial Optimization, Malaga, Spain, 11–13 April 2012; pp. 194–205.

73. Mitra, T.; Teich, J.; Thiele, L. Time-critical systems design: A survey. *IEEE Des. Test* **2018**, *35*, 8–26. [CrossRef]

74. Li, M.; Liu, W.; Yang, L.; Chen, P.; Chen, C. Chip temperature optimization for dark silicon many-core systems. *IEEE Trans. Comput.-Aided Des. Integr. Circuits Syst.* **2017**, *37*, 941–953. [CrossRef]

75. Hankendi, C.; Coskun, A.K. Scale & Cap: Scaling-aware resource management for consolidated multi-threaded applications. *ACM Trans. Des. Autom. Electron. Syst.* **2017**, *22*, 30.

76. Pourmohseni, B.; Smirnov, F.; Khdr, H.; Wildermann, S.; Teich, J.; Henkel, J. Thermally Composable Hybrid Application Mapping for Real-Time Applications in Heterogeneous Many-Core Systems. In Proceedings of the IEEE Real-Time Systems Symposium (RTSS), Hong Kong, China, 3–6 December 2019; pp. 220–232.

77. Pagani, S.; Khdr, H.; Chen, J.J.; Shafique, M.; Li, M.; Henkel, J. Thermal safe power (TSP): Efficient power budgeting for heterogeneous manycore systems in dark silicon. *IEEE Trans. Comput.* **2016**, *66*, 147–162. [CrossRef]

78. Khdr, H.; Pagani, S.; Sousa, E.; Lari, V.; Pathania, A.; Hannig, F.; Shafique, M.; Teich, J.; Henkel, J. Power density-aware resource management for heterogeneous tiled multicores. *IEEE Trans. Comput.* **2017**, *66*, 488–501. [CrossRef]

79. Fu, W.; Chen, T.; Wang, C.; Liu, L. Optimizing memory access traffic via runtime thread migration for on-chip distributed memory systems. *J. Supercomput.* **2014**, *69*, 1491–1516. [CrossRef]

80. Ge, Y.; Malani, P.; Qiu, Q. Distributed task migration for thermal management in many-core systems. In Proceedings of the Design Automation Conference (DAC), Anaheim, CA, USA, 13–18 June 2010; pp. 579–584.

81. Liu, Z.; Tan, S.X.D.; Huang, X.; Wang, H. Task migrations for distributed thermal management considering transient effects. *IEEE Trans. Large Scale Integr. Syst.* **2015**, *23*, 397–401.

82. Pourmohseni, B.; Wildermann, S.; Glaß, M.; Teich, J. Predictable run-time mapping reconfiguration for real-time applications on many-core systems. In Proceedings of the International Conference on Real-Time Networks and Systems (RTNS), Grenoble, France, 4–6 October 2017; pp. 148–157.

83. Pourmohseni, B.; Wildermann, S.; Glaß, M.; Teich, J. Hard real-time application mapping reconfiguration for NoC-based many-core systems. *Real-Time Syst.* **2019**, *55*, 433–469. [CrossRef]

84. Pourmohseni, B.; Smirnov, F.; Wildermann, S.; Teich, J. Real-time task migration for dynamic resource management in many-core systems. In *Workshop on Next Generation Real-Time Embedded Systems (NG-RES)*; Schloss Dagstuhl–Leibniz-Zentrum fuer Informatik: Dagstuhl, Germany, 2020; Volume 77, pp. 1–14.

85. Deshwal, A.; Jayakodi, N.K.; Joardar, B.K.; Doppa, J.R.; Pande, P.P. MOOS: A multi-objective design space exploration and optimization framework for NoC enabled manycore systems. *ACM Trans. Embed. Comput. Syst.* **2019**, *18*, 1–23. [CrossRef]

86. Smirnov, F.; Pourmohseni, B.; Glaß, M.; Teich, J. IGOR, get me the optimum! prioritizing important design decisions during the DSE of embedded systems. *ACM Trans. Embed. Comput. Syst.* **2019**, *18*, 1–22. [CrossRef]

87. Sagawa, M.; Aguirre, H.; Daolio, F.; Liefooghe, A.; Derbel, B.; Verel, S.; Tanaka, K. Learning variable importance to guide recombination on many-objective optimization. In Proceedings of the IIAI International Congress on Advanced Applied Informatics (IIAI-AAI), Hamamatsu, Japan, 9–13 July 2017; pp. 874–879.

88. Ito, T.; Aguirre, H.; Tanaka, K.; Liefooghe, A.; Derbel, B.; Verel, S. Estimating relevance of variables for effective recombination. In *International Conference on Evolutionary Multi-Criterion Optimization*; Springer International Publishing: Cham, Switzerland, 2019; pp. 411–423.

89. Rapp, M.; Sagi, M.; Pathania, A.; Herkersdorf, A.; Henkel, J. Power-and cache-aware task mapping with dynamic power budgeting for many-cores. *IEEE Trans. Comput.* **2020**, *69*, 1–13 [CrossRef]

90. Jin, Y. A comprehensive survey of fitness approximation in evolutionary computation. *Soft Comput.* **2005**, *9*, 3–12. [CrossRef]

91. Liu, H.Y.; Carloni, L.P. On learning-based methods for design-space exploration with high-level synthesis. In Proceedings of the Design Automation Conference (DAC), Austin, TX, USA, 2–6 June 2013; p. 50.

92. Beltrame, G.; Fossati, L.; Sciuto, D. Decision-theoretic design space exploration of multiprocessor platforms. *IEEE Trans. Comput.-Aided Des. Integr. Circuits Syst.* **2010**, *29*, 1083–1095. [CrossRef]

93. Wang, H.; Jin, Y.; Sun, C.; Doherty, J. Offline data-driven evolutionary optimization using selective surrogate ensembles. *IEEE Trans. Evol. Comput.* **2018**, *23*, 203–216. [CrossRef]

94. Wang, H.; Jin, Y.; Doherty, J. Committee-based active learning for surrogate-assisted particle swarm optimization of expensive problems. *IEEE Trans. Cybern.* **2017**, *47*, 2664–2677. [CrossRef]

95. Ge, Y.; Qiu, Q.; Wu, Q. A multi-agent framework for thermal aware task migration in many-core systems. *IEEE Trans. Large Scale Integr. Syst.* **2011**, *20*, 1758–1771. [CrossRef]

96. Rapp, M.; Pathania, A.; Mitra, T.; Henkel, J. Prediction-based task migration on S-NUCA many-cores. In Proceedings of Design, Automation and Test in Europe Conference and Exhibition (DATE), Florence, Italy, 25–29 March 2019.

97. Rapp, M.; Amrouch, H.; Wolf, M.C.; Henkel, J. Machine learning techniques to support many-core resource management: Challenges and opportunities. In Proceedings of the Workshop on Machine Learning for CAD (MLCAD), Canmore, AB, Canada, 3–4 September 2019.

98. Spieck, J.; Wildermann, S.; Teich, J. Run-time scenario-based MPSoC mapping reconfiguration using machine learning models. In Proceedings of the Workshop on Machine Learning for CAD (MLCAD), Canmore, AB, Canada, 3–4 September 2019.

99. Spieck, J.; Wildermann, S.; Schwarzer, T.; Teich, J.; Glaß, M. Data-driven scenario-based application mapping for heterogeneous many-Core systems. In Proceedings of the International Symposium on Embedded Multicore/Many-Core Systems-on-Chip (MCSoC), Singapore, 1–4 October 2019.

100. Spieck, J.; Wildermann, S.; Teich, J. Scenario-based soft real-time hybrid application mapping for MPSoCs. In Proceedings of the Design Automation Conference (DAC), San Francisco, CA, USA, 20–24 July 2020.

J. Low Power Electron. Appl. **2020**, *10*, 38

101. Zeppenfeld, J.; Bouajila, A.; Stechele, W.; Herkersdorf, A. Learning classifier tables for autonomic systems on chip. In *INFORMATIK 2008. Beherrschbare Systeme-dank Informatik. Band 2*; Gesellschaft für Informatik e.V.: Bonn, Germany, 2008.
102. Smirnov, F.; Pourmohseni, B.; Teich, J. Using learning classifier systems for the DSE of adaptive embedded systems. In Proceedings of the Design, Automation and Test in Europe Conference and Exhibition (DATE), Grenoble, France, 9–13 March 2020; pp. 957–962.
103. Urbanowicz, R.J.; Moore, J.H. Learning classifier systems: A complete introduction, review, and roadmap. *J. Artif. Evol. Appl.* **2009**. [CrossRef]

Publisher's Note: MDPI stays neutral with regard to jurisdictional claims in published maps and institutional affiliations.

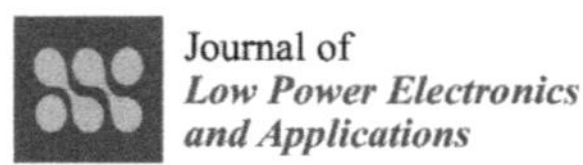

Journal of
*Low Power Electronics
and Applications*

Review

A Survey of Resource Management for Processing-In-Memory and Near-Memory Processing Architectures

Kamil Khan, Sudeep Pasricha and Ryan Gary Kim *

Department of Electrical and Computer Engineering, Colorado State University, Fort Collins, CO 80523, USA;
Kamil.Khan@colostate.edu (K.K.); Sudeep.Pasricha@colostate.edu (S.P.)
* Correspondence: Ryan.G.Kim@colostate.edu

Received: 13 August 2020; Accepted: 17 September 2020; Published: 24 September 2020

Abstract: Due to the amount of data involved in emerging deep learning and big data applications, operations related to data movement have quickly become a bottleneck. Data-centric computing (DCC), as enabled by processing-in-memory (PIM) and near-memory processing (NMP) paradigms, aims to accelerate these types of applications by moving the computation closer to the data. Over the past few years, researchers have proposed various memory architectures that enable DCC systems, such as logic layers in 3D-stacked memories or charge-sharing-based bitwise operations in dynamic random-access memory (DRAM). However, application-specific memory access patterns, power and thermal concerns, memory technology limitations, and inconsistent performance gains complicate the offloading of computation in DCC systems. Therefore, designing intelligent resource management techniques for computation offloading is vital for leveraging the potential offered by this new paradigm. In this article, we survey the major trends in managing PIM and NMP-based DCC systems and provide a review of the landscape of resource management techniques employed by system designers for such systems. Additionally, we discuss the future challenges and opportunities in DCC management.

Keywords: processing-in-memory; near-memory processing; resource management; code annotation; compiler optimizations; online heuristics; energy efficiency; 3D-stacked memories; non-volatile memories

---

1. Introduction

For the past few decades, memory performance improvements have lagged behind compute performance improvements, creating an increasing mismatch between the time to transfer data and the time to perform computations on these data (the "memory wall"). The emergence of applications that focus on processing large amounts of data, such as deep machine learning and bioinformatics, have further exacerbated this problem. It is evident that the large latencies and energies involved with moving data to the processor will present an overwhelming bottleneck in future systems. To address this issue, researchers have proposed to reduce these costly data movements by introducing data-centric computing (DCC), where some of the computations are moved in proximity to the memory architecture.

Two major paradigms of DCC have emerged in recent years: processing-in-memory (PIM) and near-memory processing (NMP). In PIM architectures, characteristics of the memory are exploited and/or small circuits are added to memory cells to perform computations. For example, [1] takes advantage of dynamic random-access memory's (DRAM) charge sharing property to perform bitwise PIM operations (e.g., AND and OR) by activating multiple rows simultaneously. These PIM operations allow computations to be done on memory where they are stored, thus eliminating most of the data movement. On the other hand, NMP architectures take advantage of existing compute substrates and integrate compute cores near the memory module. For example, modern 3D-stacked DRAM

includes a logic layer where a compute core can be integrated beneath multiple DRAM layers within the same chip [2,3]. Although the computation is carried out a little further away from memory than in PIM systems, NMP still significantly improves the latency, bandwidth, and energy consumption when compared to conventional computing architectures. For example, a commercial NMP chip called UPMEM implemented a custom core into conventional DRAM memory chips and achieved 25× better performance for genomic applications and 10× better energy consumption in an Intel x86 server compared to an Intel x86 server without UPMEM [4].

Both PIM and NMP systems have the potential to speed up application execution by reducing data movements. However, not all instructions can be simply offloaded onto the memory processor. Many PIM systems leverage memory characteristics to enable bulk bitwise operations, but other types of operations cannot be directly mapped onto the in-memory compute fabric. Even in NMP systems that utilize cores with full instruction set architecture (ISA) support, performing computation in the 3D memory stack can create high power densities and thermal challenges. In addition, if some of the data have high locality and reuse, the main processor can exploit the traditional cache hierarchies and outperform PIM/NMP systems for instructions that operate on these data. In this case, the downside of moving data is offset by the higher performance of the main processor. All of these issues make it difficult to decide which computations should be offloaded and make use of these PIM/NMP systems.

In this article, we survey the landscape of different resource management techniques that decide which computations are offloaded onto the PIM/NMP systems. These management techniques broadly rely on code annotation (programmers select the sections of code to be offloaded), compiler optimization (compiler analysis of the code), and online heuristics (rule-based online decisions). Before providing a detailed discussion of resource management for PIM/NMP systems, Section 2 discusses prior surveys in PIM/NMP. Section 3 discusses various PIM/NMP design considerations. Section 4 discusses the optimization objectives and knobs, as well as different resource management techniques utilized to manage the variety of PIM/NMP systems proposed to date. Lastly, Section 5 concludes with a discussion of challenges and future directions.

2. Prior Surveys and Scope

Different aspects of DCC systems have been covered by other surveys. Siegl et al. [5] focus on the historical evolution of DCC systems from minimally changed DRAM chips to advanced 3D-stacked chips with multiple processing elements (PEs). The authors identify the prominent drivers for this change: firstly, memory, bandwidth, and power limitations in the age of growing big data workloads make a strong case for utilizing DCC. Secondly, emerging memory technologies enable DCC; for example, 3D stacking technology allows embedding PEs closer to memory chips than ever before. The authors identify several challenges with both PIM and NMP systems such as programmability, processing speed, upgradability, and commercial feasibility.

Singh et al. [6] focus on the classification of published work based on the type of memory, PEs, interoperability, and applications. They divide their discussion into RAM and storage-based memory. The two categories are further divided based on the type of PE used, such as fixed-function, reconfigurable, and fully programmable units. The survey identifies cache coherence, virtual memory support, unified programming model, and efficient data mapping as contemporary challenges. Similarly, Ghose et al. [7] classify published work based on the level of modifications introduced into the memory chip. The two categories discussed are 2D DRAM chips with minimal changes and 3D DRAM with one or more PEs on the logic layer.

Gui et al. [8] focus on DCC in the context of graph accelerators. Apart from a general discussion on graph accelerators, the survey deals with graph accelerators in memory and compares the benefits of such systems against traditional graph accelerators that use field programmable gate arrays (FPGA) or application-specific integrated circuits (ASIC). It argues that memory-based accelerators can achieve higher bandwidth and low latency, but the performance of a graph accelerator also relies on other architectural choices and workloads.

In addition to DRAM, emerging memory technologies have been used to implement DCC architectures. Umesh et al. [9] survey the use of spintronic memory technology to design basic logic gates and arithmetic units for use in PIM operations. The literature is classified based on the type of operations performed, such as Boolean operations, addition, and multiplication. An overview of application-specific architectures is also included. Lastly, the survey highlights the higher latency and write energy with respect to static random-access memory (SRAM) and DRAM as the major challenges in large-scale adoption. Similarly, Mittal et al. [10] discuss resistive RAM (ReRAM)-based implementations of neuromorphic and PIM logical and arithmetic units.

The surveys discussed above adopt different viewpoints in presenting prominent work in the DCC domain. However, all of these surveys limit their discussion to the architecture and/or application of DCC systems and lack a discussion on the management techniques of such systems. With an increase in the architectural design complexity, we believe that the management techniques for DCC systems have become an important research area. Understandably, researchers have begun to focus on how to optimally manage DCC systems. In this survey, unlike prior surveys, we therefore focus on exploring the landscape of management techniques that control the allocation of resources according to the optimization objectives and constraints in a DCC system. We hope that this survey provides the background and the spark needed for innovations in this critical component of DCC systems.

3. Data-Centric Computing Architectures

As many modern big data applications require us to process massive datasets, large volumes of data are shared between the processor and memory subsystems. The data are too large to fit in on-chip cache hierarchies; therefore, the resulting off-chip data movement between processors (CPUs, GPUs, accelerators) and main memory leads to long execution time stalls as main memory is relatively slow to service large influx of requests. The growing disparity in memory and processor performance is referred to as the memory wall. Data movement also consumes a significant amount of energy in the system, driving the system closer to its power wall. As data processing requirements continue to increase, the cost due to each wall will make conventional computing paradigms incapable of meeting application quality-of-service goals. To alleviate this bottleneck, PEs near or within memory have emerged as a means to perform computation and reduce data movement between traditional processors and main memory. Predominantly, prior work in this domain can be classified depending on how PEs are integrated with memory.

For DCC architectures, solutions can be divided into two main categories: (1) PIM systems, which perform computations using special circuitry inside the memory module or by taking advantage of particular aspects of the memory itself, e.g., simultaneous activation of multiple DRAM rows for logical operations [1,11–25]; (2) NMP systems, which perform computations on a PE placed close to the memory module, e.g., CPU or GPU cores placed on the logic layer of 3D-stacked memory [26–42]. For the purposes of this survey, we classify systems that use logic layers in 3D-stacked memories as NMP systems, as these logic layers are essentially computational cores that are near the memory stack (directly underneath it).

3.1. Processing-In-Memory (PIM) Designs

PIM solutions proposed to date typically leverage the high internal bandwidth of DRAM dual in-line memory modules (DIMM) to accelerate computation by modifying the architecture or operation of DRAM chips to implement computations within the memory. Beyond DRAM, researchers have also demonstrated similar PIM capabilities by leveraging the unique properties of emerging non-volatile memory (NVM) technologies such as resistive RAM (ReRAM), spintronic memory, and phase-change memory (PCM). The specific implementation approaches vary widely depending on the type of memory technology used; however, the modifications are generally minimal to preserve the original function of the memory unit and meet area constraints. For instance, some DRAM-based solutions minimally change the DRAM cell architecture while relying heavily on altering memory commands

from the memory controller to enable functions such as copying a row of cells to another [16], logical operations such as AND, OR, and NOT [18,22,24], and arithmetic operations such as addition and multiplication [1,23,43]. Both DRAM and NVM-based architectures have demonstrated promising improvements for graph and database workloads [1,18,24,44] by accelerating search and update operations directly where the data are stored. In the next sections, we will briefly discuss different PIM architectures that use DRAM and NVM.

3.1.1. PIM Using DRAM

The earliest PIM implementations [11,14,15] integrated logic within DRAM. The computational capability of these in-memory accelerators can range from simply copying a DRAM row [16] to performing bulk bitwise logical operations completely inside memory [1,18,23,44]. For example, [1,16,23] use a technique called charge sharing to enable bulk AND and OR operations completely inside memory. Charge sharing is performed by the simultaneous activation of three rows called triple row activation (TRA). Two of the three rows hold the operands while the third row is initialized to zeros for a bulk AND operation or to ones for a bulk OR operation. Figure 1 shows an example TRA where cells A and B correspond to the operands (A is zero and B is one) while cell C is initialized to one to perform an OR operation ❶. The wordlines of all three cells are raised simultaneously ❷, causing the three cells to share their charge on the bitline. Since two of the three cells are in a charged state, this results in an increase in voltage on the bitline. The sense amplifier ❸ then drives the bitline to V_{DD}, which fully charges all three cells, completing the "A OR B" operation. In practice, this operation is carried out on many bitlines simultaneously to perform bulk AND or OR operations. In addition, with minimal changes to the design of the sense amplifier in the DRAM substrate, [1,44] enable NOT operations in memory, which allows the design of more useful combinational logic for arithmetic operations like addition and multiplication. Besides bitwise operations, DRAM PIM has been shown to significantly improve neural network computation inside memory. For example, by performing operations commonly found in convolutional network networks like the multiply-and-accumulate operation in memory, DRAM PIM can achieve significant speedup over conventional architectures [45].

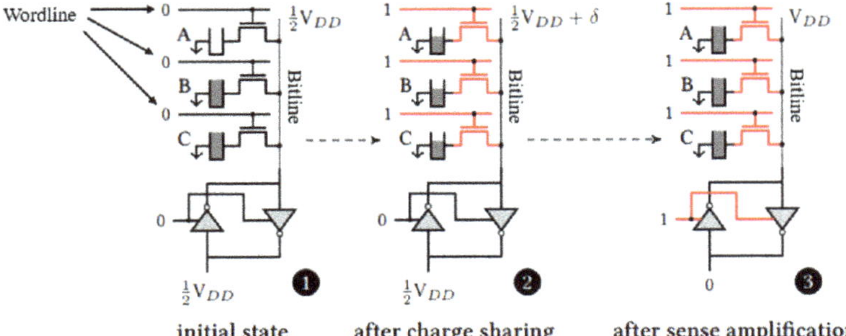

Figure 1. An example of an OR operation being performed on DRAM cells A and B using charge-sharing in triple-row activation [1].

In order to extract even more performance improvements, [12] places single instruction, multiple data (SIMD) PEs adjacent to the sense amplifiers at the cost of higher area and power per bit of memory. The inclusion of SIMD PEs inside the memory chip enables DRAM PIM to take advantage of the high internal DRAM bandwidth and minimizes the need for data to traverse power hungry off-chip links while enabling more complex computations than those afforded by TRA-based operations.

While PIM operations have been shown to outperform the host CPU (or GPU) execution, there are some technical challenges with their implementation. To enable these operations in DRAM, PIM modifications to DRAM chips are often made at the bank or sub-array level. If the source rows

containing the operands do not share the same sensing circuitry, large amounts of data may need to be internally copied between banks, increasing latency and energy consumption, necessitating the use of partitioning and data-mapping techniques. In addition, DRAM PIM operations that use TRA are normally destructive, requiring several bandwidth consuming row copy operations if the data need to be maintained [1,16]. Alternative ways to accomplish computation inside DRAM chips include using a combination of multiplexers and 3T1C (three transistors, one capacitor) DRAM cells [46,47]. While 3T1C cells allow non-destructive reads, these significantly increase the area overhead of the PIM system compared to the single transistor counterparts [1,16,44]. Moreover, DRAM is typically designed using a high threshold voltage process, which slows down PIM logic operations when PIM PEs are fabricated on the same chip [48]. These factors have resulted in relatively simple DRAM PIM designs that prevent entire applications from running entirely in memory.

3.1.2. PIM Using NVM

Charge-storage-based memory (DRAM) has been encountering challenges in efficiently storing and accurately reading data as transistor sizes shrink and operating frequencies increase. DRAM performance and energy consumption have not scaled proportionally with transistor sizes like they did for processors. Moreover, the charge storage nature of DRAM cells requires that data be periodically refreshed. Such refresh operations lead to higher memory access latency and energy consumption. Kim et al. [49] have also demonstrated that contemporary DRAM designs suffer from problems such as susceptibility to RowHammer attacks [50], which exploit the limitations of charge-based memory to induce bit-flip errors. Solutions to overcome such attacks can further increase execution time and reduce the energy efficiency of DRAM PIM designs.

On the other hand, alternative NVM memory technologies such as phase-change memory (PCM) [51–54], ReRAM [22,24,55–59], and spintronic RAM [60–66] show promise. NVM eliminates the reliance on charge memory and represents the data as cell resistance values instead. When NVM cells are read, cell resistances are compared to reference values to ascertain the value of the stored bit, i.e., if the measured resistance is within a preset range of low resistance values (R_{low}), the cell value read is a logical "1", whereas a cell resistance value within the range of high resistance values (R_{high}) is read as a logical "0". Writing data involves driving an electric current through the memory cell to change its physical properties, i.e., the material phase of the crystal in PCM, the magnetic polarity in spintronic RAM, and the atomic structure in memristors and ReRAM. The data to be stored in all cases are embedded in the resultant resistance of the memory cell, which persists even when the power supply shuts off. Thus, unlike DRAM, such memories store data in a non-volatile manner. In addition to being non-volatile, the unique properties of these memory technologies provide higher density, lower read power, and lower read latency compared to DRAM.

PIM using NVM, while based on a fundamentally different memory technology, can resemble its DRAM counterpart. For example, Li et al. [24] activate multiple rows in a resistive memory sub-array to enable bitwise logical operations. By activating multiple rows, the sense amplifiers measure the bitwise parallel resistance of the two rows on the bitlines and compare the resistance with a preset reference resistance to determine the output. Figure 2a shows how any two cells, R_1 and R_2, on the same bitline connect in parallel to produce a total resistance $R_1 \| R_2$, where $\|$ refers to the parallel resistance of two cells. Figure 2b shows how the parallel resistance formed by input resistances (R_{low} or R_{high}) measured at the sense amplifier, together with a reference value (R_{refOR}/R_{refAND}), can be used to perform the logical AND and OR operation in memory [24].

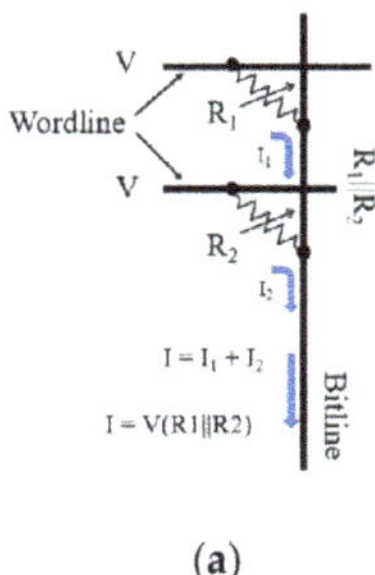

(a)

R_1	R_2	$R_1\|\|R_2$	$R_1\|\|R_2 < R_{refOR}$	$R_1\|\|R_2 < R_{refAND}$
R_{low}	R_{low}	$0.5 \cdot R_{low}$	1	1
R_{low}	R_{low}	$R_{low}\|\|R_{high}$	1	0
R_{high}	R_{low}	$R_{low}\|\|R_{high}$	1	0
R_{high}	R_{high}	$0.5 \cdot R_{high}$	0	0
$(R_{low}\|\|R_{high}) < R_{refOR} < (0.5 \cdot R_{high})$			$(0.5 \cdot R_{low}) < R_{refAND} < (R_{low}\|\|R_{high})$	

(b)

Figure 2. Example of AND and OR operations in memory that use resistance-based memory cells. (a) Activating two resistors R_1 and R_2 on the same bitline in a memristive crossbar array (MCA) results in an effective resistance of $R_1\|R_2$ on the bitline (figure adapted from [56]). This effective resistance can be compared against preset reference resistance values (R_{refOR}, R_{refAND}) to ascertain the result of the Boolean operation; (b) shows the truth table [24].

Aside from the inherent benefits of NVM technology—e.g., energy efficient reads, persistence, and latency—NVM-based PIM allows AND/OR operations on multiple rows as opposed to just two in DRAM. This is because as long as each memory cell has a large range of resistances, resistive memory can have multiple levels of resultant resistances that can represent multiple states. Another technological advantage for NVM PIM is that, unlike DRAM, reads are not destructive. This eliminates the need to copy data to special rows before operating on them, as done in DRAM-based PIM solutions.

Besides conventional logical operations, NVM PIM has also proven particularly useful for accelerating neural network (NN) computations [9,10]. Cell conductance in the rows of memristive crossbar arrays (MCA) can be used to represent synaptic weights and the input feature values can be represented by wordline voltages. Then, the current flowing through each bitline will be related to the dot product of input values and weights in a column. As this input-weight dot product is a key operation for NNs, PIM architects have achieved significant acceleration for NN computation [19] by extending traditional MCA to include in-memory multiply operations. Other applications that have been shown to benefit from NVM PIM include data search operations [67,68] and graph processing [57].

A significant drawback of using NVM for PIM is area overhead. The analog operation of NVM requires the use of DAC (digital-to-analog convertor) and ADC (analog-to-digital converter) interfaces. This results in considerable area overhead. For example, DAC and ADC converters in the implementation of a 4-layer convolutional neural network can take up to 98% of the total area and power of the PIM system [69].

In summary, PIM designs typically allow for very high internal data bandwidth, greater energy efficiency, and low area overhead by utilizing integral memory functions and components. However, to keep memory functions intact, these designs can only afford minimal modifications to the memory system, which typically inhibits the programmability and computational capability of the PEs.

3.2. Near-Memory Processing (NMP) Designs

NMP designs utilize more traditional PEs that are placed near the memory. However, as the computations are not done directly in memory arrays, PEs in NMP do not enjoy the same degree of high internal memory bandwidth available to PIM designs. NMP designs commonly adopt a 3D-stacked structure such as that provided by hybrid memory cube (HMC) [2] or high bandwidth memory (HBM) [3]. Such designs have coarser-grained offloading workloads than PIM but can use the logic layer in 3D stacks to directly perform more complex computations. In the following sections, we discuss the different PE and memory types that have been explored for NMP systems.

3.2.1. PE Types

Many different PE types have been proposed for NMP designs. The selection of PEs has a significant impact on the throughput, power, area, and types of computations that the PE can perform. For example, Figure 3 compares the throughput of different PE types in an HMC-based NMP system executing a graph workload (PageRank scatter kernel). The throughput is normalized to the maximum memory bandwidth, which is indicated by the solid line. This indicates that, while multi-threaded (MT) and SIMD cores allow flexibility, they cannot take advantage of all the available bandwidth. On the other hand, ASICs easily saturate the memory bandwidth and become memory-bound. In this particular situation, reconfigurable logic like FPGA and coarse-grained reconfigurable architecture (CGRA) may be better to balance performance and flexibility. However, in the end, this tradeoff point depends on the application, memory architecture, and available PEs. Therefore, the choice of PE is an important design consideration in NMP systems. We highlight the three types of PEs below.

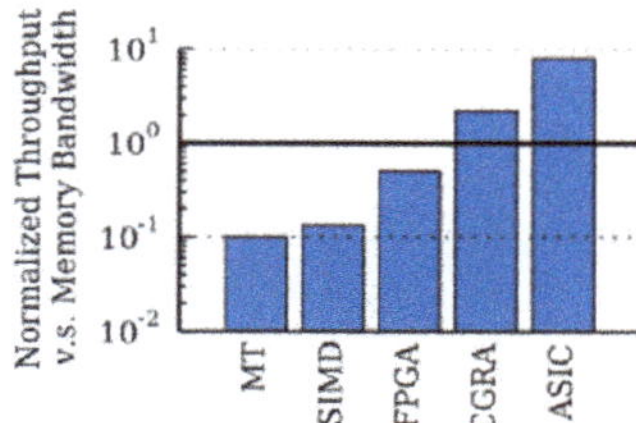

Figure 3. Normalized throughput with respect to maximum memory bandwidth for different PEs executing the PageRank scatter kernel: multi-threaded (MT) programmable core, SIMD programmable core, FPGAs, CGRAs, and fixed-function ASICs [41].

Fixed-function accelerators: Fixed-function accelerator PEs are ASICs designed with support for a reduced set of operations to speed up a particular application or task—for example, graph processing [8,32,57]. These accelerators are highly specialized in their execution, which allows them to achieve much greater throughput than general purpose processors for the same area and power budget. However, it is costly and challenging to customize an accelerator for a target workload. As these accelerators only execute a limited set of instructions, they cannot be easily retargeted to new (or new versions of existing) workloads.

Programmable logic: Programmable logic PEs can include general purpose processor cores such as CPUs [31,40,70–72], GPUs [26,27,38,73,74], and accelerated processing units (APU) [75] that can execute complex workloads. These cores are usually trimmed down (fewer computation units, less complex cache hierarchies without L2/L3 caches, or lower operating frequencies) from their conventional counterparts due to power, area, and thermal constraints. The main advantage of using these general purpose computing cores is flexibility and programmability. As opposed to fixed function accelerators, NMP systems with programmable units allow any operation normally executed on the host processor to potentially be performed near memory. However, this approach is hampered by several issues such as difficulty in maintaining cache coherence [31], implementing virtual address translation [76], and meeting power, thermal, and area constraints [29,73,77–79].

Reconfigurable logic: Reconfigurable logic PEs include CGRAs and FPGAs that can be used to dynamically change the NMP computational unit hardware [21,25,41,80]. In NMP systems, such PEs act as a compromise between the efficiency of simple fixed-function memory accelerators and the flexibility of software-programmable complex general purpose cores. However, the reconfiguration of hardware logic results in runtime overhead and additional system complexity.

3.2.2. Memory Types

The type of memory to use is an important decision for DCC system designers working with NMP systems. Different memory technologies have unique advantages that can be leveraged for particular operations. For example, write-heavy operations are more suited to DRAM than resistive memories due to the larger write latency of resistive memories. This fact and DRAM's widespread adoption has led to the creation of a commercial DRAM NMP solution UPMEM [4]. UPMEM replaces standard DIMMs, allowing it to be easily adopted in datacenters. A 3D-stacked DRAM architecture like the HMC [81] can accelerate atomic operations of the form read-modify-write (RMW) due to the memory level parallelism supported by the multi-vault, multi-layer structure shown in Figure 4. In fact, NMP designs prefer 3D-DRAM over 2D-DRAM due to several reasons. Firstly, 3D-DRAM solutions such as HMC 2.0 [81] have native support for executing simple atomic instructions. Secondly, 3D-DRAM allows for easy integration of a logic die that provides greater area for placing more complex PEs near memory than 2D counterparts. Thirdly, PEs are connected using high bandwidth through-silicon vias (TSVs) to memory arrays as opposed to off-chip links, providing much higher memory bandwidth to the PE. Fourthly, the partitioning of memory arrays into vaults enables superior memory-level parallelism.

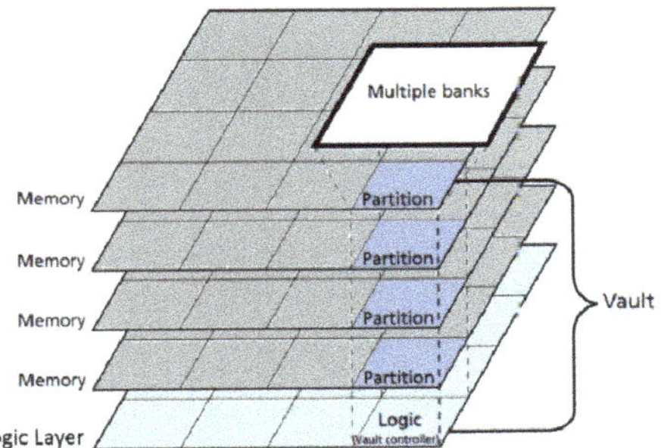

Figure 4. The architecture of hybrid memory cube (HMC) [81].

Resistive memory in its memristive crossbar array (MCA) configuration can significantly accelerate matrix multiply operations commonly found in neural network, graph, and image processing workloads [19,82–89]. Figure 5a shows a mathematical abstraction of one such operation for a single-layer perceptron implementing binary classification, where x are the inputs to the perceptron, w are the weights to the perceptron, sgn is the signum function, and y is the output of the perceptron [90]. The MCA is formed by non-volatile resistive memory cells placed at each crosspoint of a crossbar array. Figure 5b shows how the perceptron can be mapped onto an MCA. The synaptic weights of the perceptron, w, are represented by physical cell conductance values, G. Specifically, each weight value is represented by a pair of cell conductances, i.e., $w_i \propto G_i \equiv G_i^+ - G_i^-$ to represent both positive and negative weight values. The inputs, x, are represented by proportional input voltage values, V, applied to the crossbar columns. For the device shown, $V = \pm 0.2$ V. In this configuration, the weighted sum of the inputs and synaptic weights ($y = \sum_{i=0}^{9} w_i x_i$) can be obtained by reading the current at the each of the two crossbar rows ($I^+ = \sum_{i=0}^{9} G_i^+ V_i$ and $I^- = \sum_{i=0}^{9} G_i^- V_i$). Finally, the sign of the difference in the current flowing through the two rows ($sgn\left[\sum_{i=0}^{9} G_i^+ V_i - \sum_{i=0}^{9} G_i^- V_i\right]$) can be determined, which is equivalent to the original output y. In this manner, MCA is able to perform the matrix multiplication operation in just one step.

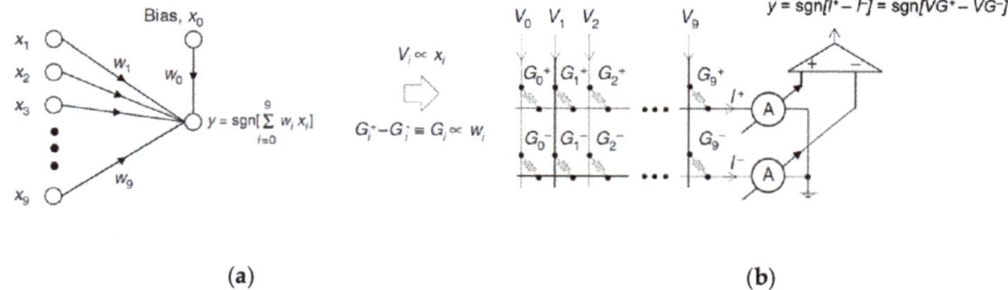

Figure 5. (**a**) A mathematical abstraction of a single-layer perceptron used for binary classification. (**b**) A memristive crossbar array implementing the weighted sum between inputs x_i (mapped to wordline voltages V_i) and weights w_i (mapped to cell conductance G_i) [90].

While the choice of a PE and memory type are important factors in determining the advantages of an NMP system over a PIM system, some features are shared by most NMP designs. Due to the external placement of the PE, NMP can afford a larger area and power budget compared to PIM designs. However, NMP also has lower maximum data bandwidth available to the PE compared to PIM systems.

3.3. Data Offloading Granularity

For both PIM and NMP systems, it is important to determine what computation will be sent (offloaded) to the memory PE. Offloading can be performed at different granularities, e.g., instructions (including small groups of instructions) [1,13,16,19,24,25,28,32,37,39,40,42,57,91,92], threads [71], Nvidia's CUDA blocks/warps [27,29], kernels [26], and applications [38,41,73,74]. Instruction-level offloading is often used with a fixed-function accelerator and PIM systems [1,13,16,19,24,25,28,29,32,37,39,42,57,92]. For example, [42] offloads atomic instructions at instruction-level granularity to a fixed-function near-memory graph accelerator. Coarser offloading granularity, such as kernel and full application, requires more complex memory PEs that can fetch and execute instructions and maintain a program counter. Consequently, coarse-grained offloading is often used in conjunction with programmable memory PEs such as CPUs [71,93] and GPUs [26,27,38,73,74].

4. Resource Management of Data-Centric Computing Systems

Although it would be tempting to offload all instructions to the PIM or NMP system and eliminate most data movements to/from the host processor, there are significant barriers to doing so. Firstly, memory chips are usually resource constrained and cannot be used to perform unrestricted computations without running into power/thermal issues. For example, 3D-DRAM NMP systems often place memory processors in the 3D memory stack that can present significant thermal problems if not properly managed. Figure 6 [29] demonstrates how the peak temperature of a 3D-stacked DRAM with a low-power GPU on the logic layer changes with the frequency of offloaded operations (PIM rate). Consistent operation at 2 op/ns will result in the DRAM module generating a thermal warning at 85 °C. Depending on the DRAM module, this can lead to a complete shutdown or degraded performance under high refresh rates. If the PIM rate is unmanaged, the DRAM cannot guarantee reliable operation due to the high amount of charge leakage under high temperature.

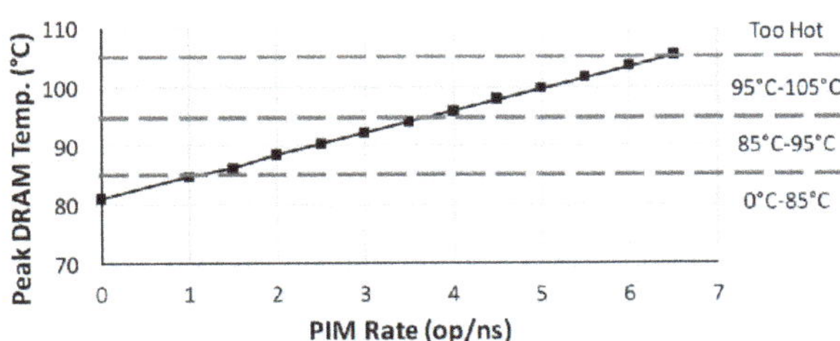

Figure 6. Peak DRAM temperatures at different workload offloading frequencies (referred to as "PIM rate" in the original work) in a 3D-DRAM NMP system [29].

Secondly, memory processors are not guaranteed to improve power and performance for all workloads. For example, workloads that exhibit spatial or temporal locality in their memory accesses are known to perform worse in PIM/NMP systems [26,28]. When executed on a host processor, these workloads can avoid excessive DRAM accesses and utilize the faster, more energy efficient on-chip cache. Only the least cache-friendly portions of an application should be offloaded to the memory processors while the cache-friendly portions are run on the host processor. Figure 7 [28] compares the speedup achieved for a "Host-Only" (entire execution on the host), "PIM-Only" (entire execution on the memory processor), and "Locality-Aware" (offloads low locality instructions) offloading approach to an NMP system for different input sizes. It should be noted that in [28], PIM refers to an HMC-based architecture that we classify as NMP. For small input sizes, all except one application suffers from severe degradation with "PIM-Only". This is because the working set is small enough to fit in the host processor's cache, thus achieving higher performance than memory-side execution. As input sizes grow from small to medium to large, the performance benefit of offloading is realized. Overall, the locality aware management policy proves to be the best of the three approaches and demonstrates the importance of correctly deciding which instructions to offload based on locality. Another useful metric to consider when making offloading decisions is the expected memory bandwidth-saving from the offloading [38,42,94].

Lastly, these DCC systems can include different types of PE and/or multiple PEs with different locations within the memory. This can make it challenging to determine which PE to select when specific instructions need to be offloaded. For example, NMP systems may have multiple PEs placed in different locations in the memory hierarchy [94] or in different vaults in 3D-based memory [26]. This will result in PEs with different memory latency and energy depending on the location of the data accessed by the computation. In these cases, it is vital that the correct instructions are offloaded to the most suitable candidate PE.

To address and mitigate the above issues in DCC systems, offloading management policies are necessary to analyze and identify instructions to offload. Like every management policy, the management of DCC systems can be divided into three main ideas: (1) defining the optimization objectives, (2) identifying optimization knobs or the parts of the system that the policy can alter to achieve its goals, and (3) defining a management policy to make the proper decisions. For the optimization objectives, performance, energy, and thermals have been targeted for PIM and NMP systems. The most common optimization knobs in DCCs include selecting offloading workloads for memory, selecting the most suitable PE in/near memory, or the timing of executing selected offloads. To implement the policy, management techniques have employed code annotation [1,13,16,19,24,25,28,31,32,37,40,57,91,95], compiler-based code analysis [27,39,40,70,92,96], and online heuristics [27–29,38,71,72,74]. Table 1 classifies prominent works based on these attributes. We further discuss the optimization objectives, optimization knobs, and management techniques in Sections 4.1–4.3, respectively.

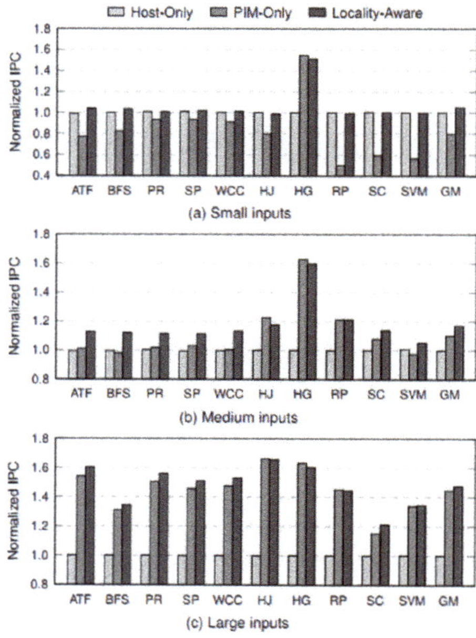

Figure 7. The speedup achieved by a Host-Only, all offloaded (PIM-Only), and low-locality offload (Locality-Aware) approaches in a DRAM NMP system for different application input sizes [28].

Table 1. Classification of resource management techniques for DCC systems (E: energy, P: performance, Pow: power, T: temperature).

Management Method	Objectives	Architecture	Offload Granularity	Work (Year)
Code Annotation	E	NMP	Instruction	[91] (2014)
	P	NMP	Group of instructions	[31] (2017)
		PIM	Instruction	[13] (1995); [24] (2016); [25] (2017)
	P/E	NMP	Instruction	[28] (2015); [32,37] (2017); [40,57] (2018)
		NMP	Group of instructions	[75] (2014); [95] (2016)
		PIM	Instruction	[16] (2013); [19] (2016); [1] (2017)
Compiler	P	NMP	Instruction	[42] (2017); [92] (2019)
		NMP	Group of instructions	[96] (2015); [70] (2017)
	P/E	NMP	Instruction	[40] (2018); [39] (2019)
		NMP	Thread	[27] (2016)
Online Heuristic	P	NMP	Thread	[71] (2018)
	P/E	NMP	Instruction	[28] (2015)
		NMP	Group of instructions	[72] (2020)
		NMP	Thread	[27] (2016); [74] (2019)
		NMP	Application	[26] (2016)
	Pow/P/E	NMP	Application	[38] (2020)
	T	NMP	Instruction	[29] (2018)

4.1. Optimization Objectives

An optimization objective is pivotal to the definition of a resource management policy. Although the direct goal of PIM/NMP systems is to reduce data movement between the host and memory, the optimization objectives are better expressed as performance improvement [1,13,16,19,22,24,25,31, 32,37–41,57,59,71–73,75,92,96], energy efficiency [38,73,91], and thermal feasibility [29] of the system.

4.1.1. Performance

DCC architectures provide excellent opportunities for improving performance by saving memory bandwidth, avoiding cache pollution, exploiting memory level parallelism, and using special purpose accelerators. Management policies are designed to take advantage of these according to the specific DCC design and workload. For example, locality-aware instruction offloading [26–28,38,42,74] involves an analysis of the cache behavior of a set of instructions in an application. This analysis is based on spatial locality (memory locations are more likely to be accessed if neighboring memory locations have been recently accessed) and temporal locality (memory locations are more likely to be accessed if recently accessed). If a section of a code's memory access pattern is found to exhibit little spatial and temporal locality, then these instructions are considered a good candidate for offloading to a memory PE. The specific method to assess locality varies across different implementations. Commonly applied methods include cache profiling [27,28,42], code analysis [26,39,42,70,96], hardware cache-hit counters [28], and heuristic or machine learning techniques [27–29,38,71,72,74].

Another source of superior performance is the use of fixed-function accelerators. While lacking programmability, these special purpose accelerators can perform specific operations many times faster than general purpose host processors. Examples include graph accelerators [8,32–36,57], stencil computation units [39], texture filtering units [37], vector processing units [92], and neuromorphic accelerators [25]. When such fixed-function accelerators are used, it is important that the policy offloads only instructions that the fixed-function accelerator is able to execute.

Finally, in the case of multiple PEs, data placement and workload scheduling become important to realizing superior performance. For example, the compiler-based offloading policy outlined in [92] selects the PE which minimizes the resulting inter-PE communication resulting from accessing data from different memory vaults during execution. Similarly, [97] uses compiler and profiling techniques to map operations to PEs that minimize data movement between memory vaults in a 3D-stacked HMC.

4.1.2. Energy Efficiency

DCC architectures have demonstrated tremendous potential for energy efficiency due to a reduction in expensive off-chip data movement, the use of energy-efficient PEs, or by using NVM instead of leakage-prone DRAM. Off-chip data movement between a host CPU and main memory is found to consume up to 63% of the total energy spent in some consumer devices [40]. The excessive energy consumption can be significantly reduced by using locality-aware offloading, as discussed in the previous section. A related metric used by some policy designs is bandwidth saving. This is calculated by estimating the change in the total number of packets that traverse the off-chip link due to an offloading decision about a set of instructions. For example, in [42], the cache-hit ratio and frequency of occurrence of an instruction is considered along with the memory bandwidth usage to decide if an instruction should be offloaded.

Another source of energy efficiency is fixed-function accelerators. Since these accelerators are designed to execute a fixed set of operations, they do not require a program counter, load-store units, or excessive number of registers and arithmetic logic units (ALU), which significantly reduces both static and dynamic energy consumption. Using NVM can eliminate the cell refresh energy consumption in DRAM, due to the inherent non-volatility in NVMs. In addition, NVMs that use an MCA architecture can better facilitate the matrix multiplication operation, requiring only one step to calculate the product of two matrices. In contrast, DRAM requires multiple loads/stores and computation steps to perform

the same operation. This advantage is exploited by management policies which map matrix vector multiplication to ReRAM crossbar arrays [19,82–89].

4.1.3. Power and Thermal Efficiency

DCC systems require power to be considered as both a constraint and an optimization objective, motivating a range of different management techniques from researchers. The memory module in particular is susceptible to adverse effects due to excessive power usage. It can lead to high temperatures, which may result in reduced performance, unreliable operation, and even thermal shutdown. Therefore, controlling the execution of offloaded workloads becomes important. Various ways in which power efficiency/feasibility is achieved using policy design include using low-power fixed-function accelerators, smart scheduling techniques for multiple PEs, and thermal-aware offloading. For example, when atomic operations are offloaded to HMC, it reduces the energy per instruction and memory stalls compared to host execution [28,29,32]. Data mapping and partitioning in irregular workloads like graph processing reduces peak power and temperature when multiple PEs are used [36]. A more reactive strategy is to use the memory module's temperature as a feedback to the offloading management policy [29] to prevent offloading in the presence of high temperatures.

4.2. Optimization Knobs

There are several knobs available to the resource management policy to achieve the goals defined in the optimization objective. Typically, the management policy identifies which operations to offload (we call the operations that end up being offloaded as offloading workloads) from a larger set of offloading candidates, i.e., all workloads that the policy determined can potentially run on a memory PE. If multiple PEs are available to execute offloading workloads, a selection must be made between them. Finally, decisions may involve timing offloads, i.e., when to offload workloads.

4.2.1. Identification of Offloading Workloads

The identification of offloading workloads concerns the selection of single instructions or groups of instructions which are suitable for execution on a memory-side processor. Depending on the computation capability of the available processor, the identification of offloading workloads can be accomplished in many ways. For example, an atomic unit for executing atomic instructions is used as the PE in [32]; then, the offloading candidates are all atomic instructions. The identification of offloading workloads is simply identifying which offloaded atomic instructions best optimizes the objectives. However, the process is not always simple, especially when the memory-side processors offer a range of computation options, each resulting in different performance and energy costs. This can happen because of variation in state variables such as available memory bandwidth, locality of memory accesses, resource contention, and the relative capabilities of memory and host processors. For example, [26] uses a regression-based predictor for identifying GPU kernels to execute on a near-memory GPU unit. Their model is trained on kernel-level analysis of memory intensity, kernel parallelism, and shared memory usage. Therefore, while single instructions for fixed-function accelerators can be easily identified as offloading workloads by the programmer or compiler, identifying offloading workloads for more complex processors in a highly dynamic system requires online techniques like heuristic or machine-learning-based methods.

4.2.2. Selection of Memory PE

A common NMP configuration places a separate PE in each memory vault of a 3D-stacked HMC module [34,92,97]. This can lead to different PEs encountering different memory latencies when accessing operand data residing in different memory vaults. In addition, the bandwidth available to intra-vault communication is much higher (360 GB/s) than that available to inter-vault communication (120 GB/s) [34]. Ideally, scheduling decisions should aim to utilize the higher intra-vault ban dwidth. An even greater scheduling challenge is posed by the use of multiple HMC modules connected by

a network [34,35]. The available bandwidth for communication between different HMC modules is even lower at 6 GB/s. Therefore, offloading decisions must consider the placement of data and the selection of PEs together to maximize performance [34,35,98]. Similarly, a reconfigurable unit may require reconfiguration before execution can begin on the offloaded workload [41].

Finally, the selection of the PE can be motivated by load balancing, especially in graph processing, where workloads are highly unstructured, e.g., when multiple HMC modules are used for graph processing acceleration. In this case, the graph is distributed across the different memory modules. Since graphs are typically irregular, this can cause irregular communication between memory modules and imbalanced load in the memory PEs [34,35]. If load distribution is not balanced by PE selection, it can lead to long waiting times for under-utilized cores. These issues influence offloading management, often requiring code annotation/analysis and runtime monitoring.

4.2.3. Timing of Offloads

While identifying offloading workloads and selecting the PE for execution are vital first steps, the actual execution is subject to the availability of adequate resources and power/thermal budget. Offloading to memory uses off-chip communication from the host processor to the memory as an essential step to convey the workload for execution. If available off-chip bandwidth is currently limited, the offloading can be halted, and the host processor can be used to perform the operations until sufficient bandwidth is available again [26,27,38,42]. Alternatively, computation can be "batched" to avoid frequent context switching [34]. Moreover, using the memory-side PE for computation incurs power and thermal cost. If it is predicted that the memory and processor system will violate power or thermal constraints while the offloading target is executed, offloading may be deferred or reduced to avoid degrading memory performance or violating thermal constraints [29]. Hence, in a resource constrained PIM/NMP system, naively offloading all offloading candidates can lead to degraded performance as well as power/thermal violations. Therefore, a management policy should be able to control not just what is being offloaded but also when and where it will be offloaded.

4.3. Management Techniques

In order to properly manage what is offloaded onto PIM or NMP systems, where it is offloaded (in the case of multiple memory PEs), and when it is offloaded, prior work has utilized one of three different strategies: (1) **code annotation**: techniques that rely on the programmers to select and determine the appropriate sections of code to offload; (2) **compiler optimization**: techniques that attempt to automatically identify what to offload during compile-time; (3) **online heuristics**: techniques that use a set of rules to determine what to offload during run-time. We discuss the existing works in each of the three categories in the following sections.

4.3.1. Code Annotation Approaches

Code annotation is a simple and cost-effective way to identify offloading workloads with the programmer's help. For these techniques to work, the programmer must manually identify offloading workloads based on their expert knowledge of the underlying execution model while ensuring the efficient use of the memory module's processing resources. Although this approach allows for greater consistency and control over the execution of memory workloads, it burdens the programmer and relies on the programmer's ability to annotate the correct instructions. This policy has been demonstrated to work well with fixed-function units [1,16,24,28,32,37] since the instructions that can be offloaded are easy to identify and the decision to offload is often guaranteed to improve the overall power and performance without violating thermal constraints. For example, [16] introduces two simple instructions for copying and zeroing a row in memory. The in-memory logic implementation of these operations is shown to be both faster and more energy-efficient, motivating the decision to offload all such operations. Code annotation has been successfully used in both PIM [1,25,98] and fixed-function NMP [28,32,40,57].

Although the most popular use of code annotation is found in PIM or fixed-function NMP, code annotation's simplicity also makes it appealing to more general architectures. This is especially true when the policy focuses more on NMP-specific optimizations such as cache-coherence, address translation, or logic reconfiguration than the system-wide optimizations outlined in Section 4.1. For example, code annotation is used in architectures with fully programmable APUs [75], CPUs [31], and CGRAs [95]. In [31], an in-order CPU core is used as a PE in 3D-stacked memory. The task of offloading is facilitated by the programmer selecting portions of code by using macros *PIM_begin* and *PIM_end*, where the main goal of the policy is to reduce coherence traffic between the host and memory PE. For more general purpose PEs and a greater number of PEs, the offloading workload becomes harder to identify with human expertise.

One major downside of code annotation approaches is that the offline nature of code annotations prevents it from adapting to the changes in the system's state, i.e., the availability of bandwidth, the cache locality of target instructions, and power and temperature constraints. To address this, code annotation has been used as a part of other management techniques like compiler-based methods [40] and online heuristics [28] that incorporate some online elements. Other variations of the technique include an extension to the C++ library for identifying offloadable operations [24,98], extensions to the host ISA for uniformity across the system [16,19,57,91], or a new ISA entirely [25]. Extending the software interface in this manner is particularly useful when the source code is not immediately executable on memory PEs or requires the programmer to repeatedly modify large sections of code. For example, a read-modify-write operation, expressed with multiple host instructions, can be condensed into a single HMC 2.0 atomic instruction by introducing a new instruction to the host processor's ISA. Similarly, library extensions provide compact functions for ease of use and readability.

- Case Study 1: Ambit—In-Memory Accelerator for Bulk Bitwise Operations Using Commodity DRAM Technology

Bulk bitwise operations like AND and OR are increasingly common in modern applications like database processing but require high memory bandwidth on conventional architectures. Ambit [1] uses a 2D-DRAM-based PIM architecture for accelerating these operations using the high internal memory bandwidth in the DRAM chip. To enable AND and OR operations in DRAM, Ambit proposes triple row activation (TRA). TRA implements bulk AND and OR operations by activating three DRAM rows simultaneously and taking advantage of DRAM's charge sharing nature (discussed in detail in Section 3.1.1). As these operations destroy the values in the three rows, data must be copied to the designated TRA rows prior to the Boolean operation if the original data are to be maintained.

To allow more flexible execution, Ambit also implements the NOT operation through a modification to the sense amplifiers shown in Figure 8. The NOT operation uses a dual-contact DRAM cell added to the sense amplifier to store the negated value of the cell in a capacitor and store it on the bitline when required. To perform the NOT operation, the value of a cell is read into the sense amplifier by activating the desired row and the bitline. Next, the n-wordline, which connects the negated cell value from the sense amplifier to the dual-contact cell capacitor is activated, which allows the capacitor to store the negated value. Finally, activating the d-wordline drives the bitline to the value stored in the dual-contact cell capacitor. AMBIT then uses techniques adapted from [16] to transfer the result to an operand row for use in computation. By using the memory components with minimal modifications, Ambit adds only 1% to the memory chip area and allows easy integration using the traditional DRAM interface.

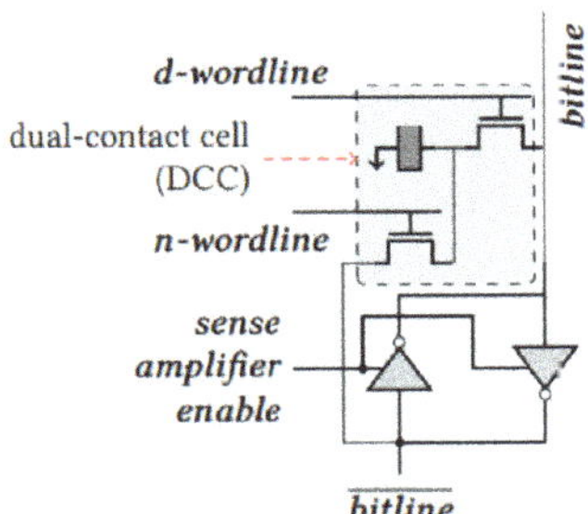

Figure 8. Ambit's implementation of the NOT operation by adding a dual-contact DRAM cell to the sense amplifier. The grey shaded region highlights the dual-contact cell [1].

To execute the Boolean operations, Ambit adds special bulk bitwise operation (bbop) instructions of the form shown in Figure 9 to the host CPU's ISA. These bbop instructions specify the type of operation (bbop), the two operand rows (src1 and src2), a destination row (dst), and the length of operation in bytes (size). Since the Ambit operations operate on entire DRAM rows, the size of the bbop operation must be a multiple of the DRAM row size. Due to Ambit's small set of operations, it is relatively easy to identify the specific code segments that can benefit from these operations. However, the authors expect that the implementers of Ambit would provide API/driver support that allows the programmer to perform code annotation to specify bitvectors that are likely to be involved in bitwise operations and have those bitvectors be placed in subarrays such that corresponding portions of each bitvector are in the same subarray. This is key to enabling TRA-based operations.

```
bbop dst, src1, [src2], size
```

Figure 9. Ambit [1] extends the ISA to include a bulk bitwise operation (bbop) instruction for computation which will be offloaded to PIM. The format of the instruction is shown.

Applications that use bulk bitwise operations show significant improvement using Ambit. Ambit accelerates a join and scan application by 6× on average for different data sizes compared to a baseline CPU execution. Ambit accelerates the Bitweaving (database column scan operations) application by 7× on average compared to the baseline, with the largest improvement witnessed for large working sets. A third application uses bitvectors, a technique used to store set data using bits to represent the presence of an element. Ambit outperforms the baseline for the application by 3× on average, with performance gains increasing with set sizes. Ambit also demonstrates better energy efficiency compared to traditional CPU execution of applications using bulk bitwise operations. While raising additional wordlines consumes extra energy, the reduction of off-chip data movement more than makes up for it, reducing energy consumption by 25.1× to 59.5×.

While the architecture of Ambit shows significant promise, its management policy is not without issues. The benefits would rely heavily on the programmer's expertise as the programmer would still need to identify and specify which bitvectors are likely to be involved in these bitwise operations. Fortunately, this technique avoids runtime overhead since all offloading decisions are part of the generated binary. However, as Ambit operates on row-wide data, the programmer is required to ensure the operand size to be a multiple or row size. Additionally, all dirty cache lines from the operand rows need to be flushed prior to Ambit operations. Similarly, cache lines from destination rows need to be invalidated. Both of these operations generate additional coherence traffic, which reduces the efficiency of this approach.

4.3.2. Compiler-Based Approaches

The main advantage of using compilers for automatic offloading over code annotation is the minimization of programmer burden. This is because the compiler can automatically identify offloading workloads [92], optimize data reuse [71], and select the PE to execute the offloading workloads [35,36]. Offloading using compilers also has the potential to outperform manual annotation by better selecting offloading workloads [30]. Like code annotation, this is an offline mechanism and avoids hardware and runtime overhead. However, similar to code annotation, this results in a fixed policy of offloads suited in particular to simple fixed-function memory accelerators which are relatively unconstrained by power and thermal limits [39,92]. While compilers can be used to predict the performance of an offloading candidate on both host and memory processors, in the case of more complex PIM and NMP designs, compiler-based offloading cannot adapt to changing bandwidth availability, memory access locality, and DRAM temperature, leading to sub-optimal performance while risking the violation of power and thermal constraints.

Compiler-based techniques have been used with both fixed-function [39,92,96] and programmable [27,40,70] memory processors. For fixed-function accelerators like vector processors [92] and stencil processors [39], the compiler's primary job is to identify instructions that can be offloaded to the memory PE and configure the operations (e.g., vector size) to match the PE's architecture. When the type of instruction is insufficient to ascertain the benefit of offloading, compiler techniques like CAIRO [42] use analytical methods to quantitatively determine the benefit of offloading by profiling the bandwidth and cache characteristics of instructions offline, at the cost of adding design (compile) time overhead.

Another advantage of using compilers is the ability to efficiently utilize NMP hardware by exploiting parallelism in memory and PEs. For example, [30] achieves 71% better floating-point utilization than hand-written assembly code using loop and memory access optimizations for some kernels. While CAIRO relies on offline profilers and [30] requires some OpenMP directives for annotation, the compiler in PRIMO [92] is designed to eliminate any reliance on third-party profilers or code annotation. Its goal is to reduce communication within multiple memory processors by the efficient scheduling of vector instructions. It can be noticed that when the memory processor is capable of a wider range of operations, compiler-based techniques have to account for aspects like bandwidth saving, cache-locality, and comparative benefit analysis with the host processor [42] while still working offline, which is not easy. Further complications arise when multiple general purpose processor options exist near memory as runtime conditions become a significant factor in exploiting the benefits of memory-side multi-core processing [27].

- Case Study 1: CAIRO

CAIRO uses an HMC-based NMP architecture to accelerate graph processing using bandwidth- and locality-aware offloading of HMC 2.0's atomic instructions using a profiler-based compiler. To enable high bandwidth and low energy, HMC stacks DRAM dies and a CMOS logic die and connects the dies using through-silicon vias (TSVs). Starting with HMC 2.0 specification, HMC has supported 18 atomic computation instructions (shown in Figure 10) on the HMC's logic layer. HMC enables a high level of memory parallelism by using a multiple vault and bank structure. Several properties of graph workloads make HMC-based NMP a perfect choice to accelerate execution. For example, graph computation involves frequent use of read-modify-write operations which can be mapped to HMC 2.0 supported instructions. When performed in HMC, the computation exploits the higher memory bandwidth and parallelism supported by HMC's architecture.

Type	Data Size	Operation	Return
Arithmetic	8/16-byte	single/dual signed add	w/ or w/o
Bitwise	8/16-byte	swap, bit write	w/ or w/o
Boolean	16-byte	AND/NAND	w/o
		OR/NOR/XOR	w/o
Comparison	8/16-byte	CAS-if equal/zero	w/
		greater/less	w/
		compare if equal w/	

Figure 10. HMC atomic instructions in HMC 2.0 [81].

Identifying offloading candidates (instructions that can be offloaded) and choosing suitable memory PEs are important compiler functions for accelerating computation. While this is the main goal with CAIRO, the authors note that additional benefits can be realized when effective cache size and bandwidth saving is considered for offloading decisions. By default, all atomic instructions are considered as offloading candidates and all data for these instructions are marked uncacheable using a cache-bypassing policy from [32]. This technique provides a simple cache coherence mechanism by never caching data that offloading candidates can modify. The underlying idea is that offloading an atomic instruction will inherently save link bandwidth when cache-bypassing is active for all atomic instructions regardless of host or memory-side execution. This is because offloading the instruction to HMC uses fewer flits than executing the instruction on the host CPU, which involves the transfer of operands over the off-chip link. Moreover, if the offloaded segment does not have data access locality, it avoids cache pollution, making more blocks available for cache-friendly data. Finally, the NMP architecture allows faster execution of atomic instructions, eliminating long stalls in the host processor.

CAIRO's first step is to identify offloading candidates that can be treated as HMC atomic instructions. Given these instructions, CAIRO then attempts to determine how much speedup can be obtained from offloading these instructions. One of the major factors in determining the speedup is the main memory bandwidth savings from offloading. Since offloading instructions frees up memory bandwidth, it is important to understand how applications can benefit from the higher available bandwidth to estimate application speedup. For applications that are limited by low memory bandwidth, increasing the available memory bandwidth leads to performance improvements (bandwidth-sensitive applications) since these applications exploit memory-level parallelism (MLP). On the other hand, compute-intensive applications are typically bandwidth-insensitive and benefit little from the increase in available memory bandwidth. To categorize their applications, the authors analyze the speedup of their applications after doubling the available bandwidth and conclude that CPU workloads and GPU workloads naturally divide into bandwidth-insensitive and bandwidth-sensitive, respectively. The authors do note that, for exceptional cases, CAIRO would need the programmer/vender to specify the application's bandwidth sensitivity.

Given the application's bandwidth sensitivity, CAIRO uses two different analytical models to estimate the potential speedup of offloading instructions. For bandwidth-insensitive applications, they model a linear relationship between the candidate instruction's miss ratio (MR), density of host atomic instructions per memory region of the application (ρ_{HA}), and speedup due to offloading the candidate instruction (SU_{tot}), as shown in Figure 11a. In other words, the decision to offload an atomic instruction depends on how often it misses in cache and how much cache is believed to be available to it. The model incorporates several machine-dependent constants, C_1, C_2, and C_3, shown in Figure 11b, that are determined using offline micro-benchmarking. When there is positive speedup ($SU_{tot} > 0$), CAIRO's compiler heuristic offloads the instruction to the HMC.

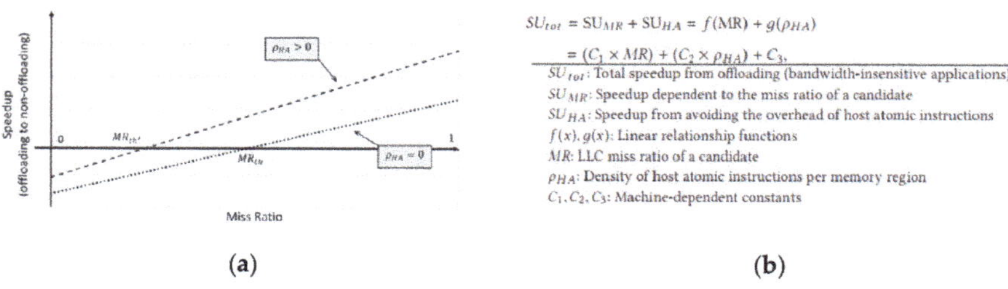

(a)   **(b)**

Figure 11. CAIRO's performance estimation for bandwidth insensitive applications. (**a**) Speedup given the host atomic instructions per memory region (ρ_{HA}) and its miss ratio. (**b**) Equation for total speedup, SU_{tot}, is the sum of speedup due to avoiding inefficient cache use SU_{MR} and speedup due to avoiding the overhead of executing atomic instructions on the host processor SU_{HA} [42].

On the other hand, bandwidth-sensitive applications can also experience additional speedup by utilizing the extra memory bandwidth due to offloading atomic instructions. Therefore, CAIRO considers bandwidth savings due to offloading in addition to the miss ratio and density of atomic instructions. As shown in Figure 12, for miss ratios higher than $MissRatio_H$, the instruction will be offloaded. Similarly, for miss ratios lower than $MissRatio_L$, the instruction will not be offloaded. However, in the region between $MissRatio_H$ and $MissRatio_L$, additional bandwidth saving analysis is performed before making the offloading decision. The speedup due to bandwidth saving, SU_{BW}, is calculated by estimating the bandwidth savings achieved, BW'_{saving} (Equations (1) and (2)). BW_{reg} and $BW_{offload}$ are estimated using hand-tuned equations based on last-level cache (LLC) hit ratios, the packet size of offloading an instruction, and the number of instructions (more details in [42]). Constants M_1 and M_2 are machine-specific and obtained using micro-benchmarking. Similar to the bandwidth-insensitive case, CAIRO's compiler heuristic offloads the instruction when the calculated speedup is greater than 0.

$$SU_{BW} = \left(M_1 \times BW'_{saving}\right) + M_2, \tag{1}$$

SU_{BW} : Speedup from bandwidth savings (bandwidth-sensitive applications)

BW'_{saving} : Normalized bandwidth savings

M_1, M_2 : Machine-dependent constants

$$BW'_{saving} = \frac{BW_{reg} - BW_{offload}}{BW_{reg}} = \frac{BW_{saving}}{BW_{reg}}, \tag{2}$$

BW_{reg} : Regular bandwidth usage (without offloading)

$BW_{offload}$: Bandwidth usage wth offloading

CAIRO's compiler heuristic allows it to consider the most important factors in offloading decisions while avoiding runtime overhead. For bandwidth-insensitive applications, the amount of speedup achieved by CAIRO increases with the cache miss ratio of the application. For a miss ratio of greater than 80%, CAIRO doubles the performance of bandwidth-insensitive CPU workloads compared to host execution. For bandwidth-sensitive applications, the amount of speedup increases with the amount of bandwidth saved by CAIRO. For bandwidth savings of more than 50%, CAIRO doubles the performance of GPU workloads. In addition, CAIRO derives energy efficiency from the reduction in data communication inherent to the HMC design. Similar to prior work [28,32], it extends the HMC instructions and ALU to support floating-point operations without violating power constraints. Despite CAIRO's performance and energy improvements, it has its drawbacks. The miss ratio and bandwidth saving analysis is machine- and application-dependent and involves considerable design time overhead. Moreover, since the heuristic does not work online, offloading decisions are based on

analytical assumptions about runtime conditions. For instance, due to its offline design, CAIRO does not consider the temperature of HMC, which can significantly impact the performance [29].

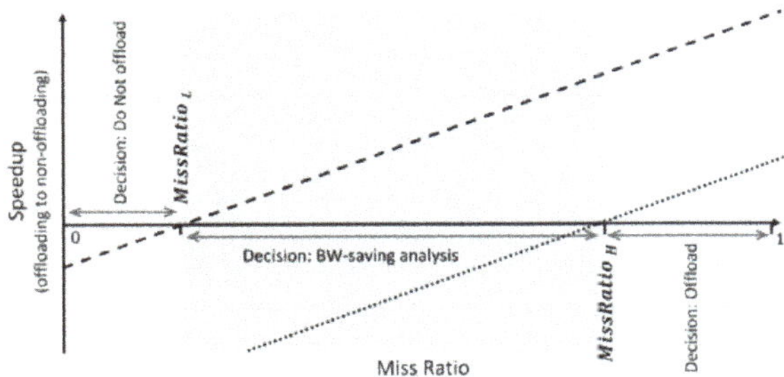

Figure 12. CAIRO's performance estimation for bandwidth insensitive applications: for miss ratios between *MissRatio_L* and *MissRatio_H*, additional bandwidth saving analysis is performed before making the decision to offload [42].

- Case Study 2: A Compiler for Automatic Selection of Suitable Processing-In-Memory Instructions (PRIMO)

NMP designs with a single PE like CAIRO can focus on offloading the most suitable sections of code to reduce off-chip data movement and accelerate execution. However, when the NMP design uses multiple PEs near memory, designers have to consider the location of data with respect to the PE executing the offloaded workload. The compiler PRIMO [92] is designed to manage such a system. It uses a host CPU connected to an 8-layer HMC unit with 32 vaults. Each of the 32 vaults has a reconfigurable vector unit (RVU) on the logic layer as a PE. The RVU uses fixed-function FUs to execute native HMC instructions extended with an Advanced Vector Extensions (AVX)-based ISA while avoiding the area overhead of programmable PEs. Each RVU has wide vector registers available to facilitate vector operations. They allow a flexible vector width with scalar operands of as little as 4 bytes and vector operands up to 256 bytes. Moreover, multiple RVUs can coordinate to operate on vector lengths of as high as 4096 bytes.

With this architecture, PRIMO has to identify not only offloading candidate code segments to convert to special NMP instructions but also which execution unit minimizes data movement within memory by exploiting internal data locality. Another way by which the compiler improves performance is the utilization of the vast number of functional units (FUs) by optimizing the vector length of offloaded instructions for the available hardware. For example, if the RVU architecture can execute instructions up to a vector size of 256 bytes, the compiler will automatically compile offloading candidate code sections with vectors larger than 256 bytes into NMP instructions, whereas operations involving smaller vector sizes are executed by the host CPU. In short, the compiler performs four functions: (1) identify operations with large vector sizes, (2) set these operations' vector size to a supported RVU vector size, (3) check for data dependencies with previously executed instructions and map instructions to the same vector unit as the previous instructions, and (4) create the binary code for execution on the NMP system. Figure 13a shows the code for a *vec-sum* kernel. Figure 13b,c compare the compiled version of the kernel for an x86 AVX-512 capable processor and an NMP architecture, respectively. The AVX-512 version performs four loads, followed by four more loads and four adds with a vector length of 64 bytes. Finally, four store operations complete an iteration of the loop. In comparison, the NMP code uses just two instructions to load the entire operand data in 256-byte registers, followed by a single 256-byte add and store operation. The NMP instruction *PIM_256B_LOAD_DWORD V_0_2048b_0 [addr]* loads 256 bytes of data starting at *addr* into register 0 of vault 0. The actual choice of the vault and RVU is made after checking for data dependencies between instructions. The use of large vector sizes allows PRIMO to exploit greater memory parallelism while the locality-aware PE minimizes data movements between vaults.

```c
for(int i=0; i<N; i++)
    c[i] = a[i] + b[i];
```

(a) C Code

```
mov rax, -16384
.LBB0_1:
vmovdqu32 zmm8, [rax+c+16576]
vmovdqu32 zmm4, [rax+c+16512]
vmovdqu32 zmm3, [rax+c+16448]
vmovdqu32 zmm0, [rax+c+16384]
vpaddd zmm9, zmm0, [rax+b+16384]
vpaddd zmm6, zmm3, [rax+b+16448]
vpaddd zmm3, zmm4, [rax+b+16512]
vpaddd zmm0, zmm8, [rax+c+16576]
vmovdqu32 [rax+a+16384], zmm0
vmovdqu32 [rax+a+16448], zmm3
vmovdqu32 [rax+a+16512], zmm6
vmovdqu32 [rax+a+16576], zmm9
add rax, 4096
jne .LBB0_1
```

(b) AVX-512 ASM Code

```
mov rax, -16384
.LBB0_1:
PIM_256B_LOAD_DWORD V_0_R2048b_0, [rax+b+16384]
PIM_256B_LOAD_DWORD V_0_R2048b_1, [rax+c+16384]
PIM_256B_VADD_DWORD V_0_R2048b_1, V_0_R2048b_0, V_0_R2048b_1
PIM_256B_STORE_DWORD [rax+a+16384], V_0_R2048b_1
add rax, 4096
jne .LBB0_1
```

(c) NMP ASM Code

Figure 13. PRIMO code generation [92].

While identifying vector instructions is common to all compiler-based management techniques, PRIMO achieves better FU utilization and minimizes internal data movement by optimizing instruction vector size and selecting PEs considering the location and vector size of the operand data. It also manages to completely automate the process and eliminate the need to annotate sections of code or run offline profilers. In addition, PRIMO performs all optimizations at design time, which allows it to eliminate runtime overhead. PRIMO achieves an average speedup of 11.8× on a set of benchmarks from the PolyBench Suite [99], performing the best with large operand vector sizes. However, such benefits come at the cost of a static policy that is unable to react to dynamic system conditions like available memory bandwidth, variations in power dissipation, and temperature. Additionally, the energy efficiency of specialized hardware and instructions need to be studied further.

4.3.3. Online Heuristic

Online heuristic-based policies use human expert knowledge of the system to make offloading decisions at runtime. To this end, additional software or hardware is tasked with monitoring and predicting the future state of the system at runtime, to inform the offloading workload identification, determining the memory PE, and timing control. The online nature of heuristic policies enables the policy to adapt to the runtime state, but it also increases its complexity and runtime overhead. The quality of decisions depends on the heuristic designer's assumptions about the system. When the assumptions are incomplete or incorrect, online heuristics will suffer from unexpected behavior.

Online heuristics are used widely when the goal is to accelerate an entire data-intensive application on general purpose memory processors [27,38,71,72,74]. In addition, online heuristics may be used in combination with code annotation and compiler-based methods [27,75]. For example, PIM-enabled instructions (PEI) [28] extends the host ISA with new PIM-enabled instructions to allow programmers to specify possible offloading candidates for PIM execution. These instructions are offloaded to PIM only if a cache hit counter at runtime suggests inefficient cache use; otherwise, the instructions are executed by the host CPU. Similarly, Transparent Offloading and Mapping (TOM) [27] requires the programmer to identify sections of CUDA code as candidate offloading blocks. The final offloading decision involves an online heuristic to determine the benefit of offloading by estimating bandwidth saving and co-locate data and computation across multiple HMC modules. Different from CAIRO, where the bandwidth saving analysis is performed entirely before compilation, TOM uses heuristic analysis of bandwidth saving both before and after compilation. It marks possible offloading candidates during a compiler pass, but the actual offloading decision is subject to runtime monitoring of dynamic system conditions like PE utilization and memory bandwidth utilization. Other methods that rely on online heuristics include [38], which uses a two-tier approach to estimating locality and the energy consumption of offloading decision, [74], which performs locality-aware data allocation and prefetching, and [29], which optimizes for thermally feasible NMP operation by throttling the frequency of issuing offloads and size of CUDA blocks offloaded to an HMC-based accelerator.

- Case Study 1: CoolPIM—Thermal-Aware Source Throttling for Efficient PIM Instruction Offloading

As noted in Section 4.1.3, 3D-stacked memory is vulnerable to thermal problems due to high power densities and insufficient cooling, especially at the bottom layers. In fact, with a passive heatsink, HMCs cannot operate at their maximum bandwidth, even without PIM and NMP (see Figure 14) [29,100]. A particular problem in these systems is that heating exacerbates charge leakage in DRAM memory cells, which demands more frequent refreshes to maintain reliable operation. Therefore, offloading in NMP, among other online runtime conditions, requires awareness of the memory system's temperature. CoolPIM [29] attempts to solve this problem by using an online-heuristic-based mechanism to control the frequency of offloading under the thermal constraints of an NMP system.

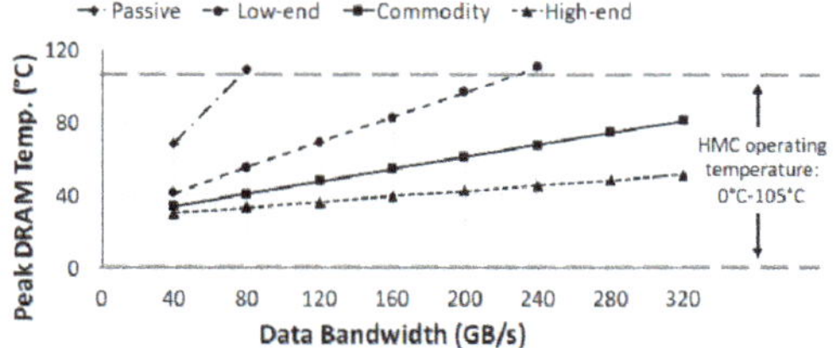

Figure 14. Peak DRAM temperature with various data bandwidth and cooling methods (passive, low-end, commodity, and high-end) [29].

CoolPIM considers a system that has a host GPU executing graph workloads with an HMC memory module capable of executing HMC 2.0 atomic instructions. In order to allow the offloading of atomic operations written in Nvidia's CUDA to HMC, the compiler is tweaked to generate an additional executable version for each CUDA block of computation. This alternative HMC-enabled version has the HMC's version of atomic instructions to be executed near-memory if the heuristic decides to offload it. All atomic instructions in HMC-enabled blocks execute on functional units on the HMC's logic layer as provided by the HMC 2.0 Specification [81]. It must be noted that the role of the compiler is not to make offloading decisions but rather to generate a set of offloading candidates that the online heuristic can select from at runtime.

The main goal of CoolPIM is to keep the HMC unit under an operational maximum temperature. The HMC module includes a thermal warning in response packets whenever the surface temperature of the module reaches 85 °C. Whenever the thermal warning is received, CoolPIM throttles the intensity of offloading by reducing the number of HMC-enabled blocks that execute on the GPU. To this end, CoolPIM introduces a software-based throttling and hardware-based throttling technique, as illustrated in Figure 15. The software-based throttling technique controls the number of HMC-enabled blocks that execute on the GPU using a thermal warning interrupt from the HMC. It implements a token-based policy where CUDA blocks request a token from the pool before starting execution. If a token is available, the block acquires it and the offloading manager executes the NMP version of the block. However, if a thermal warning is encountered, the token pool is decremented, which effectively decreases the total number of blocks that can be offloaded to the memory PE.

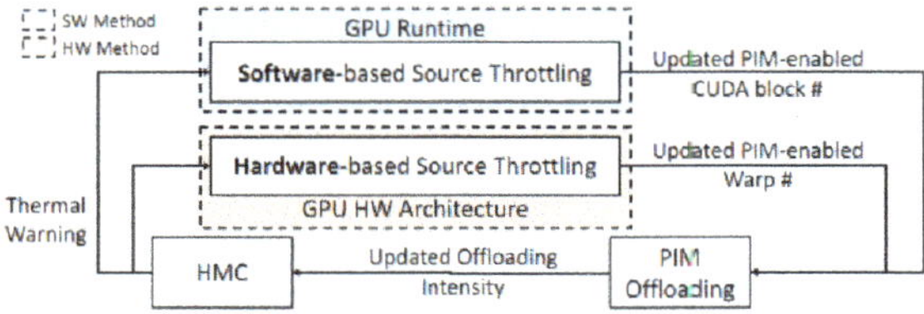

Figure 15. Overview of CoolPIM [29].

On the other hand, the hardware-based throttling technique controls the number of warps per block that are offloaded to the memory PE as a reaction to the thermal warning interrupt. Unlike the

software-based method, which requires that all warps in executing thread blocks finish before the throttling decisions take effect, the hardware-based mechanism allows the system to react faster to the temperature warning by changing the number of warps with NMP-enabled instructions during runtime. The hardware-based control is achieved by adding a hardware component to the system which can replace NMP-enabled instructions with their CUDA counterparts during the decoding process. The HMC-disabled warps, with NMP instructions replaced with CUDA instructions, will then execute entirely on the host GPU to help manage the rising HMC temperature. Figure 15 details the implementation of the policy.

The use of an online heuristic allows CoolPIM to adapt to system conditions at runtime and control temperature even when an application goes through different phases. The additional system awareness not only improves the offloading decisions to better realize the benefits of NMP but also develops a more proactive approach to management. CoolPIM improves performance by up to 40% compared to both host execution and naïve offloading while effectively managing memory temperature. While an increase in the heuristic's complexity is expected to produce better offloading decisions and performance improvement, it is necessary that runtime overhead remains manageable. While CoolPIM manages to do this, it does not consider data locality and bandwidth saving while making offloading decisions. Similarly, it does not consider energy savings except as a byproduct of minimizing the DRAM refresh rate.

- Case Study 2: Making Better Use of Processing-In-Memory through Potential-Based Task Offloading

Although works under the code annotation and compiler categories use profiling and analysis to understand the effects of offloading on key objectives, the generated results may be affected by runtime conditions related to input data, concurrent workloads, and other dynamic policies. Kim et al. [38] look at the number of $L2$ misses, memory stalls, and power in their online heuristic policy to determine offload decisions to optimize performance, energy, and power. They consider an NMP system with a host GPU and a trimmed-down GPU as the memory PE in the logic layer of an HBM unit. The GPU used as a memory PE has a lower number of streaming multiprocessors (SM) to accommodate the thermal constraints on the logic layer of the HBM. To accomplish their goals, the authors divide the heuristic into two stages: the first stage determines if the decision meets a time-energy constraint (OC_{T-E}), and if it passes, the second stage will determine if it passes a time-power constraint (OC_{T-P}). Offloading is performed only if the conditions in both stages are met. This process is represented in Figure 16.

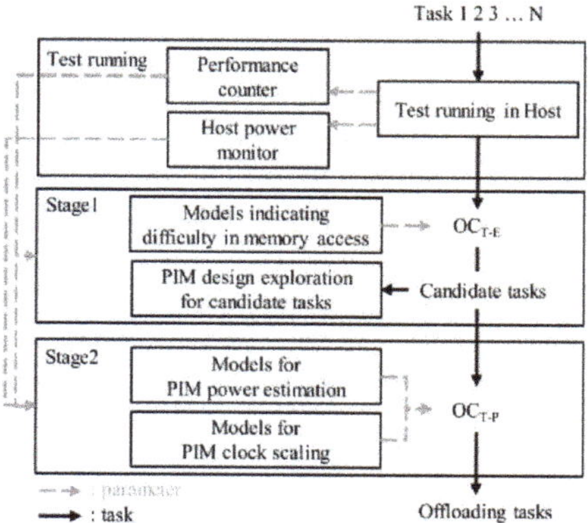

Figure 16. Overview of the offloading policy used in [38]. The offloading decision is separated into two stages. The first stage checks whether it would pass a time-energy constraint (OC_{T-E}) while the second stage checks whether it passes a time-power constraint (OC_{T-P}). If it passes both, the task is offloaded onto the memory module.

The first stage starts by profiling the application for the number of $L2$ misses per instruction ($L2MPI$) and the number of cycles L1 was stalled due to a request to $L2$ ($L1S$). The reason behind this choice is that the higher the value of each of these metrics, the slower the task will run on the host GPU and the better it will perform on the high memory bandwidth near-memory GPU. The two values are used to determine if the task should be offloaded using a simple linear model derived through empirical data from profiling (Equation (3)). If the value of this model exceeds a threshold (TH), then the task is considered as an offloading candidate based on the time-energy constraint.

$$58 \times L2MPI + 0.22 \times L1S + 0.005 > TH \tag{3}$$

If a task passes the first stage, the next stage is to determine the impact of the offloading decision on power consumption. Specifically, power is treated as a constraint with a fixed budget of 55W with a high-end-server active heat sink. The power model for estimating the task's power consumption in NMP execution is based on several assumptions. Idle power is ignored in this work. Static power is scaled by the ratio of the areas of the host and memory GPU. For the dynamic power estimation, different components of the host power are scaled down using information from profiling the application. Equation (4) shows how the relationship is modeled.

$$PIM_{SM} = \alpha \times ((ActiveSM_{Host} + L2_{Host} + ICNT_{Host}) + 0.25 \times (MC_{Host} + DRAM_{Host})) \tag{4}$$

All dynamic power components are scaled by an experimentally determined parameter α. These components include the power consumed by the active streaming multiprocessor (SM) ($ActiveSM_{Host}$), the L2 cache in the host processor ($L2_{Host}$), and the interconnect network ($ICNT_{Host}$). The DRAM and memory controller (MC) power components (MC_{Host} and $DRAM_{Host}$) are further scaled down by 0.25, which is the experimentally determined ratio between $DRAM$ and active SM power. The value of α is estimated by profiling the application. The $L2MPI$ and percentage of active SMs to idle SMs is used to infer how much the application would benefit from executing on the memory side PE. The assumption is that large $L2MPI$ values indicate inefficient use of cache and wastage of memory bandwidth. Similarly, a low percentage of active cores indicates that the application will not experience a slow-down due to the lower number of compute units in the memory PE.

Finally, another linear model based on these assumptions of power scaling is used to check if the use of dynamic voltage and frequency scaling (DVFS) will make some offloading decisions feasible within the power constraint. If all tests are passed, the task is offloaded to the memory-side GPU.

This technique comprehensively considers the energy, power, and performance aspects of offloading to power constrained NMP systems. Compared to host-only execution, the proposed technique achieves a 20.5% increase in instructions per cycle (IPC) and a 67.2% decrease in energy delay squared product (ED2P). The approach comes with additional design time and runtime overhead compared to code annotation and compiler-based methods. Additionally, it is based on strict assumptions about the system dynamics derived from profiling characteristic applications. Practically, the different system components may interact in a variety of unpredictable ways, breaking the assumption of stationarity.

5. Conclusions and Future Challenges and Opportunities

PIM and NMP architectures have the potential to significantly reduce the memory wall and enable future big data applications. However, it has been demonstrated that naïve use of these DCC systems can run into thermal issues and even reduce the performance of the overall system. This has led many to investigate offloading management techniques that are able to identify low data locality instructions or react to thermal emergencies. In this paper, we have organized these works based on the optimization objective, optimization knobs, and the type of technique they utilize. However, there are several challenges and opportunities for resource management of PIM/NMP substrates related

to generalizability, multi-objective considerations, reliability, and the application of more intelligent techniques, e.g., machine learning (ML), as discussed below:

- Due to the large variations in DCC architectures proposed to date, the management policies have been very architecture- and application-specific. For example, a policy for a near-memory graph accelerator involves the offloading of specific graph atomic operations, a policy for a stencil accelerator involves the offloading of stencil operations, and so on. This could lead to technology fragmentation, lower overall adoption, and inconsistent system improvements. As an example, an NMP system designed to exploit the high density of connections in small graph areas fails to extract significant speedup when graph connections are more uniform [33]. Future DCC architectures and resource management policies need to explore the generalizability of these systems.

- Nearly all past work has focused on the efficient use of PEs across the system, i.e., they offload tasks which minimize data movement between the host processor and memory. However, high frequency of offloading can cause the memory chip to overheat and lead to a complete shutdown. The issue is addressed reactively in [29] but more proactive and holistic resource management approaches are needed to consider both thermal- and performance-related objectives together.

- Reliability has yet to be considered in the management policies for DCC systems [59]. There is no analysis of the impact of reliability concerns in emerging NVM substrates or DRAM cells on the efficacy of PIM/NMP offloading strategies. Due to deep nanometer scaling, DRAM cell charge retention is becoming increasingly variable and unpredictable. Similarly, the use of unproven and new NVM technologies that are susceptible to disturbances during non-volatile programming brings some level of uncertainty at runtime. Resource management techniques need to be designed in a manner that is robust to these reliability issues in memory substrates when making decisions to offload to memory PEs.

- ML-based applications have exploded in recent years. ML's potential has been demonstrated for identifying offloading targets [71] using a simple regression-based model with cache performance metrics as the input. More generally, ML techniques like reinforcement learning have proven successful in improving performance by intelligently scheduling workloads on heterogenous systems [101]. As we adopt more general architectural designs, management policies will need to account for the diversity of applications and variability of processing resources. On the other hand, Internet of Things (IoT) devices have great potential to use ML for smart operation, but they lack the resources for training ML models on large datasets. Recent work [102] has shown that executing ML algorithms using near-data vector processors in IoT devices can significantly improve performance. Hence, a promising direction is to leverage the DCC approach to empower IoT devices to process data locally to improve privacy and reduce latency.

- Heterogenous manycore architectures running multi-threaded applications result in complex task mapping, load balancing, and parallelization problems due to the different PEs. Recently, complex network theory, originally inspired by social networks, has been successfully applied to the analysis of instruction and data dependency graphs and identification of "clusters" of tasks to optimally map instructions to PEs [103]. Similarly, complex network theory can be extended to include the PIM and NMP domain in order to optimize software, data orchestration, and hardware platform simultaneously.

Author Contributions: Conceptualization, K.K., S.P., and R.G.K.; investigation, K.K.; writing—original draft preparation, K.K. and R.G.K.; writing—review and editing, K.K., S.P., and R.G.K.; visualization, K.K.; supervision, S.P. and R.G.K.; project administration, S.P. and R.G.K. All authors have read and agreed to the published version of the manuscript.

Funding: This research was supported by the National Science Foundation (NSF) under grant number CCF-1813370.

Conflicts of Interest: The authors declare no conflict of interest.

References

1. Seshadri, V.; Lee, D.; Mullins, T.; Hassan, H.; Boroumand, A.; Kim, J.; Kozuch, M.A.; Mutlu, O.; Gibbons, P.B.; Mowry, T.C. Ambit: In-memory accelerator for bulk bitwise operations using commodity dram technology. In Proceedings of the 50th Annual IEEE/ACM International Symposium on Microarchitecture (MICRO), Boston, MA, USA, 14–17 October 2017; pp. 273–287. [CrossRef]
2. Jeddeloh, J.; Keeth, B. Hybrid memory cube new DRAM architecture increases density and performance. In Proceedings of the Symposium on VLSI Technology (VLSIT), Honolulu, HI, USA, 12–14 June 2012; pp. 87–88. [CrossRef]
3. Lee, D.U.; Kim, K.W.; Kim, K.W.; Kim, H.; Kim, J.Y.; Park, Y.J.; Kim, J.H.; Kim, D.S.; Park, H.B.; Shin, J.W.; et al. A 1.2V 8Gb 8-channel 128GB/s high-bandwidth memory (HBM) stacked DRAM with effective microbump I/O test methods using 29nm process and TSV. In Proceedings of the IEEE International Solid-State Circuits Conference Digest of Technical Papers (ISSCC), San Francisco, CA, USA, 9–13 February 2014; pp. 432–433. [CrossRef]
4. Devaux, F. The true processing in memory accelerator. In Proceedings of the IEEE Hot Chips 31 Symposium (HCS), Cupertino, CA, USA, 18–20 August 2019; pp. 1–24. [CrossRef]
5. Siegl, P.; Buchty, R.; Berekovic, M. Data-centric computing frontiers: A survey on processing-in-memory. In Proceedings of the Second International Symposium on Memory Systems (MEMSYS), Alexandria, VA, USA, 3–6 October 2016; pp. 295–308. [CrossRef]
6. Singh, G.; Chelini, L.; Corda, S.; Javed Awan, A.; Stuijk, S.; Jordans, R.; Corporaal, H.; Boonstra, A.-J. A review of near-memory computing architectures: Opportunities and challenges. In Proceedings of the 21st Euromicro Conference on Digital System Design (DSD), Prague, Czech Republic, 29–31 August 2018; pp. 608–617. [CrossRef]
7. Mutlu, O.; Ghose, S.; Gómez-Luna, J.; Ausavarungnirun, R. Enabling practical processing in and near memory for data-intensive computing. In Proceedings of the 56th Annual Design Automation Conference (DAC), Las Vegas, NV, USA, 2–6 June 2019; pp. 1–4. [CrossRef]
8. Gui, C.-Y.; Zheng, L.; He, B.; Liu, C.; Chen, X.-Y.; Liao, X.-F.; Jin, H. A survey on graph processing accelerators: Challenges and opportunities. *J. Comput. Sci. Technol.* **2019**, *34*, 339–371. [CrossRef]
9. Umesh, S.; Mittal, S. A survey of spintronic architectures for processing-in-memory and neural networks. *J. Syst. Archit.* **2019**, *97*, 349–372. [CrossRef]
10. Mittal, S. A survey of ReRAM-based architectures for processing-in-memory and neural networks. *Mach. Learn. Knowl. Extr.* **2018**, *1*, 75–114. [CrossRef]
11. Stone, H.S. A logic-in-memory computer. *IEEE Trans. Comput.* **1970**, *C-19*, 73–78. [CrossRef]
12. Elliott, D.G.; Stumm, M.; Snelgrove, W.M.; Cojocaru, C.; Mckenzie, R. Computational RAM: Implementing processors in memory. *IEEE Des. Test Comput.* **1999**, *16*, 32–41. [CrossRef]
13. Gokhale, M.; Holmes, B.; Iobst, K. Processing in memory: The Terasys massively parallel PIM array. *Computer* **1995**, *28*, 23–31. [CrossRef]
14. Patterson, D.; Anderson, T.; Cardwell, N.; Fromm, R.; Keeton, K.; Kozyrakis, C.; Thomas, R.; Yelick, K. A case for intelligent RAM. *IEEE Micro* **1997**, *17*, 34–44. [CrossRef]
15. Draper, J.; Chame, J.; Hall, M.; Steele, C.; Barrett, T.; LaCoss, J.; Granacki, J.; Shin, J.; Chen, C.; Kang, C.W.; et al. The architecture of the DIVA processing-in-memory chip. In Proceedings of the 16th Annual ACM International Conference on Supercomputing (ICS), New York, NY, USA, 22–26 June 2002; pp. 14–25. [CrossRef]
16. Seshadri, V.; Kim, Y.; Fallin, C.; Lee, D.; Ausavarungnirun, R.; Pekhimenko, G.; Luo, Y.; Mutlu, O.; Gibbons, P.B.; Kozuch, M.A.; et al. RowClone: Fast and energy-efficient in-DRAM bulk data copy and initialization. In Proceedings of the 46th Annual IEEE/ACM International Symposium on Microarchitecture (MICRO), Davis, CA, USA, 7–11 December 2013; pp. 185–197. [CrossRef]
17. Kvatinsky, S.; Belousov, D.; Liman, S.; Satat, G.; Wald, N.; Friedman, E.G.; Kolodny, A.; Weiser, U.C. MAGIC—Memristor-aided logic. *IEEE Trans. Circuits Syst. II Exp. Briefs* **2014**, *61*, 895–899. [CrossRef]
18. Seshadri, V.; Hsieh, K.; Boroum, A.; Lee, D.; Kozuch, M.A.; Mutlu, O.; Gibbons, P.B.; Mowry, T.C. Fast bulk bitwise AND and OR in DRAM. *IEEE Comput. Archit. Lett.* **2015**, *14*, 127–131. [CrossRef]

19. Chi, P.; Li, S.; Xu, C.; Zhang, T.; Zhao, J.; Liu, Y.; Wang, Y.; Xie, Y. PRIME: A novel processing-in-memory architecture for neural network computation in ReRAM-based main memory. In Proceedings of the ACM/IEEE 43rd Annual International Symposium on Computer Architecture (ISCA), Seoul, Korea, 18–22 June 2016; pp. 27–39. [CrossRef]

20. Jain, S.; Ranjan, A.; Roy, K.; Raghunathan, A. Computing in memory with spin-transfer torque magnetic RAM. *IEEE Trans. Very Large Scale Integr. (VLSI) Syst.* **2018**, *26*, 470–483. [CrossRef]

21. Farmahini-Farahani, A.; Ahn, J.H.; Morrow, K.; Kim, N.S. NDA: Near-DRAM acceleration architecture leveraging commodity DRAM devices and standard memory modules. In Proceedings of the IEEE 21st International Symposium on High Performance Computer Architecture (HPCA), Burlingame, CA, USA, 7–11 February 2015; pp. 283–295. [CrossRef]

22. Imani, M.; Kim, Y.; Rosing, T. MPIM: Multi-purpose in-memory processing using configurable resistive memory. In Proceedings of the 22nd Asia and South Pacific Design Automation Conference (ASP-DAC), Chiba, Japan, 16–19 January 2017; pp. 757–763. [CrossRef]

23. Gao, F.; Tziantzioulis, G.; Wentzlaff, D. ComputeDRAM: In-memory compute using off-the-shelf DRAMs. In Proceedings of the 52nd Annual IEEE/ACM International Symposium on Microarchitecture (MICRO), Columbus, OH, USA, 12–16 October 2019; pp. 100–113. [CrossRef]

24. Li, S.; Xu, C.; Zou, Q.; Zhao, J.; Lu, Y.; Xie, Y. Pinatubo: A processing-in-memory architecture for bulk bitwise operations in emerging non-volatile memories. In Proceedings of the 53rd ACM/EDAC/IEEE Design Automation Conference (DAC), Austin, TX, USA, 5–9 June 2016; pp. 1–6. [CrossRef]

25. Li, S.; Niu, D.; Malladi, K.T.; Zheng, H.; Brennan, B.; Xie, Y. DRISA: A DRAM-based reconfigurable in-situ accelerator. In Proceedings of the 50th Annual IEEE/ACM International Symposium on Microarchitecture (MICRO), Boston, MA, USA, 14–17 October 2017; pp. 288–301.

26. Pattnaik, A.; Tang, X.; Jog, A.; Kayiran, O.; Mishra, A.K.; Kandemir, M.T.; Mutlu, O.; Das, C.R. Scheduling techniques for GPU architectures with processing-in-memory capabilities. In Proceedings of the International Conference on Parallel Architecture and Compilation Techniques (PACT), Haifa, Israel, 11–15 September 2016; pp. 31–44. [CrossRef]

27. Hsieh, K.; Ebrahim, E.; Kim, G.; Chatterjee, N.; O'Connor, M.; Vijaykumar, N.; Mutlu, O.; Keckler, S.W. Transparent offloading and mapping (TOM): Enabling programmer-transparent near-data processing in GPU systems. In Proceedings of the ACM/IEEE 43rd Annual International Symposium on Computer Architecture (ISCA), Seoul, Korea, 18–22 June 2016; pp. 204–216. [CrossRef]

28. Ahn, J.; Yoo, S.; Mutlu, O.; Choi, K. PIM-enabled instructions: A low-overhead, locality-aware processing-in-memory architecture. In Proceedings of the ACM/IEEE 42nd Annual International Symposium on Computer Architecture (ISCA), Portland, OR, USA, 13–17 June 2015; pp. 336–348. [CrossRef]

29. Nai, L.; Hadidi, R.; Xiao, H.; Kim, H.; Sim, J.; Kim, H. CoolPIM: Thermal-aware source throttling for efficient PIM instruction offloading. In Proceedings of the IEEE International Parallel and Distributed Processing Symposium (IPDPS), Vancouver, BC, Canada, 21–25 May 2018; pp. 680–689. [CrossRef]

30. Nair, R.; Antao, S.F.; Bertolli, C.; Bose, P.; Brunheroto, J.R.; Chen, T.; Cher, C.-Y.; Costa, C.H.A.; Doi, J.; Evangelinos, C.; et al. Active memory cube: A processing-in-memory architecture for exascale systems. *IBM J. Res. Dev.* **2015**, *59*, 17:1–17:14. [CrossRef]

31. Boroumand, A.; Ghose, S.; Patel, M.; Hassan, H.; Lucia, B.; Hsieh, K.; Malladi, K.T.; Zheng, H.; Mutlu, O. LazyPIM: An efficient cache coherence mechanism for processing-in-memory. *IEEE Comput. Archit. Lett.* **2017**, *16*, 46–50. [CrossRef]

32. Nai, L.; Hadidi, R.; Sim, J.; Kim, H.; Kumar, P.; Kim, H. GraphPIM: Enabling instruction-level PIM offloading in graph computing frameworks. In Proceedings of the IEEE International Symposium on High Performance Computer Architecture (HPCA), Austin, TX, USA, 4–8 February 2017; pp. 457–468. [CrossRef]

33. Addisie, A.; Bertacco, V. Centaur: Hybrid processing in on/off-chip memory architecture for graph analytics. In Proceedings of the 57th Annual Design Automation Conference (DAC), Virtual DAC, San Francisco, CA, USA, 20–24 July 2020.

34. Zhuo, Y.; Wang, C.; Zhang, M.; Wang, R.; Niu, D.; Wang, Y.; Qian, X. GraphQ: Scalable PIM-based graph processing. In Proceedings of the 52nd Annual IEEE/ACM International Symposium on Microarchitecture (MICRO), Columbus, OH, USA, 12–16 October 2019; pp. 712–725. [CrossRef]

35. Dai, G.; Huang, T.; Chi, Y.; Zhao, J.; Sun, G.; Liu, Y.; Wang, Y.; Xie, Y.; Yang, H. GraphH: A processing-in-memory architecture for large-scale graph processing. *IEEE Trans. Comput. Aided Design Integr. Circuits Syst.* **2019**, *38*, 640–653. [CrossRef]

36. Zhang, M.; Zhuo, Y.; Wang, C.; Gao, M.; Wu, Y.; Chen, K.; Kozyrakis, C.; Qian, X. GraphP: Reducing communication for PIM-based graph processing with efficient data partition. In Proceedings of the IEEE International Symposium on High Performance Computer Architecture (HPCA), Vienna, Austria, 24–28 February 2018; pp. 544–557. [CrossRef]

37. Xie, C.; Song, S.L.; Wang, J.; Zhang, W.; Fu, X. Processing-in-memory enabled graphics processors for 3D rendering. In Proceedings of the IEEE International Symposium on High Performance Computer Architecture (HPCA), Austin, TX, USA, 4–8 February 2017; pp. 637–648. [CrossRef]

38. Kim, B.-H.; Rhee, C.E. Making better use of processing-in-memory through potential-based task offloading. *IEEE Access* **2020**, *8*, 61631–61641. [CrossRef]

39. Li, J.; Wang, X.; Tumeo, A.; Williams, B.; Leidel, J.D.; Chen, Y. PIMS: A lightweight processing-in-memory accelerator for stencil computations. In Proceedings of the International Symposium on Memory Systems (MemSys), Washington DC, USA, 30 September–3 October 2019; pp. 41–52. [CrossRef]

40. Boroumand, A.; Ghose, S.; Kim, Y.; Ausavarungnirun, R.; Shiu, E.; Thakur, R.; Kim, D.; Kuusela, A.; Knies, A.; Ranganathan, P.; et al. Google workloads for consumer devices: Mitigating data movement bottlenecks. In Proceedings of the 23rd International Conference on Architectural Support for Programming Languages and Operating Systems (ASPLOS), Williamsburg, VA, USA, 24–28 March 2018; pp. 316–331. [CrossRef]

41. Gao, M.; Kozyrakis, C. HRL: Efficient and flexible reconfigurable logic for near-data processing. In Proceedings of the IEEE International Symposium on High Performance Computer Architecture (HPCA), Barcelona, Spain, 12–16 March 2016; pp. 126–137. [CrossRef]

42. Hadidi, R.; Nai, L.; Kim, H.; Kim, H. CAIRO: A compiler-assisted technique for enabling instruction-level offloading of processing-in-memory. *ACM Trans. Archit. Code Optim.* **2017**, *14*, 1–25. [CrossRef]

43. Angizi, S.; Fahmi, N.A.; Zhang, W.; Fan, D. PIM-Assembler: A processing-in-memory platform for genome assembly. In Proceedings of the 57th Annual Design Automation Conference (DAC) Virtual DAC, 20–24 July 2020; p. 6.

44. Seshadri, V.; Lee, D.; Mullins, T.; Hassan, H.; Boroumand, A.; Kim, J.; Kozuch, M.A.; Mutlu, O.; Gibbons, P.B.; Mowry, T.C. Buddy-RAM: Improving the performance and efficiency of bulk bitwise operations using DRAM. *arXiv* **2016**, arXiv:1611.09988.

45. Sutradhar, P.R.; Connolly, M.; Bavikadi, S.; Pudukotai Dinakarrao, S.M.; Indovina, M.A.; Ganguly, A. pPIM: A Programmable Processor-in-Memory Architecture with Precision-Scaling for Deep Learning. *IEEE Comput. Archit. Lett.* **2020**, *19*, 118–121. [CrossRef]

46. Akerib, A.; Ehrman, E. In-Memory Computational Device 2017. U.S. Patent 9653166B2, 16 May 2017.

47. Akerib, A.; Agam, O.; Ehrman, E.; Meyassed, M. Using Storage Cells to Perform Computation. U.S. Patent 8238173B2, 7 August 2012.

48. Kim, Y.-B.; Chen, T. Assessing merged DRAM/logic technology. In Proceedings of the IEEE International Symposium on Circuits and Systems. Circuits and Systems Connecting the World (ISCAS), Atlanta, GA, USA, 15 May 1996; pp. 133–136. [CrossRef]

49. Kim, Y.; Daly, R.; Kim, J.; Fallin, C.; Lee, J.H.; Lee, D.; Wilkerson, C.; Lai, K.; Mutlu, O. Flipping bits in memory without accessing them: An experimental study of DRAM disturbance errors. In Proceedings of the ACM/IEEE 41st International Symposium on Computer Architecture (ISCA), Minneapolis, MN, USA, 14–18 June 2014; pp. 361–372. [CrossRef]

50. Mutlu, O.; Kim, J.S. RowHammer: A retrospective. *IEEE Trans. Comput. Aided Design Integr. Circuits Syst.* **2020**, *39*, 1555–1571. [CrossRef]

51. Lee, B.C.; Ipek, E.; Mutlu, O.; Burger, D. Architecting phase change memory as a scalable DRAM alternative. In Proceedings of the 36th Annual International Symposium on Computer Architecture (ISCA), Austin, TX, USA, 20–24 June 2009; pp. 2–13. [CrossRef]

52. Qureshi, M.K.; Franceschini, M.M.; Lastras-Montaño, L.A. Improving read performance of phase change memories via write cancellation and write pausing. In Proceedings of the 16th International Symposium on High-Performance Computer Architecture (HPCA), Bangalore, India, 9–14 January 2010; pp. 1–11. [CrossRef]

53. Thakkar, I.G.; Pasricha, S. DyPhase: A Dynamic Phase Change Memory Architecture with Symmetric Write Latency and Restorable Endurance. *IEEE Trans. Comput. Aided Design Integr. Circuits Syst.* **2018**, *37*, 1760–1773. [CrossRef]

54. Pan, C.; Xie, M.; Hu, J.; Chen, Y.; Yang, C. 3M-PCM: Exploiting multiple write modes MLC phase change main memory in embedded systems. In Proceedings of the International Conference on Hardware/Software Codesign and System Synthesis (CODES+ISSS), New Delhi, India, 12–17 October 2014; pp. 1–10. [CrossRef]

55. Kadetotad, D.; Xu, Z.; Mohanty, A.; Chen, P.-Y.; Lin, B.; Ye, J.; Vrudhula, S.; Yu, S.; Cao, Y.; Seo, J. Parallel architecture with resistive crosspoint array for dictionary learning acceleration. *IEEE Trans. Emerg. Sel. Top. Circuits Syst.* **2015**, *5*, 194–204. [CrossRef]

56. Shafiee, A.; Nag, A.; Muralimanohar, N.; Balasubramonian, R.; Strachan, J.P.; Hu, M.; Williams, R.S.; Srikumar, V. ISAAC: A convolutional neural network accelerator with in-situ analog arithmetic in crossbars. In Proceedings of the ACM/IEEE 43rd Annual International Symposium on Computer Architecture (ISCA), Seoul, Korea, 18–22 June 2016; pp. 14–26. [CrossRef]

57. Song, L.; Zhuo, Y.; Qian, X.; Li, H.; Chen, Y. GraphR: Accelerating graph processing using ReRAM. In Proceedings of the IEEE International Symposium on High Performance Computer Architecture (HPCA), Vienna, Austria, 24–28 February 2018; pp. 531–543. [CrossRef]

58. Kaplan, R.; Yavits, L.; Ginosar, R.; Weiser, U. A Resistive CAM Processing-in-Storage Architecture for DNA Sequence Alignment. *IEEE Micro* **2017**, *37*, 20–28. [CrossRef]

59. Imani, M.; Gupta, S.; Rosing, T. Ultra-efficient processing in-memory for data intensive applications. In Proceedings of the 54th Annual Design Automation Conference (DAC), Austin, TX, USA, 18–22 June 2017; pp. 1–6. [CrossRef]

60. Wang, Y.; Kong, P.; Yu, H. Logic-in-memory based big-data computing by nonvolatile domain-wall nanowire devices. In Proceedings of the 13th Non-Volatile Memory Technology Symposium (NVMTS), Minneapolis, MN, USA, 12–14 August 2013; pp. 1–6. [CrossRef]

61. Butzen, P.F.; Slimani, M.; Wang, Y.; Cai, H.; Naviner, L.A.B. Reliable majority voter based on spin transfer torque magnetic tunnel junction device. *Electron. Lett.* **2016**, *52*, 47–49. [CrossRef]

62. Kang, W.; Chang, L.; Wang, Z.; Zhao, W. In-memory processing paradigm for bitwise logic operations in STT-MRAM. In Proceedings of the IEEE International Magnetics Conference (INTERMAG), Dublin, Ireland, 24–28 April 2017. [CrossRef]

63. Fan, D.; Angizi, S.; He, Z. In-memory computing with spintronic devices. In Proceedings of the IEEE Computer Society Annual Symposium on VLSI (ISVLSI), Bochum, Germany, 3–5 July 2017; pp. 683–688. [CrossRef]

64. Fan, D.; He, Z.; Angizi, S. Leveraging spintronic devices for ultra-low power in-memory computing: Logic and neural network. In Proceedings of the IEEE 60th International Midwest Symposium on Circuits and Systems (MWSCAS), Boston, MA, USA, 6–9 August 2017; pp. 1109–1112. [CrossRef]

65. Parveen, F.; He, Z.; Angizi, S.; Fan, D. HielM: Highly flexible in-memory computing using STT MRAM. In Proceedings of the 23rd Asia and South Pacific Design Automation Conference (ASP-DAC), Jeju, Korea, 22–25 January 2018; pp. 361–366. [CrossRef]

66. Bhosale, S.; Pasricha, S. SLAM: High performance and energy efficient hybrid last level cache architecture for multicore embedded systems. In Proceedings of the IEEE International Conference on Embedded Software and Systems (ICESS), Las Vegas, NV, USA, 2–3 June 2019; pp. 1–7. [CrossRef]

67. Imani, M.; Peroni, D.; Rosing, T. Nvalt: Nonvolatile approximate lookup table for GPU acceleration. *IEEE Embed. Syst. Lett.* **2018**, *10*, 14–17. [CrossRef]

68. Imani, M.; Gupta, S.; Arredondo, A.; Rosing, T. Efficient query processing in crossbar memory. In Proceedings of the IEEE/ACM International Symposium on Low Power Electronics and Design (ISLPED), Taipei, Taiwan, 24–26 July 2017; pp. 1–6. [CrossRef]

69. Xia, L.; Tang, T.; Huangfu, W.; Cheng, M.; Yin, X.; Li, B.; Wang, Y.; Yang, H. Switched by input: Power efficient structure for RRAM-based convolutional neural network. In Proceedings of the 2016 53nd ACM/EDAC/IEEE Design Automation Conference (DAC), Austin, TX, USA, 5–9 June 2016; pp. 1–6. [CrossRef]

70. Tang, X.; Kislal, O.; Kandemir, M.; Karakoy, M. Data movement aware computation partitioning. In Proceedings of the 50th Annual IEEE/ACM International Symposium on Microarchitecture (MICRO), Boston, MA, USA, 14–17 October 2017; pp. 730–744. [CrossRef]

71. Tsai, P.-A.; Chen, C.; Sanchez, D. Adaptive scheduling for systems with asymmetric memory hierarchies. In Proceedings of the 51st Annual IEEE/ACM International Symposium on Microarchitecture (MICRO), Fukuoka, Japan, 20–24 October 2018; pp. 641–654. [CrossRef]

72. Lockerman, E.; Feldmann, A.; Bakhshalipour, M.; Stanescu, A.; Gupta, S.; Sanchez, D.; Beckmann, N. Livia: Data-centric computing throughout the memory hierarchy. In Proceedings of the Twenty-Fifth International Conference on Architectural Support for Programming Languages and Operating Systems (ASPLOS), Lausanne, Switzerland, 16–20 March 2020; pp. 417–433. [CrossRef]

73. Wen, W.; Yang, J.; Zhang, Y. Optimizing power efficiency for 3D stacked GPU-in-memory architecture. *Microprocess. Microsyst.* **2017**, *49*, 44–53. [CrossRef]

74. Choi, J.; Kim, B.; Jeon, J.-Y.; Lee, H.-J.; Lim, E.; Rhee, C.E. POSTER: GPU based near data processing for image processing with pattern aware data allocation and prefetching. In Proceedings of the 28th International Conference on Parallel Architectures and Compilation Techniques (PACT), Seattle, WA, USA, 23–26 September 2019; pp. 469–470. [CrossRef]

75. Zhang, D.; Jayasena, N.; Lyashevsky, A.; Greathouse, J.L.; Xu, L.; Ignatowski, M. TOP-PIM: Throughput-oriented programmable processing in memory. In Proceedings of the 23rd International Symposium on High-Performance Parallel and Distributed Computing (HPDC), Vancouver, BC, Canada, 23–27 June 2014; pp. 85–98. [CrossRef]

76. Hsieh, K.; Khan, S.; Vijaykumar, N.; Chang, K.K.; Boroumand, A.; Ghose, S.; Mutlu, O. Accelerating pointer chasing in 3D-stacked memory: Challenges, mechanisms, evaluation. In Proceedings of the IEEE 34th International Conference on Computer Design (ICCD), Scottsdale, AZ, USA, 2–5 October 2016; pp. 25–32. [CrossRef]

77. Scrbak, M.; Greathouse, J.L.; Jayasena, N.; Kavi, K. DVFS space exploration in power constrained processing-in-memory systems. In Proceedings of the 30th International Conference on Architecture of Computing Systems (ARCS), Vienna, Austria, 3–6 April 2017; pp. 221–233. [CrossRef]

78. Eckert, Y.; Jayasena, N.; Loh, G.H. Thermal feasibility of die-stacked processing in memory. In Proceedings of the 2nd Workshop on Near-Data Processing (WoNDP), Cambridge, UK, 14 December 2014.

79. Nai, L.; Hadidi, R.; Xiao, H.; Kim, H.; Sim, J.; Kim, H. Thermal-aware processing-in-memory instruction offloading. *J. Parallel Distrib. Comput.* **2019**, *130*, 193–207. [CrossRef]

80. Gokhale, M.; Lloyd, S.; Hajas, C. Near memory data structure rearrangement. In Proceedings of the International Symposium on Memory Systems (MEMSYS), Washington DC, USA, 5–8 October 2015; pp. 283–290. [CrossRef]

81. Hybrid Memory Cube Specification 2.1. 2 February 2017. Available online: https://web.archive.org/web/20170202004433/; http://hybridmemorycube.org/files/SiteDownloads/HMC-30G-VSR_HMCC_Specification_Rev2.1_20151105.pdf (accessed on 12 August 2020).

82. Ankit, A.; Sengupta, A.; Panda, P.; Roy, K. RESPARC: A reconfigurable and energy-efficient architecture with memristive crossbars for deep spiking neural networks. In Proceedings of the 54th Annual Design Automation Conference (DAC), Austin, TX, USA, 18–22 June 2017; pp. 1–6. [CrossRef]

83. Mittal, S.; Vetter, J.S. AYUSH: A technique for extending lifetime of SRAM-NVM hybrid caches. *IEEE Comput. Archit. Lett.* **2015**, *14*, 115–118. [CrossRef]

84. Tang, S.; Yin, S.; Zheng, S.; Ouyang, P.; Tu, F.; Yao, L.; Wu, J.; Cheng, W.; Liu, L.; Wei, S. AEPE: An area and power efficient RRAM crossbar-based accelerator for deep CNNs. In Proceedings of the IEEE 6th Non-Volatile Memory Systems and Applications Symposium (NVMSA), Hsinchu, Taiwan, 16–18 August 2017; pp. 1–6. [CrossRef]

85. Zha, Y.; Li, J. IMEC: A fully morphable in-memory computing fabric enabled by resistive crossbar. *IEEE Comput. Archit. Lett.* **2017**, *16*, 123–126. [CrossRef]

86. Zidan, M.A.; Jeong, Y.; Shin, J.H.; Du, C.; Zhang, Z.; Lu, W.D. Field-programmable crossbar array (FPCA) for reconfigurable computing. *IEEE Trans. Multi-Scale Comput. Syst.* **2018**, *4*, 698–710. [CrossRef]

87. Zheng, L.; Zhao, J.; Huang, Y.; Wang, Q.; Zeng, Z.; Xue, J.; Liao, X.; Jin, H. Spara: An energy-efficient ReRAM-based accelerator for sparse graph analytics applications. In Proceedings of the IEEE International Parallel and Distributed Processing Symposium (IPDPS), New Orleans, LA, USA, 18–22 May 2020; pp. 696–707. [CrossRef]

88. Challapalle, N.; Rampalli, S.; Jao, N.; Ramanathan, A.; Sampson, J.; Narayanan, V. FARM: A flexible accelerator for recurrent and memory augmented neural networks. *J. Signal. Process. Syst.* **2020**, *92*, 1–15. [CrossRef]
89. Abunahla, H.; Halawani, Y.; Alazzam, A.; Mohammad, B. NeuroMem: Analog graphene-based resistive memory for artificial neural networks. *Sci. Rep.* **2020**, *10*, 9473. [CrossRef]
90. Alibart, F.; Zamanidoost, E.; Strukov, D.B. Pattern classification by memristive crossbar circuits using ex situ and in situ training. *Nat. Commun.* **2013**, *4*, 2072. [CrossRef] [PubMed]
91. Guo, Q.; Alachiotis, N.; Akin, B.; Sadi, F.; Xu, G.; Low, T.M.; Pileggi, L.; Hoe, J.C.; Franchetti, F. 3D-stacked memory-side acceleration: Accelerator and system design. In Proceedings of the Workshop on Near-Data Processing (WoNDP), Cambridge, UK, 14 December 2014.
92. Ahmed, H.; Santos, P.C.; Lima, J.P.C.; Moura, R.F.; Alves, M.A.Z.; Beck, A.C.S.; Carro, L. A compiler for automatic selection of suitable processing-in-memory instructions. In Proceedings of the Design, Automation Test in Europe Conference Exhibition (DATE), Florence, Italy, 25–29 March 2019; pp. 564–569. [CrossRef]
93. Corda, S.; Singh, G.; Awan, A.J.; Jordans, R.; Corporaal, H. Platform independent software analysis for near memory computing. In Proceedings of the 22nd Euromicro Conference on Digital System Design (DSD), Kallithea, Greece, 28–30 August 2019; pp. 606–609. [CrossRef]
94. Pattnaik, A.; Tang, X.; Kayiran, O.; Jog, A.; Mishra, A.; Kandemir, M.T.; Sivasubramaniam, A.; Das, C.R. Opportunistic computing in GPU architectures. In Proceedings of the ACM/IEEE 46th Annual International Symposium on Computer Architecture (ISCA), Phoenix, AZ, USA, 22–26 June 2019; pp. 210–223. [CrossRef]
95. Asghari-Moghaddam, H.; Son, Y.H.; Ahn, J.H.; Kim, N.S. Chameleon: Versatile and practical near-DRAM acceleration architecture for large memory systems. In Proceedings of the 49th Annual IEEE/ACM International Symposium on Microarchitecture (MICRO), Taipei, Taiwan, 15–19 October 2016; pp. 1–13. [CrossRef]
96. Sura, Z.; O'Brien, K.; Nair, R.; Jacob, A.; Chen, T.; Rosenburg, B.; Sallenave, O.; Bertolli, C.; Antao, S.; Brunheroto, J.; et al. Data access optimization in a processing-in-memory system. In Proceedings of the 12th ACM International Conference on Computing Frontiers (CF), Ischia, Italy, 18–21 May 2015; pp. 1–8. [CrossRef]
97. Xiao, Y.; Nazarian, S.; Bogdan, P. Prometheus: Processing-in-memory heterogeneous architecture design from a multi-layer network theoretic strategy. In Proceedings of the Design, Automation Test in Europe Conference Exhibition (DATE), Dresden, Germany, 19–23 March 2018; pp. 1387–1392. [CrossRef]
98. Ahn, J.; Hong, S.; Yoo, S.; Mutlu, O.; Choi, K. A scalable processing-in-memory accelerator for parallel graph processing. In Proceedings of the ACM/IEEE 42nd Annual International Symposium on Computer Architecture (ISCA), Portland, OR, USA, 13–17 June 2015; pp. 105–117. [CrossRef]
99. Pouchet, L.-N.; Yuki, T. PolyBench/C 4.1. SourceForge. Available online: http://polybench.sourceforge.net/ (accessed on 12 August 2020).
100. Kim, Y.; Song, Y.H. Analysis of thermal behavior for 3D integration of DRAM. In Proceedings of the 18th IEEE International Symposium on Consumer Electronics (ISCE), JeJu Island, Korea, 22–25 June 2014; pp. 1–2. [CrossRef]
101. Ipek, E.; Mutlu, O.; Martínez, J.F.; Caruana, R. Self-Optimizing Memory Controllers: A Reinforcement Learning Approach. In Proceedings of the International Symposium on Computer Architecture, Beijing, China, 21–25 June 2008; pp. 39–50. [CrossRef]
102. Santos, P.C.; de Lima, J.P.C.; de Moura, R.F.; Ahmed, H.; Alves, M.A.Z.; Beck, A.C.S.; Carro, L. Exploring IoT platform with technologically agnostic processing-in-memory framework. In Proceedings of the Workshop on INTelligent Embedded Systems Architectures and Applications (INTESA), Turin, Italy, 4 October 2018; pp. 1–6. [CrossRef]
103. Xiao, Y.; Xue, Y.; Nazarian, S.; Bogdan, P. A load balancing inspired optimization framework for exascale multicore systems: A complex networks approach. In Proceedings of the IEEE/ACM International Conference on Computer-Aided Design (ICCAD), 13–16 November 2017; pp. 217–224. [CrossRef]

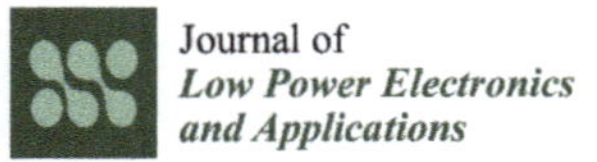

Journal of
*Low Power Electronics
and Applications*

Article

Reliability-Aware Resource Management in Multi-/Many-Core Systems: A Perspective Paper

Siva Satyendra Sahoo *,†, **Behnaz Ranjbar** *,† **and Akash Kumar** *

CFAED, Technische Universität Dresden, 01062 Dresden, Germany
* Correspondence: siva_satyendra.sahoo@tu-dresden.de (S.S.S.); behnaz.ranjbar@tu-dresden.de (B.R.); akash.kumar@tu-dresden.de (A.K.)
† These authors contributed equally to this work.

Abstract: With the advancement of technology scaling, multi/many-core platforms are getting more attention in embedded systems due to the ever-increasing performance requirements and power efficiency. This feature size scaling, along with architectural innovations, has dramatically exacerbated the rate of manufacturing defects and physical fault-rates. As a result, in addition to providing high parallelism, such hardware platforms have introduced increasing unreliability into the system. Such systems need to be well designed to ensure long-term and application-specific reliability, especially in mixed-criticality systems, where incorrect execution of applications may cause catastrophic consequences. However, the optimal allocation of applications/tasks on multi/many-core platforms is an increasingly complex problem. Therefore, reliability-aware resource management is crucial while ensuring the application-specific Quality-of-Service (QoS) requirements and optimizing other system-level performance goals. This article presents a survey of recent works that focus on reliability-aware resource management in multi-/many-core systems. We first present an overview of reliability in electronic systems, associated fault models and the various system models used in related research. Then, we present recent published articles primarily focusing on aspects such as application-specific reliability optimization, mixed-criticality awareness, and hardware resource heterogeneity. To underscore the techniques' differences, we classify them based on the design space exploration. In the end, we briefly discuss the upcoming trends and open challenges within the domain of reliability-aware resource management for future research.

Keywords: multi/many-core platforms; reliability; resource management; mixed-criticality

Citation: Sahoo, S.S.; Ranjbar, B.; Kumar, A. Reliability-Aware Resource Management in Multi-/Many-Core Systems: A Perspective Paper. *J. Low Power Electron. Appl.* **2021**, *11*, 7. https://doi.org/10.3390/jlpea11010007

Academic Editor: Amit Kumar Singh

Received: 17 December 2020
Accepted: 19 January 2021
Published: 25 January 2021

1. Introduction

From the colossus machines of 1943 [1], to the modern Internet of Thing (IoT) devices, there has been a massive growth in the variety of applications that use electronic computing systems. Every sector of our day-to-day lives—Consumer products, Telecommunication, Education, Agriculture, Healthcare, Automobiles, Military defence etc.—usually involves some form of information processing on electronic systems. While the scale of such information processing platforms can vary from small energy-harvesting IoT nodes to large data centres, the overall computing performance requirements for each of the application areas has undoubtedly increased in the last two decades. Prior to the 2000s, the major semiconductor manufacturers would answer the need for increasing performance requirements by technology scaling and micro-architectural enhancements only. Methods such as deep pipelining, increased cache sizes and complex dynamic Instruction Level Parallelism (ILP) were the primary tools as long as Dennard scaling could be achieved [2]. However, as shown in Figure 1a, the quest for higher clock frequency at lower technology nodes resulted in high power density and heat dissipation beyond the capacity of inexpensive cooling methods. This phenomenon, is often referred to as the power-wall [3]. Further, the continued technology scaling and micro-architectural innovations did not necessarily translate to faster memory technologies. As a result, the increasing gap between the

compute clock frequency and the memory access frequency, referred to as the memory-wall, did not provide much benefits even at very high computation speeds. As a result, as shown in Figure 1b, the rate of improvement in single thread performance reduced considerably [4]. Further, the data dependencies of each application imposes implicit limits to the extent to which the application can exploit ILP. Traditional approaches to exploiting ILP such as deeper pipelines and branch prediction can exacerbate the power density problem while increasing the time and energy wastage due to branch mispredictions.

The cumulative effect of the three walls—memory, power and ILP—has led to diminishing returns from the efforts to provide performance scaling by increasing clock frequency of single core systems. As shown in Figure 1b, since 2005, the semiconductor industry has adopted the on-chip integration of multiple cores and processors as the weapon-of-choice for satisfying increasing computation complexity. In addition to using the ILP for each core, the multi-/many-core systems allow the software developer to exploit any form of thread-, task- and data-level parallelism in the application. Further, the technology scaling allows for additional application-specific cores to be integrated on the chip. The heterogeneity of the cores could be targeted for generic objectives such as power-performance trade-offs (e.g., big.LITTLE from ARM [5]) or for some specific computations (e.g., the CELL Broadband Engine [6]). However, the quest for increasing the number of on-chip transistors through technology scaling and improving performance of each core through architectural innovations have also increased the reliability issues across all electronic systems. Increased unreliability—either in terms of computation errors or the reduced lifetime of systems—has led to the increasingly complex problem of ensuring the reliable execution of applications on increasingly unreliable hardware [7].

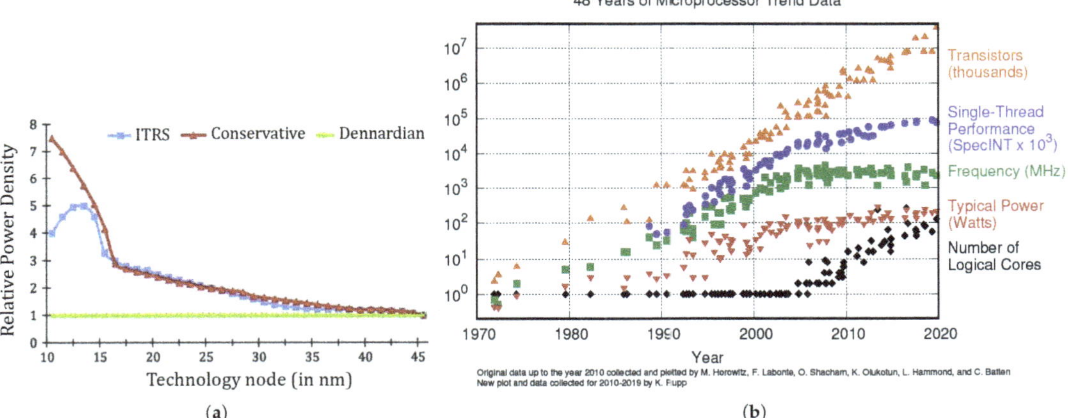

Figure 1. Technology scaling and its impact. (**a**) Power density trends. (**b**) Impact of cheaper transistors [4].

1.1. Need of Reliability in Multi/Many-Core Systems

The increasing unreliability of electronic systems can be explained using the bath-tub curves shown in Figure 2. The reliability-specific life-cycle of typical electronic systems is characterized by three types of failures. The infant mortality, caused by premature failure of weak components as a result of manufacturing defects and burn-in testing, the constant failures due to random faults and the wearout-based faults due to aging. The solid curve in the figure shows the net effect of all three factors. With aggressive technology scaling, the rate of manufacturing defects has increased, resulting in higher infant mortality and higher susceptibility to aging-related faults. Similarly, the architectural innovations such as deeper pipelines and supply voltage scaling have reduced the clocking-window masking and electrical masking, respectively [8], thus increasing the constant failure rate due to random faults. Further, the parallel processing of multiple cores can increase the power

dissipation resulting in accelerated aging. The net result of all these factors is the increased failure rate as shown by the dashed bath-tub curve in Figure 2. However, each of the these factors can manifest as degradation of different system-level performance metrics and the priority of each such metric may vary with each application.

For instance, in real-time systems the timeliness of execution has the highest priority. Similarly, in financial and scientific computations, the accuracy of calculations would be more important than the execution time. Additionally, in systems such as consumer products and space missions, extended lifetime of the system may have higher priority. Further, in a system executing multiple applications, each of the application may have varying criticality w.r.t. each reliability-related performance metric. In this scenario, the ideal solution would be to design a custom hardware platform for each application. However, from a market perspective, every application may not warrant the development of a custom hardware platform. Hence, standard multi-/many-core systems should be used for ensuring application-specific Quality of Service (QoS) requirements—both reliability-related and otherwise.

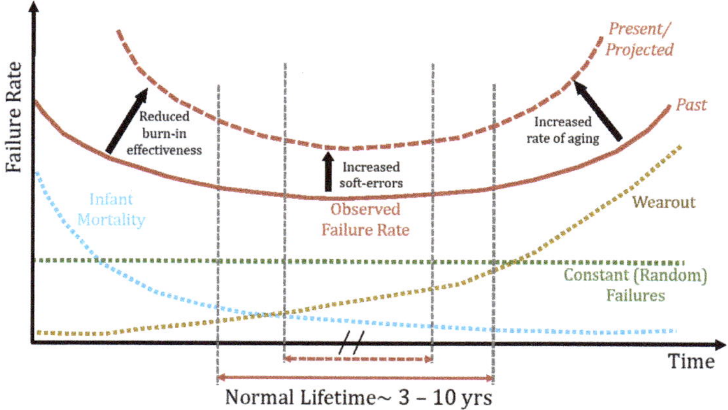

Figure 2. Increasing unreliability in electronic systems.

1.2. Reliability-Aware Resource Management in Multi-/Many-Core Systems

Given the variety of applications executed on multi-/many-core systems, optimal allocation of on-chip resources is an increasingly complex problem. We define this reliability-aware resource management problem as below:

Definition 1. *Reliability-aware resource management refers to the appropriate allocation of on-chip resources—computation, communication and memory—to applications/tasks executing on a multi-/many-core system while ensuring application-specific QoS and optimizing other system-level performance goals.*

The increasing unreliability in electronic systems has also resulted in a large body of research being devoted to ensuring reliable execution of applications. The research works related to solving the problem stated in Def. Figure 1 usually focus on one or more of following aspects:

1. Application Specificity: The varying priority among QoS metrics across different applications presents both a scope for application-specific optimization as well as challenges for ensuring application specific constraints.
2. Mixed criticality: Scheduling tasks with different criticality levels on a common platform is challenging, in which executing the tasks must be guaranteed in terms of both safety and real-time aspects to prevent the probability of failure and, consequently, catastrophic consequences.

3. Resource Heterogeneity: The heterogeneity of cores provides a scope for leveraging the availability of custom hardware implementations. Similarly, the availability of reconfigurable logic provides the scope for implementing accelerators on standard hardware platforms. However, such heterogeneity also introduces additional complexity for ensuring optimal resource sharing.

4. Design Space Exploration: All the above aspects introduce additional degrees of freedom in the design space. Consequently, the Design Space Exploration (DSE) for the joint optimization across all these aspects can result in an exponential increase in complexity.

In this article, we provide a survey of some of the more recent research works that explore these aspects. The rest of the article is organised as follows. We provide a brief overview of the relevant background and the taxonomy that is used through the rest of the article in Section 2. A generic system model and the generic problem statement along with the classification of the approaches is presented in Sections 3 and 4, respectively. We provide a detailed survey of related works across Sections 5–7. We present a brief discussion of emerging approaches to reliability management in Section 8. Finally, we conclude the article with a summary in Section 9.

2. Background and Taxonomy for Reliability Management Methodologies

The research works surveyed in this article aim to improve one or more aspects of reliability using different techniques of resource management under varying scenarios. The scenario can vary depending upon the executing application(s) and the resources available on the hardware platform. Similarly, the resource management could be aimed at maximizing different types of reliability and may be achieved by implementing various DSE methods. Figure 3 shows the taxonomy for the various terms that will be used frequently in the rest of the article for reviewing the related works. The tree-like structure in Figure 3 is used to categorize and show the relationships among these different aspects addressed by the related works discussed in this article. The figure also serves as a checklist for determining the scope and the assumptions used in each reviewed article. The colour coding of the rectangular boxes relates to the major aspects of reliability management as discussed in Section 1.2. The orange boxes show the terms related to reliability-types and causes. The gray and green boxes correspond to application and architecture scenarios, respectively. The blue boxes show the terms related to DSE. The current and the next section provide the background of and relationship (shown as arrows) among the terms shown in Figure 3.

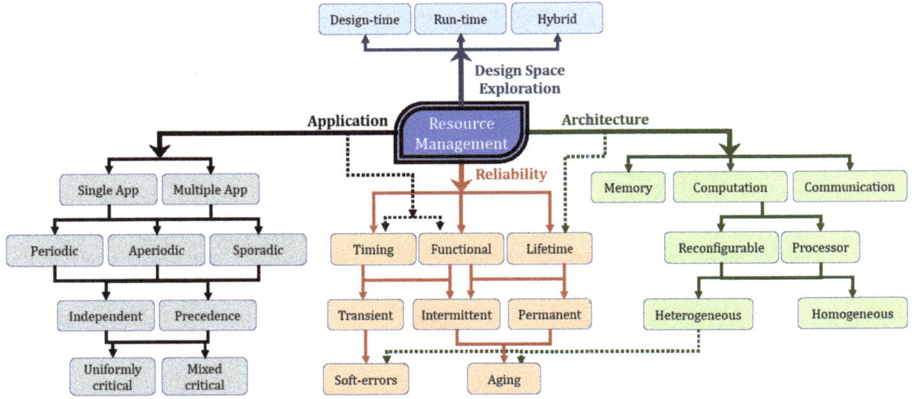

Figure 3. Taxonomy: The terms used in the article are categorized under four aspects of reliability-aware resource management—Application, Architecture, Reliability and Design Space Exploration.

2.1. Reliability in Electronic Systems

The rise in fault-rates in electronic systems has led to adverse effects on system performance in more than one way. Direct effects include application failures in terms of incorrect functionality and inability of the system to complete execution within the specified timing constraints. Similarly, indirect effects include over-designing for fault-mitigation—e.g., the high power dissipation of Triple Modular Redundancy (TMR)—leading to accelerated aging in the system and multiple re-executions of some critical tasks—for reducing chances of error—resulting in deadline violations. Given the variations in application-specific requirements across different application domains, the reliability-related QoS can be categorized as discussed next.

2.1.1. Lifetime Reliability

The expected operational life of the system can be characterized by its lifetime reliability. Depending upon whether the system is repairable, and the cost of such repairs, metrics such as Mean Time To Failure (MTTF), Mean Time To Crash (MTTC) and Availability can be used to characterize the system's lifetime reliability. MTTF refers to the expected time to the first observed failure in the system. In healthcare applications and consumer electronics, the need for predictable and extended MTTF can be the primary objective. Similarly, MTTC refers to the expected operational time for the point at which the system does not have sufficient resources for ensuring the expected behavior and is usually applicable for repairable systems. In applications with long mission times such as space exploration, repairing the failure mechanism is used to extend the MTTC. However, repair-time plays a critical role in high-availability applications such as automated control of power generation.

The reduced lifetime in electronic systems usually occurs due to the aging caused by electrical stress [9]. Most research works focus on improving the lifetime reliability by one/more of the following methods—reducing continuous computation on the processing elements, reducing the power dissipation, improving heat dissipation, reducing computations involved in executing the application etc. The availability of multiple processing elements enables improving the lifetime reliability by utilizing such techniques more effectively. For instance, the electrical stress on a single processing element can be reduced by using different resources for different execution instances. However, selecting the appropriate optimization technique along with the optimal configuration poses a significant research problem.

2.1.2. Timing Reliability

The performance of the system in terms of the expected behavior concerning the timeliness of execution completion can be expressed as its timing reliability. It is used only in terms of real-time systems and depending upon the criticality of the application, can be expressed in terms of Worst-case Execution Time (WCET), Mean Time between Failures (MTBF), Probability of Completion and Average Makespan [10–12]. WCET is usually used in hard real-time systems such as pacemakers and automobile safety features where any instance of missing a deadline can have fatal consequences. MTBF, frequently used in the context of repairable systems, can also be used for expressing the timing reliability in firm real-time systems such as manufacturing assembly lines, where infrequent failures can be tolerated, provided sufficient availability is ensured. Average makespan and probability of completion are usually used in soft real-time systems such as streaming devices and gaming consoles where frequent deadline misses can be tolerated as long as they do not affect user experience.

As more and more applications that require fast reaction times implement some level of automaton, for example autonomous driving, timing reliability can be a prime QoS metric for a large number of systems. Usual methods for improving timing reliability include faster execution, isolation of critical computation and communication etc. The spatial parallelism in many/multi-core systems can be used to exploit the inherent parallelism in an application and reduce the execution latency. However, distributing the computation

across multiple processing elements introduces design complexities in terms of isolation and communication latency.

2.1.3. Functional Reliability

With the rising constant failure rates during the normal life of the system, the chances of such failures manifesting as incorrect computations has also increased. Hence, in applications that require high levels of computational accuracy such as financial transactions in point-of-sales systems or ATMs, scientific applications, the corresponding QoS can be expressed in terms of functional reliability. It concerns the correctness of the results computed by a system operating in a fault-inducing environment. The functional reliability can be quantified by the probability of no physical fault-induced errors occurring during application execution or the MTBF. Improving the functional reliability usually involves implementing one/more among–Spatial, Temporal and Informational redundancy. In many/muti-core systems, additional processing elements can be used to implement spatial redundancy effectively without considerable overheads on the execution latency. However, the increased power dissipation overheads and reduced spatial parallelism for computation resulting from this approach can adversely affect the other reliability metrics.

2.2. Fault Model

The reliability-specific events in a system can be classified as one of–failure, error and fault [13]. An application failure refers to an event where the service delivered by the system deviates from the expected service defined by the application requirements. An error refers to the deviation of the system from a correct service state to an erroneous one. Faults refer to the adjudged or hypothesized cause of the error. The physical faults in a system can be further classified into the following types, based on their frequency and persistence of occurrence.

1. *Transient faults* occur at a particular time, remain in the system for some period and then disappear. Such faults are initially dormant but can become active at any time. Examples of such faults occur in hardware components which have an adverse reaction to some external interference, such as electrical fields or radioactivity.
2. *Intermittent faults* show up in systems from time to time due to some inherent design issue or aging. An example is a hardware component that is heat sensitive—it works for some time, stops working, cools down and then starts to work again.
3. *Permanent faults* such as a broken wire or a software design error show a more persistent behavior than intermittent faults—start at a particular time and remain in the system until they are repaired.

The reliability-aware resource management presented in this article concerns the mitigation of physical faults only. Such physical faults, if unmasked, lead to errors which in turn may lead to application failures. The manifestation of all types of physical faults can be studied under two broad categories—Soft-errors and Aging. The effects of soft-errors caused by transient faults are considered for functional reliability analysis. Similarly, the aging-related intermittent and permanent faults are used in the analysis for lifetime reliability. Timing reliability issues occur usually due to aging-induced slower execution and additional computations that are employed to mitigate soft-errors.

2.2.1. Soft-Errors

Soft-error refer to non-reproducible hardware malfunctions caused by transient faults. The additional charge induced by external interference (such as alpha particles from radioactive impurities in chip packaging materials [14], and neutrons generated by cosmic radiation's interaction with the earth's atmosphere [15]) can sometimes (when $> Q_{crit}$) lead to changing the logic value of the affected nodes in the system. While in memory elements the changed value is retained until the next refresh, in combinational circuits the computations are affected only if the wrong value is latched by a memory element. The probability of such computational errors is reduced by either of the three masking

effects—Logical masking, Electrical masking and Latching-window masking. In the quest for faster faster systems, although logical masking has remained unchanged, the deeper pipelines and aggressive voltage scaling have reduced latching-window masking and electrical masking, respectively, leading to increased Soft Error Rate (SER).

2.2.2. Aging

The term aging broadly refers to the degradation of semiconductor devices due to continued electrical stress that may lead to timing failures and reduced operational life of the integrated circuits (ICs). The primary physical phenomena causing aging are listed next [16].

1. Bias Temperature Instability (BTI) results in an increase in the threshold voltage, V_{th} due to the accumulation of charge in the dielectric material of the transistors [17]. The use of high-k dielectrics in lower technology nodes has resulted in an increased contribution of BTI to aging.
2. Hot Carrier Injection (HCI) occurs when charge carriers with higher energy than the average stray out of the conductive channel between the source and drain and get trapped in the insulating dielectric [18]. Eventually it leads to building up electric charge within the dielectric layer, increasing the voltage needed to turn the transistor on.
3. Time Dependent Dielectric Breakdown (TDDB) comes into play when a voltage applied to the gate creates electrically active defects within the dielectric, known as traps, that can join and form an outright short circuit between the gate and the current channel. Unlike the other aging mechanisms, which cause a gradual decline in performance, the breakdown of the dielectric can lead to the catastrophic failure of the transistor, causing a malfunction in the circuit.
4. Electromigration (EM) [19] occurs when a surge of current knocks metal atoms loose and causes them to drift along with the flow of electrons. The thinning of the metal increases the resistance of the connection, sometimes to the point that it can become an open circuit. Similarly, the accumulation of the drifted material can lead to electrical shorts.

3. System Model

The research works reviewed in this article, in general, aim at improving the reliability of the system. These improvements could be aimed at mitigating one or more types of faults across a subset of all the components of the system. In order to evaluate the impact of the proposed improvements, each of the works makes certain assumptions regarding the system components. We provide an overview of the system components under two models—Architecture and Application.

3.1. Architecture Model

The architecture model encapsulates all the features and the assumptions regarding the hardware platform. Figure 4 shows the components of a typical architecture model. The fault mechanisms discussed in Section 2.2 can affect different aspects of processing-computation, memory and communication.

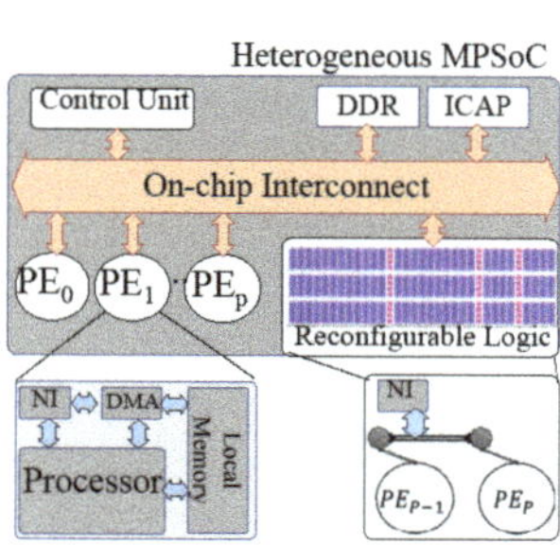 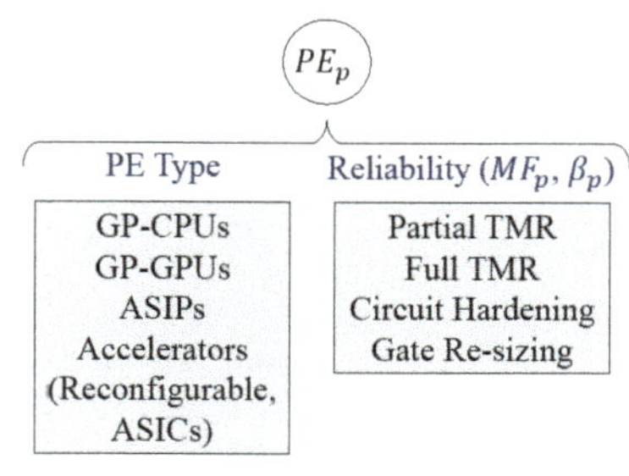

Figure 4. Architecture model.

3.1.1. Computation

As shown in Figure 4 we will use the term Processing Element (PE) to refer to each core/processor/accelerator implemented on the reconfigurable logic of the architecture. Soft-errors and the stuck-at faults due to aging directly affect the factional reliability. Similarly aging-related delays can result in timing violations. A more indirect impact of the timing and lifetime reliability can be observed as a result of fault-mitigation measures implemented for reducing errors in computation. Spatial redundancy measures such as TMR and Dual Modular Redundancy (DMR) result in higher power dissipation, thereby accelerating aging. Similarly, temporal and information-redundancy based methods introduce additional computations for processing the same workload and can led to degradation of both timing and lifetime reliability. The architecture may be homogeneous or heterogeneous w.r.t. the extent of spatial redundancy and other hardening techniques implemented in each of the PEs. Other types of heterogeneity may be due to the result of varying implementations for low-power design (ex. Big-Little from ARM), ISAs, ASIPs, etc.

3.1.2. Communication

As shown in Figure 4, we use the term on-chip interconnect to denote the set of communication-related components on the hardware architecture. Although the implementation of the on-chip interconnect may vary among bus-based, etc., with the rising number of PEs, Network-on-Chip (NoC)-based communication is being increasingly used in multi-/many-core systems [20]. Reliability of the on-chip interconnects involves ensuring error-free inter PE communication by using a combination of different types of redundancy methods across multiple layers of the Open Systems Interconnect (OSI) model. A detailed account of various reliability issues in on-chip interconnects and their mitigation can be found in [21,22]. Resource management of communication elements usually involves allocating links and routers to communicating tasks. Depending upon the reliability requirements and the criticality, the allocation algorithm may choose to provide different types of redundancy methods.

3.1.3. Memory

Similar to computation, soft-errors and stuck-at faults may result in incorrect values being read from the memory elements on the hardware platform, leading to reduced functional reliability. Information redundancy in the form of additional bits for Error Checking and Correcting (ECC) is commonly used for both Static Random Access Memory (SRAM)-based caches and Dynamic Random Access Memory (DRAM)-based main memory. Hamming [23] or Hsiao [24] code based Single-bit-Error-Correcting (SEC) and Double-bit-Error-Detecting (DED) codes are usually sufficient for most systems. More robust methods like Double-bit-Error-Correcting (DEC) and Triple-bit-Error-Detecting (TED) codes can be used for higher resilience against random bit errors. Storage overhead and power are the cost factors associated with design of a resilient memory system. Flexible error protection methods can enable adaptation to system-level requirements, both at design-time as well

as run-time. ECC granularity and fault-coverage provide tunable parameters, while the memory controller acts as the tuning knob for varying error protection levels based on system requirements.

3.2. Application Model

In general, one application or multiple applications within a system can be executed on a platform according to its mission. These applications consist of different tasks whose properties are dependency, periodicity, execution time, and criticality. We introduce each property of these tasks briefly.

3.2.1. Task Dependencies

Tasks within an application can be dependent or independent to each other. Independent tasks mean that there is no precedence or communication among them, while the dependency represents the flow of data between tasks and induces a partial order on the task set. Such precedence relations are usually described through a Directed Acyclic Graph (DAG), where tasks are represented by nodes, and precedence relations by arrows. Figure 5a represents a combination of dependent (as DAG) and independent tasks. Synchronous Data Flow Graph (SDFG) is another common application representation that models cyclic dependency. This task model is used in streaming multimedia applications, in which support for pipe-lined execution are needed [12,25,26]. Figure 5b shows an example of SDFG for the H.263 encoder application [25,26], in which nodes are called actors. Each actor is executed by reading data from its inputs and writing the results as a token on the output port.

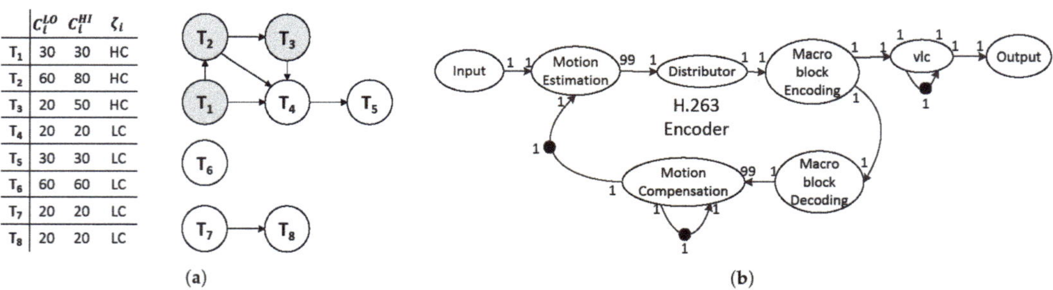

	C_i^{LO}	C_i^{HI}	ζ_i
T_1	30	30	HC
T_2	60	80	HC
T_3	20	50	HC
T_4	20	20	LC
T_5	30	30	LC
T_6	60	60	LC
T_7	20	20	LC
T_8	20	20	LC

(a)

(b)

Figure 5. Examples of task dependency. (**a**) DAG of UAV application [27]. (**b**) SDFG of H.263 encoder application [26].

3.2.2. Application/Tasks' Periodicity

Another property of tasks is their execution periodicity (i.e., the time of activation to be executed) that can be periodic, aperiodic, or sporadic. A periodic task consists of an infinite sequence of jobs, that are regularly activated at each period (i.e., the time to complete one iteration). An aperiodic task also consists of an infinite sequence of jobs, but their activations are not regularly interleaved. Sporadic tasks are aperiodic tasks where consecutive jobs are separated by minimum initiation time interval.

3.2.3. Application/Tasks' Criticality

In general, all tasks running on a common platform, may not be equally critical (i.e., not uniform criticality) for doing a correct service. Avionics, automotive and medical devices are examples of these systems. Indeed, due to the various safety demand for tasks, within an application, tasks can have different reliability requirements and criticality levels [28]. The systems with different criticality tasks are called mixed-criticality. In this regard, a set of industrial standards, e.g., DO-178B [29], has been introduced with five levels of safety, i.e., A, B, C, D, and E, (A and E provide the highest and the lowest levels of safety, respectively). A failure occurring in tasks with different criticality levels has a

different impact on the system, which are shown in Table 1. The Probability-of-Failure-per-Hour (PFH) values (adopted by safety standards for tasks' safety measurement) have been determined for all the criticality levels to guarantee system safety.

In these mixed-criticality systems, the correct execution of tasks with higher criticality levels and QoS of tasks with lower criticality levels are considered, especially for service-oriented systems. Furthermore, mixed-criticality tasks may be real-time, i.e., the high-criticality tasks must be executed correctly before their deadlines to not cause catastrophic consequences. For tasks with lower-criticality levels, guaranteeing the deadlines is commonly regarded as a QoS parameter. Therefore, to ensure the correct execution of high-criticality tasks in any situation and the minimum QoS of low-criticality tasks while the system computation time is beneficially utilized, different WCET estimations, pessimistic (C_i^{HI}) and optimistic (C_i^{LO}), are used. As a result, different operational modes, according to the number of WCET estimations are considered for these systems. Figure 5a shows an example of mixed-criticality applications, Unmanned Air Vehicle (UAV), where some tasks are dependent on other tasks. In this application, T_1, T_2 and T_3 are the high-criticality (HC) tasks which are responsible for the collision avoidance, navigation, and stability of the system. Failure in the execution of these tasks may lead to system failure and cause irreparable damage to the system. Besides, low-criticality (LC) tasks ($\{T_4, \ldots, T_8\}$ in the figure) are responsible for recording sensors data, GPS coordination, and video transmissions, to help the system carry out its mission successfully. Furthermore, as shown in this figure, since the correct execution of high-criticality tasks is crucial, two different WCETs are computed for them.

Table 1. DO-178B safety requirement [29].

x	A	B	C	D	E
PFH_x	$<10^{-9}$	$<10^{-7}$	$<10^{-5}$	$\geq 10^{-5}$	-
Failure Condition	Catastrophic	Hazardous	Major	Minor	No Effect

From the mixed-criticality system operation perspective, the system starts execution in the low-criticality mode in which all tasks are expected to finish their execution before their optimistic WCET (C_i^{LO}). If the execution of at least one high-criticality task exceeds its C_i^{LO}, while the correct output of the task is not ready, the system switches to the high-criticality mode and all task are expected to finish their execution before their pessimistic WCET (C_i^{HI}). Hence, since the requested demand for the system computation time is increased in this mode, some low-criticality tasks may be dropped to guarantee the correct execution of high-criticality tasks.

4. Reliability Management in Multi/Many-Core Systems

4.1. Problem Statement

The related research problem involves the optimization of the allocation of the available hardware resources to the computation, communication and storage requirements of an application(s) while satisfying the reliability and criticality constraints. We formulate an optimisation problem that serves as a generic framework, and the research works mentioned in this article are discussed with reference to it.

Application: Each application is represented as a tuple $G_g(\mathbb{T}_g, \mathbb{E}_g)$ containing the set of nodes $\mathbb{T}_g$ representing tasks/actors, and a set of directed edges $\mathbb{E}_g$, representing the precedence and the communication between the tasks. Similarly, a scenario denoting the applications executing in parallel can be represented by a set of applications $S_s = \{G_1, G_2, \ldots\}$.

Architecture: Similar to an application, a mesh NoC-based hardware platform can be represented by a tuple $H(\mathbb{SW}, \mathbb{C})$. $\mathbb{SW}$ is the set of switches and $\mathbb{C}$ represents the connections among the switches. Each switch $Sw_s \in \mathbb{SW}$ can be attached to one or more

cores. Assuming P_s cores are connected to Sw_s, the total number of cores in $H(\mathbb{SW}, \mathbb{C})$ is $|\mathbb{P}| = \sum |P_s|_{Sw_s \in \mathbb{SW}}$, where the set $\mathbb{P}$ denotes the collection of cores.

Allocation: Resource management involves timely allocation of appropriate hardware resources for all the tasks and communication in the scenario. We represent the resource allocation by two sets—(1) Task allocation: $T_{alloc} = \{(T_t, P_t, St_t), \forall T_t \in \mathbb{T}\}$, where $\mathbb{T}$ is the set of all tasks in the scenario, and (2) Edge allocation: $E_{alloc} = \{(E_e, C_e, St_e), \forall E_t \in \mathbb{E}\}$, where $\mathbb{E}$ is the set of all edges in the scenario. For a feasible solution, the sets T_{alloc} and E_{alloc} must satisfy certain constraints. The constraints may be of the following types:

- Precedence constraints: Any task must start execution only after all its preceding tasks and incoming communication is complete.
- Scheduling constraints: Properties such as each task/actor assigned to a single core, not more than a single task executing on a single core at a time etc. must be satisfied.
- Criticality constraints: PFH and reliability requirements should be corresponded with the criticality levels.

Performance Estimation: Varying the resource allocation results in varying system-level performance. Different methods of estimating these performance metrics—both analytical and empirical—as a function of T_{alloc} and E_{alloc} have been used across the works discussed in this article. We use the notation shown in Equation (1) to denote these methods. The estimated performance is used during the decision-making for reliability resource management. While the priority of each metric varies with application, Equation (2) shows the generic optimization problem. The terms $w_{\langle m \rangle}$ determine the application-specific priority of the system-level metrics ($\langle m \rangle$). Similarly, the terms $c_{\langle m \rangle}$ are used to denote the existence of constraints due to application-specific QoS requirements. The terms $\mathbb{T}_{alloc}$ and $\mathbb{E}_{alloc}$ represent the set of all possible task and edge allocations, respectively.

$$\text{System-level Performance Estimation: Timing Reliability: } \mathcal{T}_{sys} = funcR_T(T_{alloc}, E_{alloc})$$
$$\text{Functional Reliability: } \mathcal{F}_{sys} = funcR_F(T_{alloc}, E_{alloc}); \text{Lifetime Reliability: } \mathcal{L}_{sys} = funcR_L(T_{alloc}, E_{alloc}) \quad (1)$$
$$\text{Power Dissipation: } \mathcal{W}_{sys} = funcP(T_{alloc}, E_{alloc}); \text{Energy Consumption: } \mathcal{J}_{sys} = funcE(T_{alloc}, E_{alloc})$$

$$\underset{\forall T_{alloc} \in \mathbb{T}_{alloc}, \forall E_{alloc} \in \mathbb{E}_{alloc}}{\text{minimize}} \quad \{w_{\mathcal{T}}\mathcal{T}_{sys}, w_{\mathcal{F}}\mathcal{F}_{sys}, w_{\mathcal{L}}\mathcal{L}_{sys}, w_{\mathcal{W}}\mathcal{W}_{sys}, w_{\mathcal{J}}\mathcal{J}_{sys}\}$$
$$\text{s.t., } \mathcal{T}_{sys} \geq c_{\mathcal{T}} \, \mathcal{T}_{SPEC}; \quad (2)$$
$$\mathcal{F}_{sys} \geq c_{\mathcal{F}} \, \mathcal{F}_{SPEC}; \, \mathcal{L}_{sys} \geq c_{\mathcal{L}} \, \mathcal{L}_{SPEC};$$
$$\mathcal{J}_{sys} \leq c_{\mathcal{J}} \, \mathcal{J}_{SPEC}; \, \mathcal{W}_{sys} \leq c_{\mathcal{W}} \, \mathcal{W}_{SPEC};$$

4.2. Classification of Solution Approaches

The research articles discussed in this survey employ various methods for finding the optimal resource allocation. Further, each article focuses on optimizing a subset of the system-level performance metrics shown in Equation (2). In addition to providing an overview of the various approaches and prioritization (among different metrics) presented in each of the articles, we also classify them based on the criterion shown in Figure 6. Some of the criteria, such as fault-types, application model, architecture model, and criticality have been covered in the earlier sections. Additionally the articles can be classified on the experimental evaluation methodology. While some works use analytical methods to estimate the effectiveness of their proposed methods, others use simulation based approaches. In most of the cases real-world applications, in terms of standard benchmark suites have been used for the experiments. Along with reliability, other system level metrics such as power dissipation, energy consumption, throughput and thermal limits have been reported from the experiments as well. Further, some works—especially for reconfigurable systems—use resource utilization as an optimization objective.

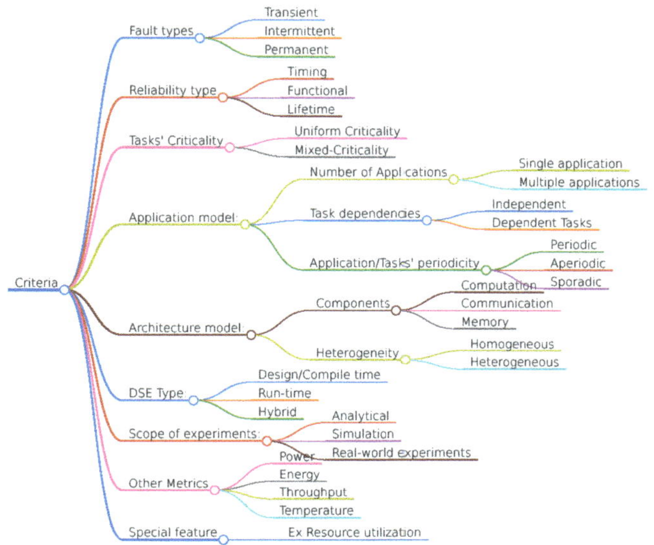

Figure 6. Classification criterion.

The DSE approach used in the research works can be categorized under (1) Design-/compile-time, (2) Run-time, and (3) Hybrid. While in most cases the methods used in each article can be classified under one of the three types, in some cases more than one of the approaches have also been used—typically for different objectives.

- Design-/compile-time: In this approach all the design decisions and the related optimizations are performed before the system is deployed. As shown in Figure 7, the related analysis is made under the design-time assumptions regarding both the application workload and the system's hardware resource availability. This approach allows the designers to generate highly optimized solutions. However, it also limits the adaptability of the system to dynamic operating conditions.
- Run-time: This approach involves implementing all resource allocation decisions only after the deployment of the system. This allows the system to adapt to varying operating conditions—both external and internal. External variations might include changing external radiation, changing workload resulting in varying QoS requirements etc. Internal variations include the changing performance/availability of the cores due to aging, as shown in Figure 7, low energy availability in mobile systems etc. In the run-time DSE approach the dynamic adaptability comes at the cost of result quality. Since all the resource management decisions are determined at run-time, it may result in sub-optimal solutions due to computation and availability constraints.
- Hybrid: The hybrid approach attempts to combine the best of both design-/compile-time and run-time approaches. It usually involves analysing most of the possible run-time operating conditions, finding the optimal solution for each condition of design time and storing the optimal solution to be used for run-time adaptation. As shown in Figure 7, the design-time analysis could involve determining the possible scenarios and determining the optimal solution for each scenario for varying core availability that might change during run-time due to aging and the resulting physical faults.

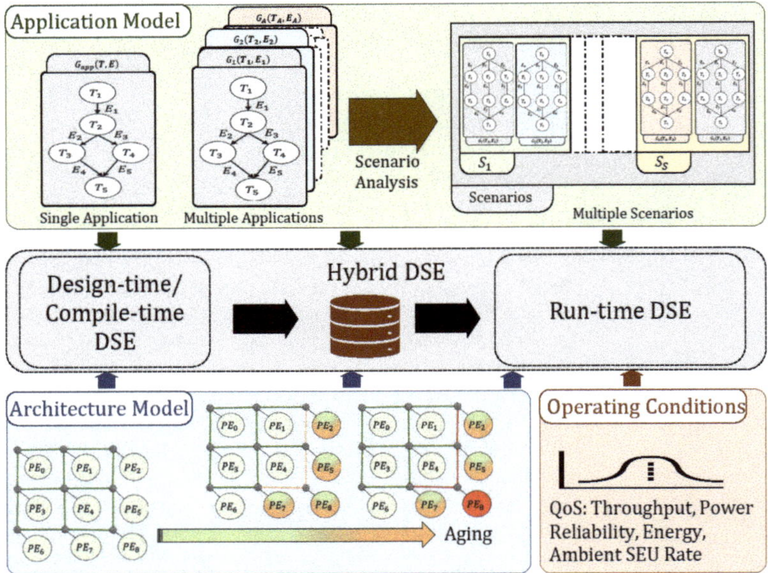

Figure 7. DSE approaches: Design-time/Compile-time, Run-time and Hybrid.

5. Lifetime Reliability Management in Multi/Many-Core Processors

Resource management for improving lifetime reliability may involve direct approaches such as allocating spare cores in case of faults or indirect approaches such as wear-leveling and thermal management to delay the onset of faults. We summarize the related works under three categories; design-time approaches (D.T), run-time approaches (R.T), and hybrid (H), which we discuss in detail as follows. Table 2 lists the works in lifetime reliability management of these three categories for both uniform and mixed-criticality task models in detail. In this table, we also present the considered criteria in the state-of-the-arts, such as fault model (Transient, Intermittent, permanent), system model (Components and Heterogeneity), and application model (Dependency-Independent (I), Dependent (D), Periodicity-Periodic (P), Sporadic (S), Aperiodic (A)), that have been explained in detail, in Section 2.2, Section 3.1, and Section 3.2, respectively.

5.1. Design-Time Strategies

A purely design-time DSE approach to improving the system's lifetime reliability usually involves significant analytical estimation of workload stress. Furthermore, the scope of design-time analysis is limited by the assumptions of complete knowledge of the applications that will be executed on the hardware platform. Similarly, the estimation usually employs assumptions about the variability of the execution time of the various tasks in the application. The case for resource allocation explicitly targeting lifetime improvements is presented by [30]. Hartman er al. showed the improvement in system lifetime when task-mapping was optimized directly for reducing aging rather than trying to reduce the thermal stress in the system. The authors used an Ant Colony Optimization (ACO)-based design-time optimization of lifetime and compared the results with randomized task-mapping and task-mapping optimized for reducing the temperature using Simulated Annealing (SA). The authors attribute the improvements in the ACO-based approach to the negligence of the temperature optimization approach to factors such as supply voltage, current density and the aging-related interaction between the application and the architecture. Ref. [31] improved upon the lifetime optimization approach with task-mapping by determining the appropriate allocation of computation, communication and memory resources during the design of an NoC-based system for an application. Specifically, the au-

thors have proposed the search for the optimal allocation of execution slack, storage slack and communication architecture during the hardware platform design. The authors extended this approach in [32] to perform joint optimization of system lifetime and the yield. While the optimization for lifetime involves slack distribution such that the system can survive most wear-out induced failures, the yield optimization involves surviving the most defect-induced failures in the system. The optimization methodologies presented in [30–32] use system-level simulation for estimating the lifetime as a result of the design decisions. In [33], Ma et al. presented a Multi-Armed Bandit (MAB)-based exploration for allocating less simulations for sampling of weaker solutions. The authors compared their approach against regular Monte-Carlo Simulations (MCS) in the optimization of lifetime-aware Multi-Processor System-on-Chip (MPSoC) design and report similar quality of results as MCS with up to $5.26\times$ fewer samples. A more analytical approach, proposed in [34], for the estimation of system-level lifetime reliability in multi-core systems have found use in multiple research works that rely on design-time optimization for the DSE problem. In [35], Das et al., used the average execution time of each task, and the corresponding wear-out due to EM, to estimate the systems MTTF for varying task-mapping configurations and different Dynamic Voltage and Frequncy Scaling (DVFS) modes. The net aging effect on each core has been modelled as the average across all the tasks mapped to the core in every period (of periodic applications). The aging estimation methodology used in [35] is based on the techniques presented by [34]. Further, [26] use similar lifetime estimation approach to present a DSE methodology for showing the trade-off between permanent and transient fault-tolerance. Specifically, the authors explore the effect of temporal redundancy on the EM-related wear-out failures. Figure 8a shows the results from [26], depicting the effect of increasing number of checkpoints in the tasks of an application, on the average (expected) execution time and transient- and permanent-fault reliability. A more indirect approach of improving system lifetime by reducing the core temperatures is presented in [36], where the authors explore both the temporal and spatial effects of task-mapping on the peak temperature of a core. Recently, [37] have proposed a more generic DSE framework for joint optimization across varying types of redundancy methods and multiple design objectives. The authors presented the benefits of using a cross-layer optimization approach and proposed improved Multi-Objective Evolutionary Algorithms (MOEA)-based search methods for the large design space. Most design-time DSE works for lifetime reliability use the wear-out estimation for electromigration. Such EM-related wear-out effects are particularly disruptive for on-chip communication components. To this end, [38] proposed a dual physical channel switch architecture that was designed to improve the system's lifetime in the presence of permanent faults. The design-time analysis and optimization for mixed-criticality systems introduces the additional complexity of considering multiple criticality levels across the tasks. Related research for mixed criticality systems are discussed next.

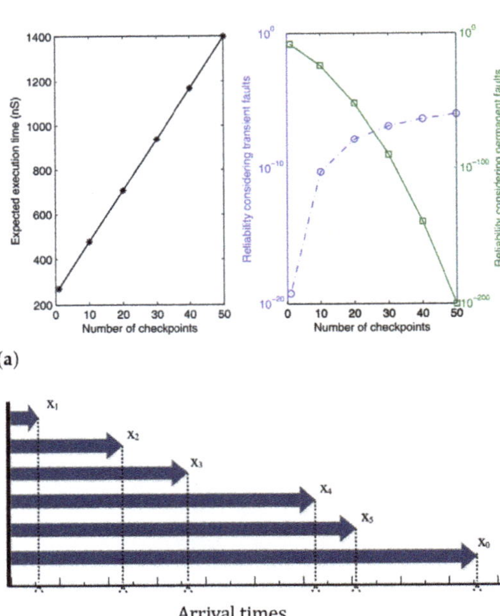

(**a**)

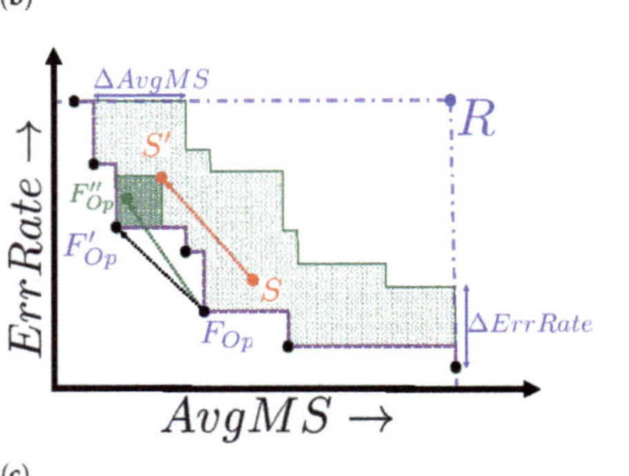

$$U\_val = \frac{\sum x_i/n - x_0/2}{x_0\sqrt{1/(12n)}}$$

(**b**)

(**c**)

Figure 8. Design space exploration for lifetime reliability. (**a**) Design-time DSE results in [26]. (**b**) Aging estimation for run-time DSE [39]. (**c**) Hybrid DSE [40].

Table 2. Summary of state-of-the-art approaches in lifetime reliability aware resource management.

	Fault Model			App. Model			System Model				Imp.		DSE	Technique
	Transient	Intermittent	Permanent	Criticality	Dependency	Periodicity	Communication	Computation	Memory	Heterogeneity	Real Board	Real App.		
Hartman'10 [30]	×	×	✓	×	D	P	×	✓	×	Het.	×	✓	D.T	Task Mapping
Meyer'10 [31]	×	×	✓	×	D	P	✓	✓	✓	Het.	×	✓	D.T	Slack Allocation
Meyer'14 [32]	×	×	✓	×	D	P	✓	✓	✓	Het.	×	✓	D.T	Slack Allocation, Yield
Ma'17 [33]	×	×	✓	×	D	P	✓	✓	×	Het.	×	✓	D.T	MAB Simulation
Das'14 [35]	✓	×	✓	×	D	P	×	✓	×	Hom.	×	✓	D.T	Task Mapping
Das'13 [26]	✓	×	✓	×	D	P	×	✓	×	Het.	×	✓	D.T	Task Mapping
Das'14 [36]	×	×	✓	×	D	P	×	✓	×	Hom.	×	✓	D.T	Task Mapping
Sahoo'20 [37]	✓	×	✓	×	D	P	×	✓	×	Het.	×	×	D.T	Task Mapping
Kakoee'11 [38]	✓	×	✓	×	×	×	✓	×	×	×	×	✓	D(R).T	Hardware Redundancy
Hartman'12 [41]	×	×	✓	×	D	P	✓	✓	×	Het.	×	✓	R.T	Task (Re)-Mapping
Duque'15 [42]	×	✓	✓	×	D	P	×	✓	×	Hom.	×	✓	R.T	Task (Re)-Mapping
Sahoo'16 [39]	×	✓	✓	×	D	P	×	✓	×	Hom.	×	✓	R.T	Task (Re)-Mapping
Rathore'19 [43]	×	×	✓	×	D	P	×	✓	×	Hom.	×	✓	R.T	Task (Re)-Mapping
Venkataraman'15 [44]	×	×	✓	×	D	P	×	✓	×	Hom.	×	×	R.T	Hardware Migration
Wang'19 [45]	×	×	✓	×	D	P	×	✓	×	Hom.	×	✓	R.T	Task (Re)-Mapping
Haghbayan'16 [46]	×	×	✓	×	D	P	×	✓	×	Hom.	×	✓	R.T	Task (Re)-Mapping
Haghbayan'17 [47]	×	×	✓	×	D	P	×	✓	×	Hom.	×	✓	R.T	Task (Re)-Mapping
Rathore'18 [48]	×	×	✓	×	D	P	×	✓	×	Hom.	×	✓	R.T	Task (Re)-Mapping
Raparti'17 [49]	×	×	✓	×	D	P	✓	✓	×	Hom.	×	✓	R.T	(3D) Task Mapping
Bauer'15 [50]	✓	✓	✓	×	D	P	✓	✓	×	Het.	×	✓	R.T	Multi-layer
Das'13 [51]	×	×	✓	×	D	P	✓	✓	×	Hom.	×	✓	H	Task (Re)-Mapping
Das'16 [52]	×	×	✓	×	D	P	✓	✓	×	Hom.	×	✓	H	Task (Re)-Mapping
Das'13 [53]	×	✓	✓	×	D	P	×	✓	×	Hom.	×	✓	H	Task (Re)-Mapping
Bolchini'13 [54]	×	×	✓	×	D	P	✓	✓	×	Hom.	×	✓	H	Task (Re)-Mapping
Namazi'19 [55]	✓	×	✓	×	D	P	✓	✓	×	Hom.	×	✓	H	Task (Re)-Mapping
Kriebel'16 [56]	✓	×	✓	×	D	P	✓	✓	×	Hom.	×	✓	H	Task (Re)-Mapping
Nahar'15 [57]	✓	×	✓	×	D	P	✓	✓	×	Hom.	×	✓	H	Redundancy
Sahoo'19 [40]	✓	×	✓	×	D	P	×	✓	×	Het.	×	×	H	Task (Re)-Mapping
Axer'11 [58]	✓	×	✓	✓	I	P	×	✓	×	Hom.	×	×	D.T	Redundancy
Pathan'17 [59]	✓	×	✓	✓	I	S	×	✓	×	Hom.	×	×	D.T	Redundancy
Safari'20 [60]	✓	×	✓	✓	D	P	×	✓	×	Hom.	×	×	D.T	Redundancy
Saraswat'09 [61]	✓	×	✓	✓	I	P	✓	✓	✓	Het.	×	✓	R.T	Task Re-Mapping
Liu'13 [62]	×	×	✓	✓	I	P	×	✓	×	Hom.	×	×	R.T	Task Re-Mapping
Ranjbar'19 [63]	×	×	×	✓	D	P	×	✓	×	Hom.	×	✓	R.T	Thermal Management
Iacovelli'18 [64]	×	×	✓	✓	I	P	×	✓	×	Hom.	×	×	H	Task (Re)-Mapping
Bayati'16 [65]	×	×	✓	✓	I	S	×	✓	×	Hom.	×	×	H	Task (Re)-Mapping

Using redundancy in multi-core systems is one of the most useful techniques to manage permanent faults and consequently, lifetime reliability at design-time, used in [58–60]. Ref. [58] have presented a reliability analysis to tolerate soft errors by using checkpointing and redundancy techniques. The application is modeled as a set of independent tasks that consist of fault-tolerant tasks, which are more critical, and non-fault-tolerant tasks. In this paper, the lifetime reliability is represented by MTTF, and to improve the reliability, in addition to checkpointing, redundancy is used that each task is mapped two times perhaps on the same core (time redundancy) or different cores (hardware redundancy). To evaluate their method and show the improvement of lifetime reliability, the results are compared to a reference Monte-Carlo simulation. The work in [59] has presented a resource-efficient scheduling algorithm for independent safety-critical sporadic tasks. In this algorithm, first, the number of sufficient backups (multiple instances) for each task is determined to guarantee timing and lifetime reliability. Then, to reduce the processing

resources consumption, the number of active backups for tasks are determined and the other backups are counted as passive. Hence, active backups can execute in parallel with primary tasks, while the passive backups would be executed in the case of occurring faults in active backups. In the end, the effectiveness of their proposed method is shown by using an example application.

In [60], researchers have presented an approach for dependent mixed-criticality tasks running on homogeneous multi-core processors, in which parallelism and redundancy policy have been applied for fault-tolerance. They enhance the reliability at design-time by using spare cores and tolerate both transient and permanent faults. Hence, they just improve the reliability of tasks with higher criticality by running task replica on the spare core. To allocate tasks on cores, first, all tasks are mapped to primary cores, and if the deadline of at least one high-criticality task is missed, the parallelism policy is used. It means the low-criticality tasks that can be scheduled concurrently are re-mapped one by one to spare cores until high-criticality tasks' deadlines are met. If there is still a high-criticality task whose deadline is missed, the reduction policy is used in which the QoS of low-criticality tasks is aggravated by reducing the worst-case execution time of low-criticality tasks on primary cores. To enhance reliability, they schedule the replicas of high-criticality tasks in the spare core by postponing the execution of replicas as much as possible. After allocating tasks on primary and spare cores, the free slack in each core is used to minimize the energy consumption by changing the V-f levels, such that the reliability constraints of criticality tasks would not be violated. Eventually, the presented method has been evaluated in simulation by using some random task graphs.

5.2. Run-Time Strategies

Run-time optimization of system lifetime usually involves periodically assessing the aging-level of each core and other components in the hardware platform and modifying the workload distribution in order to achieve wear-leveling. Wear-out estimation using special hardware structures and corresponding task-remapping to improve the system lifetime was proposed by [41]. The authors in [41] assumed the presence of wear sensors in every PE in the architecture and showed that the usage of such sensors instead of temperature sensors can improve the lifetime of a system by up to 14.6% compared to a thermal optimization approach. A heuristic scoring system, based on the weighted score from the wear sensors of the various architecture elements, is used for evaluating the candidate mapping solutions at run-time and perform the task-remapping accordingly. There have been more recent works that leverage the frequency of the occurrence of intermittent faults for wear-out estimation instead of using special hardware structures such as those used by [41]. For instance, ref. [42] presented a averaging window-based approach, that computes the average number of intermittent faults observed in each core over a fixed number of past execution cycles (moving window), to determine the wear-out level in each core. They ranked the cores in terms of their wear-out level to remap the tasks of an application accordingly. However, the evaluation in [42] involved experiments with the Infant Mortality stage of the system lifecycle (Figure 2), which primarily represents the initial failures due to manufacturing defects and burn-in tests, and not the wear-out region that represents the aging of electronic components. Ref. [39] improved upon this approach by using the Centroid test [66] over the arrival times of the intermittent faults in each core, as shown in Figure 8b. The centroid test presents a better statistical estimate of the wear-out of each core compared to a moving window approach. While the moving window approach suffers from a form of short-term memory, the centroid test is more reflective of the trends of aging seen by any arbitrary PE. Using this approach, considerable improvements over [42] were reported in [39]. Both [39,42] used the partial repairable nature of aging mechanism such as Negative Bias Temperature Instability (NBTI) to improve the lifetime both in terms of MTTF and MTTC. However, both [39,42] present the results with known workloads and do not consider the optimization for new application execution requests. Ref. [43] presented an aging-aware run-time task-remapping methodology that can cater to

new application requests in addition to using a Reinforcement Learning (RL)-based online adaptation method for ensuring performance and improving lifetime. The authors present a learning based approach to keep track of the interactions between the tasks and cores to learn the process variations and the aging effects in the many-/multi-core system in order to determine the optimal operating frequency for each core. The authors show their approach to be scalable with the number of cores in the architecture. A novel approach to reducing the overheads during task-remapping due to permanent or intermittent faults was presented by [44]. The authors proposed a hardware-based task-migration technique that could reduce the re-mapping latency by up to 6.5×. A host of such run-time estimation and wear-leveling across multiple layers and different architectures are presented in [50]. Most works focused towards improving system lifetime treat the performance metrics such as throughput and latency as constraints and the lifetime as the optimization objective. However, ref. [45] presented a resource management approach that aims at optimizing the throughput under user-specified system lifetime constraints. The authors presented a run-time DSE methodology that implements a borrowing strategy to differentiate the resource allocation for compute- and communication-intensive applications. Specifically, the proposed method relaxes the short-term lifetime reliability constraints to improve throughput for communication-intensive applications while ensuring long-term lifetime of the system.

Most of the articles discussed in this section do not show the scalability of their proposed methods for many number of cores in the architecture. The scaling performance of these methods is especially important in the dark silicon era [67]. Dark Silicon refers to the phenomenon where, with each technology generation, the thermal and power limits of the system reduces the fraction of transistors that can operate at maximum frequency. Therefore, appropriate run-time resource management becomes crucial for enabling the useful integration of a large number of cores in the architecture. In [46], Haghbayan et al. present a methodology for lifetime-aware run-time task-mapping in many-core systems for the dark silicon era. The proposed technique involved running a long-term reliability analysis unit to track the aging of the cores along with a short-term re-mapping unit. The re-mapping unit utilizes the information from the analysis unit to provide longer recovery times to highly aging cores in the system. In [47], Haghbayan et al. extended this approach for the joint optimization of performance and lifetime reliability. Additionally, they adopted a hierarchical approach to task re-mapping where the first stage determined the appropriate region in the hardware and the second stage determined the appropriate PEs in the selected region to be used for mapping the application. A similar hierarchical approach to lifetime-aware run-time mapping of applications to many-core systems is presented by [48], where the authors intersperse the selected region with dark cores to maintain the thermal limits and reduce accelerated aging. An aging-aware run-time task-mapping for 3D NoC-based systems is presented in [49]. In this work, Raparti et al. consider the additional aging effects faced due to the 3D nature of the hardware platform. The higher current densities and limited number of power pins in such structures warrant special considerations for the EM-related aging of the power delivery network along with the aging of the PEs in the system.

Next, we discuss some related research in the context of mixed criticality systems. As we mentioned, one of the techniques to guarantee lifetime reliability is using task re-mapping at run-time. As shown in Table 2, Refs. [61,62] have used this technique that we explain each work in detail. Ref. [61] proposed a heuristic to manage the transient and permanent faults for mixed-criticality systems that tasks can be hard or soft (a task is said to be hard if missing its deadline may cause catastrophic consequences and, also a task is said to be soft if missing the deadline cause a performance degradation [68]). In this heuristic, checkpointing and roll-back recovery is used to tolerate transient faults and also task re-mapping to tolerate permanent faults. Meeting the deadlines of hard tasks and maximizing the QoS of soft tasks through task re-mapping are the targets of the heuristic. In the case of a permanent failure for a core, first the hard tasks and then soft tasks re-

mapped to other healthy cores. Besides, researchers in [62] have proposed an algorithm for independent periodic mixed-criticality tasks to minimize the number of task re-mapping, while the most critical applications continue to meet their deadlines in homogeneous multi-core systems. When a core fails due to the permanent faults, first high-criticality tasks are re-mapped to other healthy cores, then, low-criticality tasks running on the processor may be re-mapped to other processors or even dropped to increase performance. Indeed, the algorithm makes a trade-off between the number of task re-allocations and the system's performance. The efficiency of the algorithm has been evaluated through simulation with a random task generation.

On the other hand, in [63,69], online peak power, and thermal management heuristic has been proposed, in which the re-mapping technique is used in the case of available dynamic slack to re-map a ready task from the hot core to a core with a lower temperature to manage the system's maximum temperature which improves the lifetime reliability. In addition, researchers have also proposed an approach to assign available dynamic slack to an appropriate task among k look-ahead tasks, which has more impact on system power and maximum temperature and reduce the V-f levels. The proposed method has been evaluated for dependent periodic mixed-criticality tasks (both real-life and random task set generation) in simulation. Figure 9 depicts an example of the thermal hotspot mitigation result of the system based on the proposed method in [63] and a state-of-the-art [27]). As shown, the proposed method can help in balancing the difference in temperature between the cores, which results in lifetime reliability improvement in the long-term.

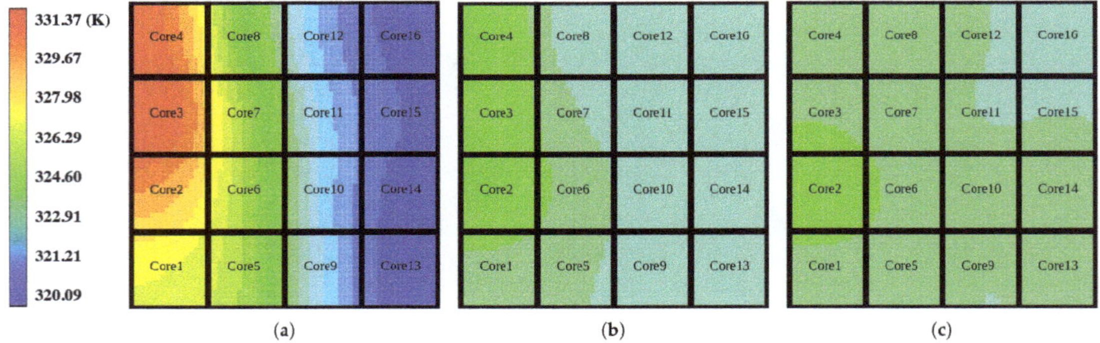

Figure 9. Temperature profiles of different approaches. (**a**) [27]. (**b**) [63], k = 1. (**c**) [63], k = 4.

5.3. Hybrid Strategies

The hybrid DSE approach to improving lifetime reliability usually involves searching and storing multiple system configurations for various fault/aging-scenarios at design-/compile-time that the system can be reconfigured to during run-time. Correspondingly, ref. [51] presented a methodology for reliability-driven task-mapping for MPSoCs that could be used for multiple applications as well. Their compile-time analysis used convex optimization to find the optimal task-mapping for different system states resulting from permanent faults to one or more cores. The run-time optimization in [51] involved considering the aging of the NoC for determining the appropriate task-mapping to be selected dynamically. The experimental evaluation in [51], involved testing the proposed methods for both DAG and SDFG representation of synthetic and real-world applications. Ref. [52] used a similar approach with the added design objective of energy consumption and the consideration of the thermal effect of adjoining cores on the reliability of any arbitrary core. In another similar work, ref. [53] presented hybrid DSE approaches that considered the occurrence of both permanent and intermittent faults during design-time analysis. The run-time optimization in [53] takes into account the energy consumption of task-remapping during the selection of the appropriate dynamic system configuration.

Similarly, Ref. [54] presented a task-remapping methodology for mitigation of core aging and reduction of communication energy. In addition to selecting and storing a set of Pareto-front points obtained from design-time optimization for performance and reliability, they proposed a fast heuristics-based run-time task-remapping for reducing energy consumption. A similar two-pronged approach to improve system lifetime is proposed in [55]. Here, the authors use the offline optimization to counter the effect of transient faults and the online re-mapping is aimed at migration-reduced adaptation to permanent faults. Similarly, in [56], Kriebel et al. propose an aging-aware resource management in the context of Redundant Multi-Threading (RMT). RMT allows the mitigation of soft-errors by executing copies of a task as two parallel threads. However, RMT can also lead to more aging of the core due to higher utilization. Kriebel et al. store multiple compiled versions of the application exhibiting varying vulnerability to soft-errors, aging and execution time. At run-time, the appropriate version is selected to be executed on an appropriate PE, along with the decision to enable/disable RMT, depending upon the vulnerability and aging-impact of the application. All of the works implementing run-time adaptation to failing PEs implement some form of task-migration. However, ref. [57] proposed a spatial redundancy based task-mapping approach that aims to remove the task-migration overhead completely by replicating the execution of tasks. In Nahar and Meyer [57], provide tolerance to both transient and permanent faults by ensuring that no single failure affects more than one copy of the redundant task. The authors report considerable improvements in the fault-tolerant lifetime of the system compared to traditional spatial redundancy-based methods such as TMR and DMR.

Most of the works in hybrid DSE store either the complete set or a subset of the Pareto-front points that are used during run-time. Recently [40] proposed a methodology where additional non-Pareto points are also stored by the system. These additional points were useful in providing configurations that resulted in lower reconfiguration time at the cost of slightly sub-optimal performance. Figure 8c shows the rationale behind this approach. In the figure, the system has to reconfigure due to changing requirements ($S \rightarrow S'$) at run-time. If only Pareto-front points were stored, it would result in the system switching from F_{Op} to F'_{Op} However there might be non-dominating points (such as F''_{Op}) that satisfy the new requirements while costing lower reconfiguration than F'_{Op}. Thereby, storing additional points within $\Delta ErrRate$ and $\Delta AvgMS$ around the Pareto-front points may result in better dynamic adaptation.

There are a few works [64,65], in which the researchers have taken advantage of both run-time and design-time phases to improve the lifetime reliability in mixed-criticality systems. In [64], researchers have considered an independent periodic task model that tasks with different criticality levels run on a homogeneous multi-core processor. The algorithm has two steps, task partitioning between cores and core utilization optimization. So, tasks are sorted in decreasing order of criticality and then assigned to the cores with the least load allocated. Indeed, the design-time algorithm adapts the resource shortage at run-time. Now, in run-time phase, if a permanent failure happens in a core, the algorithm must re-map tasks to other cores. In the case of not having enough space on the remaining cores, the heuristic drops first tasks with the least criticality and utility. In addition, the work in [65] proposed a Mixed Integer Linear Programming (MILP) based design space exploration process to design a reliable mixed-criticality system. At design-time, tasks are mapped, and the task schedulability in each core is tested by MILP. At run-time, to support the tasks running on cores from permanent faults, the re-mapping technique is used in run-time phase. Therefore, each high-criticality task would be run in two processors, primary and backup. When the primary processor fails, the high-criticality tasks on the failed processor are re-mapped to the predefined backup processor, and all low-criticality tasks assigned in the primary failed processor are dropped. Hence, this algorithm just supports a single processor failure.

5.4. Critique and Perspectives

Almost all the works discussed for improving lifetime reliability use one or more from a very limited set of techniques. These techniques involve either a reactive re-mapping of tasks in the event of a fault, or, pro-active distribution of workload for reducing the electrical stress on the individual cores to achieve wear-leveling. Most articles typically use an additional design objective along with system lifetime to present their novel methodology. While this does result in solving a problem of higher complexity, it does not necessarily contribute to improving system lifetime. Similarly most works listed in Table 2 ignore the reliability of communication and memory structures of the hardware platform. With increasing usage of NoCs for many-/multi-core systems, the reliability of communication elements should find more focus in research. Further, with emerging technologies shifting the focus to memory systems—for both storage and computation—reliability of memory elements needs more research. Finally, the evaluation methodology for lifetime reliability optimizations should include real world benchmarks that include more relevant applications from the domain of machine learning, computer vision etc. Further, lifetime reliability optimization for mixed criticality applications needs more direct approaches to improve system lifetime compared to the more prevalent indirect approach of thermal management.

6. Timing and Functional Reliability Management in Multi/Many-Core Processors

In this section, we aim to study the state-of-the-art works, which manage the timing or functional reliability in multi/many-core processors. In general, most papers have employed redundancy techniques, such as timing, hardware and information to guarantee the reliability in the systems. In the following, we present the existing criticality-aware and non-criticality-aware approaches in three categories of design-time, run-time, and hybrid strategies. Table 3 lists the works based on the criteria in detail. In the end of this section, we discuss about the critique and perspectives on the presented research works.

6.1. Design-Time Strategies

Some of works discussed under the lifetime reliability optimization earlier also analyze for functional and/or timing reliability as a design objective. For instance, ref. [26] used checkpointing with rollback recovery for mitigating the effect of transient faults on functional reliability. The analysis in [26] is aimed at determining the appropriate number of checkpoints in each constituent task of the application for providing sufficient functional correctness. Similarly in [35], Das et al. included the impact of DVFS on soft-error rate while varying the number of replications for each task. In a similar approach, ref. [37] proposed a methodology for an early stage evaluation of the impact of using multiple types of redundancies on the application's timing and functional reliability. In their analysis, ref. [37] integrated the effect of DVFS, imperfect fault-mitigation and implicit fault-masking across multiple layers. Figure 10a,b show the effect of DVFS and varying implicit masking on the average execution time and probability of error of a single task, respectively. Ref. [37] used a Markov Chain-based model to estimate the average execution time and the probability of error. A similar modelling approach for estimating the probability of task completion within a deadline was also presented recently by [70]. A task-mapping and priority assignment for similar deadline miss ratio-constrained systems has been proposed by [71]. In [71], the authors provide a design-time optimization for a heterogeneous architecture and use the stochastic execution time of tasks to optimize the task-mapping. A collection of methods for improving the fractional and timing reliability by using various mitigation methods across different layers, and with varying resource requirements, can be found in [50,72,73]. Similarly, research works targeting improved functional and timing reliability of on-chip communication include [38,74–76].

Table 3. Summary of state-of-the-art approaches in timing/functional reliability aware resource management.

	Fault Model			App. Model			System Model				Imp.		DSE	Technique
	Transient	Intermittent	Permanent	Criticality	Dependency	Periodicity	Communication	Computation	Memory	Heterogeneity	Real Board	Real App.		
Manolache'08 [71]	✗	✗	✗	✗	D	P	✓	✓	✗	Het.	✗	✓	D.T	Task Mapping
Das'14 [35]	✓	✗	✓	✗	D	P	✗	✓	✗	Hom.	✗	✓	D.T	Task-Replication
Bauer'15 [50]	✓	✓	✓	✗	D	P	✓	✓	✗	Het.	✗	✓	R.T	Hardware blackundancy
Das'13 [26]	✓	✗	✓	✗	D	P	✗	✓	✗	Het.	✗	✓	D.T	Checkpointing
Sahoo'20 [37]	✓	✗	✓	✗	D	P	✗	✓	✗	Het.	✗	✗	D.T	Cross-layer Redundancy
Frantz'07 [74]	✓	✗	✗	✗	✗	✗	✓	✗	✗	✗	✗	✓	D.T	HW/SW Redundancy
Lehtonen'07 [75]	✓	✓	✓	✗	✗	✗	✓	✗	✗	✗	✗	✓	D.T	Fault-spec. Opt.
Vitkovskiy'10 [76]	✗	✗	✓	✗	✗	✗	✓	✗	✗	✗	✗	✓	D(R).T	Latency Reduction
Kakoee'11 [38]	✓	✗	✓	✗	✗	✗	✓	✗	✗	✗	✗	✓	D(R).T	Hardware Redundancy
Duque'15 [42]	✗	✓	✓	✗	D	P	✗	✓	✗	Hom.	✗	✓	R.T	Task (Re)-Mapping
Wells'08 [77]	✗	✓	✗	✗	I	S	✗	✓	✗	Hom.	✗	✓	R.T	Hardware Redundancy
Lehtonen'10 [78]	✗	✗	✓	✗	✗	✗	✓	✗	✗	✗	✗	✓	R.T	Online Testing
Yu'10 [79]	✓	✗	✓	✗	✗	✗	✓	✗	✗	✗	✗	✓	R.T	Hardware Redundancy
Rehman'16 [80]	✓	✗	✗	✗	D	A	✗	✓	✗	Hom.	✗	✓	H	Cross-layer
Weichslgartner'18 [81]	✗	✗	✗	✗	D	A	✓	✓	✓	Het.	✗	✓	H	Task Mapping
Sahoo'19 [40]	✓	✗	✓	✗	D	P	✗	✓	✗	Het.	✗	✗	H	Cross-layer Redundancy
Pourmohseni'19 [82]	✗	✗	✗	✗	D	P	✓	✓	✗	Het.	✗	✓	H	Task Mapping
Pathan'17 [59]	✓	✗	✓	✓	I	S	✗	✓	✗	Hom.	✗	✗	D.T	Hardware Redundancy
Safari'20 [60]	✓	✗	✓	✓	D	P	✗	✓	✗	Hom.	✗	✗	D.T	Hardware Redundancy
Safari'19 [83]	✓	✗	✗	✓	I	P	✗	✓	✗	Hom.	✗	✗	D.T	Hardware Redundancy
Rambo'17 [84]	✓	✗	✗	✓	I	P	✓	✓	✗	Hom.	✗	✗	D.T	Hardware Redundancy
Bolchini'13 [85]	✓	✗	✗	✓	D	P	✓	✓	✗	Het.	✗	✗	D.T	Hardware Redundancy
Kang'14 [86]	✓	✗	✗	✓	D	P	✓	✓	✗	Hom.	✗	✓	D.T	Hardware Redundancy
Kang'14a [87]	✓	✗	✗	✓	D	P	✓	✓	✗	Het.	✗	✓	D.T	Hardware Redundancy
Choi'18 [88]	✓	✗	✗	✓	D	P/S	✗	✓	✗	Hom.	✗	✓	D.T	Hardware Redundancy
Jiang'18 [89]	✓	✗	✗	✓	D	P	✓	✓	✗	Hom.	✗	✗	D.T	Hardware Redundancy
Zeng'16 [90]	✓	✗	✗	✓	I	S	✗	✓	✗	Hom.	✗	✓	D.T	Hardware Redundancy
Caplan'17 [91]	✓	✗	✗	✓	I	S	✗	✓	✗	Hom.	✗	✗	D.T	Hardware Redundancy
Axer'11 [58]	✓	✗	✓	✓	I	P	✗	✓	✗	Hom.	✗	✗	D.T	Timing Redundancy
Saraswat'09 [61]	✓	✗	✓	✓	I	P	✓	✓	✓	Het.	✗	✓	D.T	Timing Redundancy
Saraswat'10 [92]	✓	✗	✗	✓	I	P	✓	✓	✓	Het.	✗	✗	D.T	Timing Redundancy
Bagheri'14 [93]	✓	✗	✗	✓	D	-	✓	✓	✗	Hom.	✗	✗	D.T	Timing Redundancy
Kajmakovic'19 [94]	✓	✗	✗	✓	✗	✗	✗	✗	✓	✗	✓	✗	D.T	Information Redundancy
Liu'19 [95]	✓	✗	✗	✓	D	P	✓	✓	✗	Het.	✗	✗	D.T	Rel.-Aware Mapping
Thekkilakattil'14 [96]	✓	✗	✗	✓	I	P	✗	✓	✗	Hom.	✗	✗	H	Mapping & Redundancy
Koc'19 [97]	✗	✗	✗	✓	D	P	✗	✓	✗	Het.	✗	✗	H	Mapping & Redundancy

Next, we study the prior works that managed the timing or functional reliability of mixed-criticality systems in the design-time phase. Most of the papers that exploited the multi/many-core processors achieve reliability improvement by using redundancy technique, such as hardware redundancy (using replica, which is Active, Passive, or Hybrid), timing redundancy using re-execution after error detecting and check-pointing with roll-back recovery, and information redundancy by using the addition of redundant information to data to mitigate soft error. A summary of the literature based on their exploiting techniques are detailed as follows.

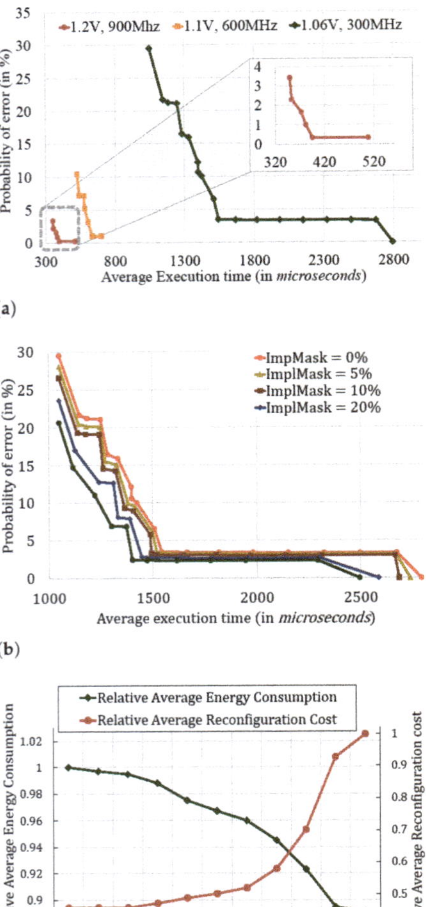

(a)

(b)

(c)

Figure 10. DSE for timing and functional reliability. (**a**) Reliability and DVFS [37]. (**b**) Implicit masking [37]. (**c**) Hybrid DSE results [40].

6.1.1. Multi-Core Platform Used for Spatial and Temporal Redundancy

Two types of redundancy, which are used in the most papers [59,60,83–91] for tolerating faults and consequently, improving reliability are timing and hardware. Some papers have used the feature of multi-core platforms and applied just the hardware redundancy, i.e., replication [59,60,83–85]. As we mentioned in Section 5, authors in [59,60] have used backup redundancy to improve timing and lifetime reliabilities, while the deadline of tasks is guaranteed. In [83], a scheme has been presented to minimize energy, guarantee reliability by computing the optimum number of replication for each criticality task, and maximize QoS when the systems switch to the high-criticality mode. To map tasks on cores, first high-criticality tasks and their replicas and then, low-criticality tasks are mapped on cores that are selected based on worst-fit decreasing and first-fit decreasing. Safari et al. claimed that worst-fit decreasing is the best policy from the energy-awareness perspective. The effect of this algorithm is validated through simulation based on random task generation. Besides, ref. [84] have presented a replica-aware co-scheduling method for mixed-criticality systems by exploiting cross-layer fault tolerance mechanisms. This

method has supported network-on-chip communication delay and replication management overheads. In addition, ref. [85] have presented a methodology to map and schedule tasks on heterogeneous multi-core processors and optimize overall performance. In this methodology, different fault management techniques (Fault Detection/Tolerance) are exploited on different portions of the task graph. Indeed, for some parts of a task graph that needed to be tolerated against faults, different reliability improvement techniques such as replication would be exploited based on architecture features.

On the other hand, there are some works [86–91] that considered both hardware and timing (re-execution) redundancies to tolerate faults before tasks' deadlines and improve timing reliability. In [86,87], an offline heuristic for mapping optimization has been proposed for dependable tasks with different reliability requirements and tolerated transient faults and, consequently, guaranteed the tasks' reliability before their deadlines. The number of re-execution or replication for each task is defined based on its criticality. Besides, ref. [88] have presented a framework to find an optimal mapping of dependent mixed-criticality tasks on multi-core processors, while the QoS is increased in the faulty state and also the power consumption in the normal and faulty states are minimized. To tolerate the transient faults based on the probability distribution of fault occurrences, re-execution and replication are used, and also, the algorithm is designed to endure the maximum number of faults. Choi et al. use the genetic algorithm to find the optimum mapping while all objectives are guaranteed. To investigate the fault tolerance techniques on message communication between dependent tasks, ref. [89] have proposed an optimization to map tasks, minimize the scheduling length and application security vulnerability by considering fault-tolerant constraints. Two techniques of re-execution and active replication are used to tolerate faults of tasks.

Besides, ref. [90] find the optimum number of replication and re-execution for each criticality task to guarantee the reliability of tasks based on the DO-178B safety requirements, in which PFH is defined as their metric. In [91], the authors have used this metric to guarantee the reliability of high-criticality tasks in the case of fault occurrence in different situations. In that paper, an efficient mapping and scheduling algorithm based on the genetic algorithm is proposed on heterogeneous multi-core platforms, while transient faults are tolerated by using on-demand redundancy. In on-demand redundancy, three types of Dual Modular Redundancy, Triple Modular Redundancy, and Passive Replication are supported. In this algorithm, all low-criticality tasks must be executed in a normal situation. If the system is overloaded or a fault occurs, these tasks can be dropped to guarantee the correct execution of high-criticality tasks. Hence, the QoS of low-criticality tasks is one of the objectives of this algorithm that would be maximized.

6.1.2. Timing Redundancy with Check-Pointing and Rollback Recovery

The authors in [58,61,92,93] have improved the reliability by using Check-pointing and rollback recovery for mixed-criticality systems in multi-core processors. Using the Check-pointing technique helps to tolerate transient faults and guarantee reliability. At design-time, authors find the optimum number of check-points during the execution of tasks, such that the deadline of tasks would be guaranteed. In the case of faults occurring during run-time phase, the faulty task is recovered from the previous check-point and continues its execution. In [92], a Tabu search-based approach for task mapping is presented, in which the deadlines of the hard tasks (higher criticality tasks) are guaranteed, even in the case of transient faults, and the QoS for the soft tasks (tasks with lower criticality) is maximized. To improve the timing reliability of tasks with higher criticality, check-pointing with rollback recovery is used. The optimum number of check-points is calculated by considering the overheads of establishing the check-point, error detection, and recovery, while the task deadlines are guaranteed. The proposed algorithm has been evaluated with several random and real-life benchmarks. In [93], a framework of dependable NoC-based multiprocessor is designed for mixed-criticality DAG task models, in which the inter-task communication has been considered. Bagheri and Jervan [93] guarantee the deadline of just

high-criticality tasks even in the presence of transient faults. Transient faults are tolerated by check-pointing, and also, its timing overhead is considered as part of WCET.

6.1.3. Information Redundancy, Mitigating Soft Errors by Using Parity

From the perspective of reliability improvement of memories, ref. [94] have considered soft errors and used redundant parity bit to detect and recover the data utilizing the parity bit. Considering mixed-criticality for memories allows us to improve the system reliability and increase the protection of more criticality memory parts. In this work, faulty data would be corrected in the minimum time to maximize the performance and minimize the run-time memory overhead. Indeed, this parity approach to detect soft errors is explored to use in the design-time phase. In the case of detecting fault at run-time, faulty data recover by copying healthy data. Hence, the Kajmakovic et al. have not used any additional hardware components to tolerate faults.

6.1.4. Task Reliability-Aware Mapping

Another technique to guarantee the reliability of tasks is finding an efficient mapping of mixed-criticality tasks on heterogeneous multi-core platforms to ensure the timing reliability and minimize the probability of fault occurrence. In [95], a heuristic has been proposed that the reliability requirements are satisfied, and also the deadline miss ratio of high-criticality tasks is minimized. In this heuristic, the authors find the optimum mapping and scheduling of tasks with different criticality and reliability requirements on multi-cores with different reliability levels to reduce the probability of transient fault occurrence. The proposed heuristic efficiency has been validated through simulation and shows an outstanding reduction in the deadline miss ratio.

6.2. Run-Time Strategies

Run-time adaptation for improving functional and timing reliability usually involves varying the redundancy levels and using faster cores, respectively. While achieving better functional reliability with a purely run-time approach may be achieved by avoiding faulty cores and/or replicating task execution (both spatially and/or temporally), guaranteeing/improving timing reliability requires much more analysis and is better achieved with a hybrid approach. For instance, multiple research works involve detection of permanent faults in cores and re-mapping tasks to healthier cores [39,41,42]. Similarly, ref. [77] proposed multiple mitigation approaches for intermittent faults including pausing execution, using spare cores and avoiding faulty cores to allow for self-repair. Similarly, ref. [50] proposed multiple resource management methods that include both proactive (avoiding hot-spots) and reactive (online testing and error detection) methods. Some of the approaches proposed in [50] also include using TMR for improving functional reliability. Similar redundancy based improvement of on-chip communication by using information encoding and spare wires are proposed by [75]. A novel methodology for online testing, detection and bypassing of permanent faults in NoCs is presented in [78].

From the run-time criticality-aware reliability management perspective, ref. [98] have recently investigated MC systems. In this paper, authors have presented the Information Processing Factory (IPF) paradigm to achieve long-term dependability for MC systems, where a 5-layer hierarchical organization has been introduced. IPF is introduced as a self-aware and self-organizing system used to manage resources by decomposing approach, planning, and confining them during run-time. The IPF can also detect and predict potential hazards and handle these upcoming risks in different layers. As a result, in this introduced framework, the requirements of safety-critical functions are always met at run-time. In the end, the proposed method efficiency has been evaluated by showing achieving the reliability levels (functional reliability) and MTTF as lifetime reliability.

6.3. Hybrid Strategies

The design-/compile-time analysis for improving functional reliability involves determining the different levels of error tolerance provided by varying levels of redundancy. For timing reliability, the analogous stage involves finding the various resource allocation configurations that ensure the timing requirements of the application(s). Ref. [80] proposed a cross-layer reliability approach in single-processor systems, where multiple executables for each task, with varying execution time and error probabilities, were generated at compile time. The run-time allocation involved dynamically selecting the appropriate version of each task, depending upon the available slack w.r.t. the deadline. Ref. [81] also tackled the problem of achieving predictable execution time in MPSoCs with a hybrid approach. The authors presented a compile time DSE methodology for generating clustered tasks and constraint graphs based on the timing requirements of the application (modelled as a DAG). The run-time management involved mapping the clustered tasks and edges on available cores and interconnects of the hardware platform. A similar approach to mapping of hard real-time applications on many-core system using a hybrid approach was presented by [82]. A novel methodology for using hybrid DSE to adapt to varying QoS requirements was proposed by [40]. In addition to ensuring timing and functional reliability, the run-time process proposed in [40] allowed the user to dynamically select the priority of energy consumption and reconfiguration cost during run-time adaptation. Figure 10c shows the different trade-offs obtained for a sample application by varying the user-controlled parameter p_{RC}.

From the perspective of criticality-aware strategies, some papers have improved the timing reliability for mixed-criticality tasks in both design-time and run-time phases [96,97]. For example, ref. [96] have presented an approach to map and schedule tasks with different criticality levels on multi-core processors in which the timing constraints of high-criticality tasks are guaranteed at design-time even in the case of fault occurrence and also, the flexibility for the low-criticality tasks are ensured. To tolerate the faults and improve reliability, the timing redundancy technique, re-execution, is used for high-criticality tasks at design-time and low-criticality tasks at run-time. Besides, the work [97] has focused on finding the best mapping of mixed-criticality tasks to minimize execution latency by considering the reliability of both system and high-criticality tasks at design-time. At run-time, when the system switches to the high-criticality mode, high-criticality tasks are re-mapped to the highly reliable cores to be executed before their deadlines and, if possible, low-criticality tasks are scheduled without exceeding the minimum latency.

6.4. Critique and Perspectives

Most works in this category have evaluated their proposed approaches in simulation, which can be seen in Table 3. Although simulation can validate the proposed method good enough, it is not sufficient due to some reasons such as overheads at run-time, like communication, and fault detection and tolerance; then the proposed method may not be applicable in some cases. Researchers in [69] have discussed that if the timing overheads, such as the delay for changing the V-f levels of cores, are not considered while scheduling the tasks, it may cause deadline violation and consequently, catastrophic consequences may happen. Therefore, evaluating the proposed methods based on real applications on a real platform is needed, which has not been considered in most existing works.

Besides, as mentioned in this section, most of the works have used redundancy techniques to guarantee timing or functional reliability. Since the system may execute safely without fault occurrence, these techniques such as hardware and timing redundancy may waste the system's resources. Therefore, investigating the run-time approaches is needed to efficiently use the computational resources and optimize the other objectives.

In addition, intermittent faults are one of the common faults in embedded systems. The reason for these faults can be inherent design issue or unstable hardware. Due to the behavior of repeated conditions of causing faults, errors may occur. Investigating the

intermittent faults can help the system be optimized more efficiently at run-time, which has not been considered in previous works.

From the criticality-aware reliability management perspective, most of the existing works have guaranteed the timing reliability only for the tasks with higher criticality tasks in any system operational modes. However, in some mixed-criticality embedded systems, such as avionics, low-criticality tasks, are mission-critical, and guaranteeing the reliability requirement of both low- and high-criticality tasks are crucial. Most of the existing approaches cannot be applied to these systems; thus, new studies to guarantee each criticality level's reliability in each operational mode are required in multi-core mixed-criticality systems.

As can be illustrated from Table 3, most of the works have not considered the communication and memory of multi/many-core processors and only focused on guaranteeing the reliability in computational parts. However, the memories and data sharing can be the bottleneck to guarantee the reliability of applications in multi-core platforms. Designing the system for reliability management and analyzing it at run-time by having a comprehensive view on the whole system resources are required in multi/many-core platforms.

7. Reliability Management in Reconfigurable Architectures

Some of the related works discussed till now assume the availability of reconfigurable logic on the hardware platform and tackle the related problem of resource management by allocating accelerators to a select subset of tasks of an application (hardware/software partitioning). Therefore the DSE methods presented in works such as [26,40,50] can be directly used. However, in this section we survey the works that assume the complete architecture comprising of reconfigurable hardware logic, specifically for Field Programmable Gate Array (FPGA)s.

Improving functional reliability in FPGAs has mostly been focused on improving the reliability of the configuration bits. Traditional methods methods such as ECC [99], scrubbing [100] and hardware checkpointing [101] have been used to provide protection from transient faults in the configuration memory. Additionally, circuit design methods that involve TMR has also been employed for enabling the usage of FPGAs in high-radiation environments [102]. However, there has been a growing trend of lifetime reliability improvement in FPGA-based systems. For instance, ref. [103] proposed various phenomenon-specific methods, tailored for each failure mechanism, to mitigate aging in FPGAs. Similarly, ref. [104] proposed multiple generic electrical stress hot-spot reduction techniques for FPGAs. Ref. [105] presented a stress-aware run-time wear-leveling approach that leverages Dynamic Partial Reconfiguration (DPR) in FPGA-based systems. In [106], Zhang et al. used module diversification, to generate multiple accelerator designs with spatially varying aging effects. These diverse modules were used to leverage DPR by periodically swapping accelerators that use different CLBs of the FPGA fabric. A novel approach combining module diversification and dynamic adaptation to varying aging-effects at run-time was proposed by [107]. Similarly, ref. [108] presented a reliability-aware floorplanning methodology along with delay-based aging estimation and run-time reconfiguration. A joint mitigation methodology using DPR, aimed at both soft errors and permanent faults in FPGAs was proposed by [109]. The authors presented reconfiguration as a solution to both types of faults, which can be a costly approach. Almost all the research works employing DPR assume using homogeneous Partially Reconfigurable Region (PRR) (comprising of equivalent amount of FPGA resources). However, as shown in Figure 11, Sahoo et al. [110,111] proposed a hardware/hardware partitioning methodology that allows using application specific heterogeneous PRRs, that provided the scope for improving both the latency (average makespan) and reliability (MTTF) in DPR-based systems.

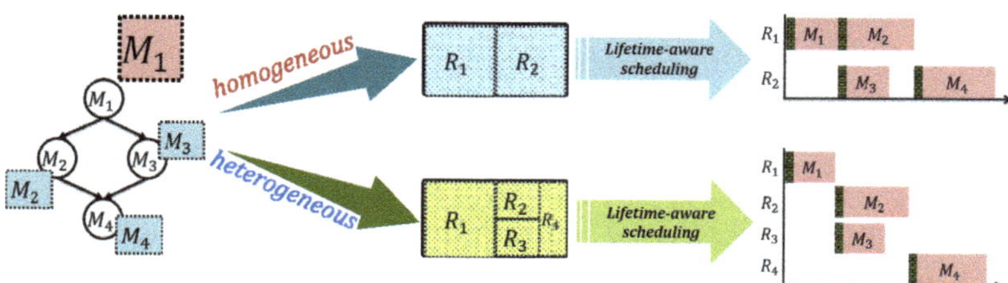

Figure 11. Reliability-aware HW/HW partitioning [110].

A few papers such as [112–116], have addressed reconfigurable processors in a system with different criticality tasks to improve timing reliability. In [112,113], Santos, et al. have proposed a new efficient scrubbing mechanism to increase the system's reliability by considering the criticality and timing of the hardware task execution. Hence, scrubbing mechanism takes advantages of FPGA reconfiguration, and verify the reconfiguration periodically to tolerate faults like Single Event Upset (SEU) in the Static-RAM (SRAM) [112,117]. Researchers in [112] have proposed a static heuristic to schedule the tasks based on their criticality level, running on reconfigurable embedded systems to maximize each task's reliability. In addition, they have presented a dynamic scrubbing mechanism in [113] to have high reliability by using the windows and fixed priority scheduling. They also consider the reliability as well as the criticality level for the tasks. They have claimed that they significantly reduce the amount of memory required to store the scrubbing schedule. Ref. [114] have presented an efficient resource management mechanism and then architecture to provide fault tolerance in the context of a time-triggered NoC-based mixed-criticality system that invokes reconfiguration. Their method establishes the fault-recovery and efficient resource utilization in Mixed-Criticality Networks-on-Chip (MCNoCs) by monitoring the resource requests and reconfiguring them based on a recovery strategy. Besides, ref. [115] have proposed a reliability driven scheduling approach for mixed-criticality tasks by handling periodic, aperiodic, and sporadic tasks on FPGAs against hardware trojan horse attacks. In this approach, redundancy is used to increase reliability on task criticality level, offline, and then attempt to prevent faulty data propagation in the run-time phase. Ref. [116] have detected an error by using the Secure Hash Algorithm and corrected them by using parity based two-dimensional erasure code, while the performance is reduced, which consists of time error detection and correction. This method has taken the execution period and criticality into account to correct faulty data.

8. Upcoming Trends and Open Challenges

In this section, we briefly address the upcoming trends and challenges relevant to reliability-aware resource management in multi/many-core systems.

- Cross-layer Reliability: Most of the research articles discussed in this survey employ/select different redundancy-based methods to improve the system's reliability. Similarly, there is an increasing trend of using multiple layers of the system stack in the design for reliability [73,118]. This is unlike the traditional approach of mitigating each fault-mechanism at the hardware layer and providing a fault-free abstraction to the other layers. Although this phenomenon-based approach makes the design process simpler for the non-hardware layers, the high cost of hardware-based fault-mitigation can make this approach infeasible for resource-constrained systems. In contrast, the cross-layer approach involves multiple layers sharing the fault-mitigation activities during run-time [119]. Similarly, various methods of leveraging at cross-layer reliability at design-time have been proposed [50].
One of the major advantages of the cross-layer approach is the inherent suitability for application-specific optimizations. Since the overheads of fault-tolerance varies with the type of redundancy being used, application-specific tolerances to degradation in some form of reliability can be used to improve other reliability metrics.

Further, with the cross-layer approach, the implicit masking of multiple layers can be used to provide low-cost fault-tolerance [120]. However, the joint optimization across multiple layers increases the design space considerably. Recently there have been multiple works that try to provide efficient DSE for cross-layer reliability for various system-level design tasks such as task mapping [37], hardware-hardware partitioning [121], run-time adaptation [40], hardware design [72] etc. However, most of the works assume rather simplistic reliability models such as the one shown in Figure 12a where each layer is limited to a specific type of redundancy [37,40,121]. A more holistic approach to the design of cross layer reliability is necessary for more realistic reliability models that integrate multiple reliability methods at each layer. As shown in Figure 12b having reliability interfaces, similar to those used for functionality and performance, can enable far better DSE than current state-of-the-art works. An interface for functional reliability was proposed by [80] that used Architectural Vulnerability Factor (AVF), Instruction Vulnerability Index (IVI) and Function Vulnerability Index (FVI) for characterising different implementations of an embedded processor, instruction set and function libraries, respectively. However, similar interfaces for timing and lifetime reliability need to be developed for designing efficient cross-layer reliability.

- Self-Aware System Design: In general, design-time approaches are applied to optimize resource usage and guarantee the reliability in the worst-case scenarios. However, due to the various run-time behaviors of applications and fault occurrence, we cannot efficiently manage the reliability, especially lifetime reliability and system utilization. Therefore, run-time system monitoring and optimization are essential to control and have a reliable operation of applications, especially mixed-criticality applications, and efficient resource management of multi/many-core platforms [122]. IPF paradigm is recently used to manage the system dynamically, according to the changes in system and workload [98,123]. This self-aware paradigm improves the reliability and resource utilization by combining different techniques in different hierarchical layers. As a result, the online optimization based on the current state and variations in applications and system by monitoring the hardware and software components, and using the IPF to conquer the complexity is essential, especially for safety-critical systems.

- Reliable Communication and Data Sharing: Safety and dependability are critical issues in designing the mixed-criticality systems on multi/many-core platforms, in which data are shared between concurrent execution of tasks with different criticality [124]. The strict control of data (critical and non-critical), communication, sharing, and storage in such systems for safety assurance, e.g., in medical devices, is crucial. Most state-of-the-art works have concentrated on the reliability management of tasks in processors of multi/many-core systems regardless of safe data sharing among communication and memories. As a result, safe mixed-criticality system design considering all system resources, like communications, and memory access, and processors are needed.

In addition to the trends discussed above many new approaches to computing have emerged over the past few years. These emerging technologies have brought forward novel opportunities and challenges to reliability-aware resource management. For example, Approximate Computing (AxC) has emerged as a new computing paradigm that offers the promise of low-power and faster execution [125]. While AxC might hold promise for improving timing and lifetime reliability, the deliberate introduction of computational inaccuracies requires careful design for dependable systems. Similarly, the introduction of post-CMOS transistors for next-generation computing requires a thorough reliability analysis of emerging devices. Finally, we are already witnessing AI-based resource management in multi-/many-core systems for both design-time optimization and run-time adaptation to varying operating conditions [40,126,127].

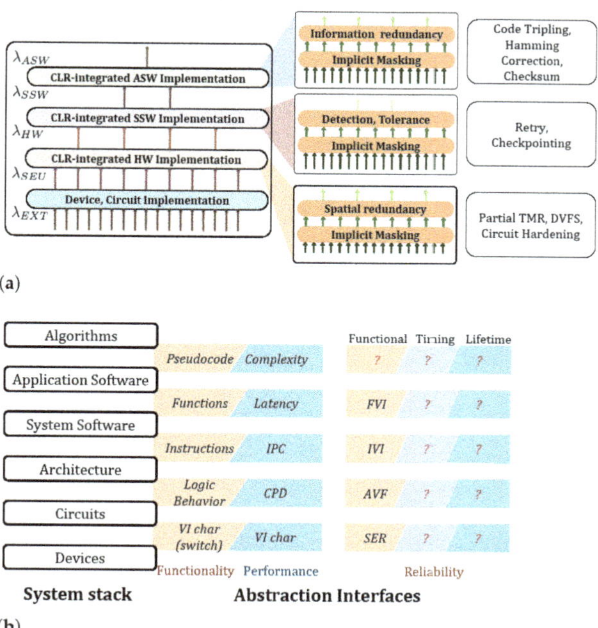

(a)

(b)

Figure 12. Cross-layer design approach to reliability. (**a**) Redundancy methods across layers. (**b**) Interfaces for cross-layer design approach.

9. Conclusions

With increasing susceptibility of modern electronic systems to physical faults, reliability-aware resource management is a topical research problem. While technology scaling and architectural innovations such as 3D integration allows us to integrate more and more cores in a system, extracting reliable performance from increasingly unreliable semiconductor heightens the need for resource management in multi-/many-core systems. To this end, this article presents our perspective on the related research. To begin, we have provided a detailed overview of the various phenomena that have contributed to the growing need for reliability in modern electronic systems. Then, we provided a taxonomy along with a detailed background of the various aspects of reliability improvements that are explored in related research works. Specifically we looked at the different types of reliability—lifetime, timing and functional (the different fault models and the system model) application and architecture. We formulated a generic problem statement for reliability-aware resource management that lets us determine the scope and classify the methods adopted in each of the related works. A survey of the related works is then presented, categorized under the type of DSE approach—design/compile-time, run-time and hybrid–adopted for each type of reliability. We have also presented a brief survey related research works targeted for FPGA-based systems. The presented survey focuses on the application-specific reliability, mixed-criticality awareness and hardware resource heterogeneity. In the end, we have provided a brief discussion on the upcoming trends in reliability-aware resource management and the challenges therein, to encourage further research in this topic.

Author Contributions: Conceptualization, A.K.; writing—original draft preparation, S.S.S. and B.R.; writing—review and editing, S.S.S., B.R. and A.K.; supervision, A.K.; project administration, A.K.; funding acquisition, A.K. All authors have read and agreed to the published version of the manuscript.

About the Editors

Amit Kumar Singh (Associate Professor) is working in the School of Computer Science and Electronic Engineering at the University of Essex, UK. He received a B.Tech. degree from IIT, Dhanbad, India, in 2006, and a Ph.D. degree from Nanyang Technological University (NTU), Singapore, in 2013. He has over five years of post-doctoral research experience at several reputed universities. His current research interests are design and optimization of multi-core-based computing systems with a focus on performance, energy, temperature, reliability and security. He has published over 100 papers in reputed journals/conferences and received several best paper awards, e.g., IEEE TC February 2018 Featured Paper, ICCES 2017, ISORC 2016, PDP 2015, HiPEAC 2013 and GLSVLSI 2014 runner-up. He has served on the TPC of IEEE/ACM conferences such as DAC, DATE, CASES and CODES+ISSS.

Amlan Ganguly is an Associate Professor and Department Head of Computer Engineering at the Rochester Institute of Technology, NY, USA. He received his B.Tech. in Electronics and Electrical Communication Engineering at the Indian Institute of Technology and MS and Ph.D. degrees from Washington State University, USA. His research interests are in energy-efficient and high-performance multi-core processors, data centers, novel non-von Neumann architectures such as in-memory processing and hardware security. He has published over 100 peer-reviewed journal and conference articles. He is in the Technical Program Co-Chair of the International Green and Sustainable Computing conference (IGSC, 2020) and the ACM Workshop on Network-on-Chip Architectures (NoCArc, 2020). He is a member of the Technical Program Committee of several highly regarded conferences. He is an Associate Editor of the Elsevier journal *Sustainable Computing: Informatics and Systems* (SUSCOM) and MDPI's *Journal of Low Power Electronics and Applications* (JLPEA).

Funding: This work is supported in part by the German Research Foundation (DFG) within the Cluster of Excellence Center for Advancing Electronics Dresden (CFAED) at the Technische Universität Dresden.

Conflicts of Interest: The authors declare no conflict of interest.

Abbreviations

The following abbreviations are used in this manuscript:

ACO	Ant Colony Optimization
AVF	Architectural Vulnerability Factor
AxC	Approximate Computing
BTI	Bias Temperature Instability
DAG	Directed Acyclic Graph
DEC	Double-bit-Error-Correcting
DED	Double-bit-Error-Detecting
DMR	Dual Modular Redundancy
DPR	Dynamic Partial Reconfiguration
DRAM	Dynamic Random Access Memory
DSE	Design Space Exploration
DVFS	Dynamic Voltage and Frequncy Scaling
ECC	Error Checking and Correcting
EM	Electromigration
FPGA	Field Programmable Gate Array
FVI	Function Vulnerability Index
HCI	Hot Carrier Injection
ICs	integrated circuits
ILP	Instruction Level Parallelism
IoT	Internet of Thing
IPF	Information Processing Factory
IVI	Instruction Vulnerability Index
MAB	Multi-Armed Bandit
MCS	Monte-Carlo Simulations
MILP	Mixed Integer Linear Programming
MOEA	Multi-Objective Evolutionary Algorithms
MPSoC	Multi-Processor System-on-Chip
MTBF	Mean Time between Failures
MTTC	Mean Time To Crash
MTTF	Mean Time To Failure
NBTI	Negative Bias Temperature Instability
NoC	Network-on-Chip
PE	Processing Element
PFH	Probability-of-Failure-per-Hour
PRR	Partially Reconfigurable Region
QoS	Quality of Service
RL	Reinforcement Learning
RMT	Redundant Multi-Threading
SA	Simulated Annealing
SDFG	Synchronous Data Flow Graph
SEC	Single-bit-Error-Correcting
SER	Soft Error Rate
SEU	Single Event Upset
SRAM	Static Random Access Memory
TDDB	Time Dependent Dielectric Breakdown
TED	Triple-bit-Error-Detecting
TMR	Triple Modular Redundancy
WCET	Worst-case Execution Time

References

1. Coombs, A.W.M. The Making of Colossus. *Ann. Hist. Comput.* **1983**, *5*, 253–259. [CrossRef]
2. Dennard, R.H.; Gaensslen, F.H.; Rideout, V.L.; Bassous, E.; LeBlanc, A.R. Design of ion-implanted MOSFET's with very small physical dimensions. *IEEE J. Solid State Circuits* **1974**, *9*, 256–268. [CrossRef]
3. Patterson, D.A.; Hennessy, J.L. *Computer Organization and Design ARM Edition: The Hardware Software Interface*; Morgan Kaufmann: Burlington, MA, USA, 2016.
4. Rupp, K. 42 Years of Microprocessor Trend Data. Available online: https://www.karlrupp.net/2018/02/42-years-of-microprocessor-trend-data/ (accessed on 12 December 2020).
5. ARM. big. LITTLE Technology: The Future of Mobile. Available online: https://img.hexus.net/v2/press_releases/arm/big.LITTLE.Whitepaper.pdf (accessed on 12 December 2020).
6. Chen, T.; Raghavan, R.; Dale, J.N.; Iwata, E. Cell Broadband Engine Architecture and its first implementation–A performance view. *IBM J. Res. Dev.* **2007**, *51*, 559–572. [CrossRef]
7. Borkar, S. Designing reliable systems from unreliable components: The challenges of transistor variability and degradation. *IEEE Micro* **2005**, *25*, 10–16. [CrossRef]
8. Shivakumar, P.; Kistler, M.; Keckler, S.W.; Burger, D.; Alvisi, L. Modeling the effect of technology trends on the soft error rate of combinational logic. In Proceedings of the International Conference on Dependable Systems and Networks, Washington, DC, USA, 23–26 June 2002; pp. 389–398. [CrossRef]
9. Nightingale, E.B.; Douceur, J.R.; Orgovan, V. Cycles, Cells and Platters: An Empirical Analysisof Hardware Failures on a Million Consumer PCs. In Proceedings of the Sixth Conference on Computer Systems, Salzburg, Austria, 10–13 April 2011; ACM: New York, NY, USA, 2011; pp. 343–356. [CrossRef]
10. Liu, J.W.S. *Real-Time Systems*; Prentice Hall: Upper Saddle River, NJ, USA, 2000.
11. Halang, W.A.; Gumzej, R.; Colnaric, M.; Druzovec, M. Measuring the performance of real-time systems. *Real Time Syst.* **2000**, *18*, 59–68. [CrossRef]
12. Das, A.K.; Kumar, A.; Veeravalli, B.; Catthoor, F. *Reliable and Energy Efficient Streaming Multiprocessor Systems*; Springer: Berlin/Heidelberg, Germany, 2018.
13. Avizienis, A.; Laprie, J.; Randell, B.; Landwehr, C. Basic concepts and taxonomy of dependable and secure computing. *IEEE Trans. Dependable Secur. Comput.* **2004**, *1*, 11–33. [CrossRef]
14. May, T.C.; Woods, M.H. Alpha-particle-induced soft errors in dynamic memories. *IEEE Trans. Electron Devices* **1979**, *26*, 2–9. [CrossRef]
15. Ziegler, J.F.; Lanford, W.A. Effect of Cosmic Rays on Computer Memories. *Science* **1979**, *206*, 776–788. [CrossRef]
16. Keane, J.; Kim, C.H. An odomoeter for CPUs. *IEEE Spectr.* **2011**, *48*, 28–33. [CrossRef]
17. Zhang, J.F.; Eccleston, W. Positive bias temperature instability in MOSFETs. *IEEE Trans. Electron Devices* **1998**, *45*, 116–124. [CrossRef]
18. Takeda, E.; Suzuki, N.; Hagiwara, T. Device performance degradation to hot-carrier injection at energies below the Si-SiO$_2$ energy barrier. In Proceedings of the 1983 International Electron Devices Meeting, Washington, DC, USA, 5–7 December 1983; pp. 396–399. [CrossRef]
19. Black, J.R. Electromigration: A brief survey and some recent results. *IEEE Trans. Electron Devices* **1969**, *16*, 338–347. [CrossRef]
20. Benini, L.; De Micheli, G. Networks on chips: A new SoC paradigm. *Computer* **2002**, *35*, 70–78. [CrossRef]
21. Radetzki, M.; Feng, C.; Zhao, X.; Jantsch, A. Methods for Fault Tolerance in Networks-on-chip. *ACM Comput. Surv.* **2013**, *46*, 8:1–8:38. [CrossRef]
22. Postman, J.; Chiang, P. A Survey Addressing On-Chip Interconnect: Energy and Reliability Considerations. Availabe online: https://www.hindawi.com/journals/isrn/2012/916259/ (accessed on 12 December 2020).
23. Hamming, R.W. Error detecting and error correcting codes. *Bell Syst. Tech. J.* **1950**, *29*, 147–160. [CrossRef]
24. Hsiao, M.Y. A Class of Optimal Minimum Odd-weight-column SEC-DED Codes. *IBM J. Res. Dev.* **1970**, *14*, 395–401. [CrossRef]
25. Das, A.; Kumar, A. Fault-aware task re-mapping for throughput constrained multimedia applications on NoC-based MPSoCs. In Proceedings of the IEEE International Symposium on Rapid System Prototyping (RSP), Tampere, Finland, 11–12 October 2012.
26. Das, A.; Kumar, A.; Veeravalli, B. Aging-aware hardware-software task partitioning for reliable reconfigurable multiprocessor systems. In Proceedings of the International Conference on Compilers, Architectures and Synthesis for Embedded Systems (CASES), Montreal, QC, Canada, 29 September–4 October 2013; pp. 1–10. [CrossRef]
27. Medina, R.; Borde, E.; Pautet, L. Availability enhancement and analysis for mixed-criticality systems on multi-core. In Proceedings of the Design, Automation Test in Europe Conference & Exhibition (DATE), Dresden, Germany, 19–23 March 2018.
28. Ranjbar, B.; Safaei, B.; Ejlali, A.; Kumar, A. FANTOM: Fault Tolerant Task-Drop Aware Scheduling for Mixed-Criticality Systems. *IEEE Access* **2020**, *8*, 187232–187248. [CrossRef]
29. Johnson, L.A. DO-178B, Software considerations in airborne systems and equipment certification. *Crosstalk Oct.* **1998**, *199*, 11–20.

30. Hartman, A.S.; Thomas, D.E.; Meyer, B.H. A Case for Lifetime-Aware Task Mapping in Embedded Chip Multiprocessors. In Proceedings of the IEEE/ACM/IFIP International Conference on Hardware/Software Codesign and System Synthesis (CODES), CODES/ISSS '10, Scottsdale, AZ, USA, 24–29 October 2010; Association for Computing Machinery: New York, NY, USA, 2010; pp. 145–154. [CrossRef]
31. Meyer, B.H.; Hartman, A.S.; Thomas, D.E. Cost-effective slack allocation for lifetime improvement in NoC-based MPSoCs. In Proceedings of the Design, Automation Test in Europe Conference & Exhibition (DATE), Dresden, Germany, 8–12 March 2010; pp. 1596–1601. [CrossRef]
32. Meyer, B.H.; Hartman, A.S.; Thomas, D.E. Cost-Effective Lifetime and Yield Optimization for NoC-Based MPSoCs. *ACM Trans. Des. Autom. Electron. Syst.* **2014**, *19*. [CrossRef]
33. Ma, C.; Mahajan, A.; Meyer, B.H. Multi-armed bandits for efficient lifetime estimation in MPSoC design. In Proceedings of the Design, Automation Test in Europe Conference & Exhibition (DATE), Lausanne, Switzerland, 27–31 March 2017; pp. 1540–1545. [CrossRef]
34. Xiang, Y.; Chantem, T.; Dick, R.P.; Hu, X.S.; Shang, L. System-level reliability modeling for MPSoCs. In Proceedings of the IEEE/ACM/IFIP International Conference on Hardware/Software Codesign and System Synthesis (CODES), Scottsdale, AZ, USA, 24–29 October 2010; pp. 297–306.
35. Das, A.; Kumar, A.; Veeravalli, B.; Bolchini, C.; Miele, A. Combined DVFS and Mapping Exploration for Lifetime and Soft-Error Susceptibility Improvement in MPSoCs. In Proceedings of the Design, Automation Test in Europe Conference & Exhibition (DATE), Dresden, Germany, 24–28 March 2014; European Design and Automation Association: Leuven, Belgium, 2014.
36. Das, A.; Kumar, A.; Veeravalli, B. Temperature aware energy-reliability trade-offs for mapping of throughput-constrained applications on multimedia MPSoCs. In Proceedings of the Design, Automation Test in Europe Conference & Exhibition (DATE), Dresden, Germany, 24–28 March 2014; pp. 1–6. [CrossRef]
37. Sahoo, S.S.; Veeravalli, B.; Kumar, A. CL(R)Early: An Early-stage DSE Methodology for Cross-Layer Reliability-aware Heterogeneous Embedded Systems. In Proceedings of the Design Automation Conference (DAC), San Francisco, CA, USA, 19–22 July 2020; pp. 1–6. [CrossRef]
38. Kakoee, M.R.; Bertacco, V.; Benini, L. ReliNoC: A reliable network for priority-based on-chip communication. In Proceedings of the Design, Automation Test in Europe Conference & Exhibition (DATE), Grenoble, France, 14–18 March 2011; pp. 1–6. [CrossRef]
39. Sahoo, S.S.; Kumar, A.; Veeravalli, B. Design and evaluation of reliability-oriented task re-mapping in MPSoCs using time-series analysis of intermittent faults. In Proceedings of the Design, Automation Test in Europe Conference & Exhibition (DATE), Dresden, Germany, 14–18 March 2016; pp. 798–803.
40. Sahoo, S.S.; Veeravalli, B.; Kumar, A. A Hybrid Agent-Based Design Methodology for Dynamic Cross-Layer Reliability in Heterogeneous Embedded Systems. In Proceedings of the Design Automation Conference (DAC), Las Vegas, NV, USA, 2–6 June 2019; Association for Computing Machinery: New York, NY, USA, 2019. [CrossRef]
41. Hartman, A.S.; Thomas, D.E. Lifetime Improvement through Runtime Wear-Based Task Mapping. In Proceedings of the IEEE/ACM/IFIP International Conference on Hardware/Software Codesign and System Synthesis (CODES), CODES + ISSS '12, Tampere, Finland, 7–12 October 2012; Association for Computing Machinery: New York, NY, USA, 2012; pp. 13–22. [CrossRef]
42. Duque, L.A.R.; Diaz, J.M.M.; Yang, C. Improving MPSoC reliability through adapting runtime task schedule based on time-correlated fault behavior. In Proceedings of the Design, Automation Test in Europe Conference & Exhibition (DATE), Grenoble, France, 9–13 March 2015; pp. 818–823.
43. Rathore, V.; Chaturvedi, V.; Singh, A.K.; Srikanthan, T.; Shafique, M. Life Guard: A Reinforcement Learning-Based Task Mapping Strategy for Performance-Centric Aging Management. In Proceedings of the Design Automation Conference (DAC), Las Vegas, NA, USA, 2–6 June 2019; pp. 1–6.
44. Venkataraman, S.; Santos, R.; Kumar, A.; Kuijsten, J. Hardware task migration module for improved fault tolerance and predictability. In Proceedings of the International Conference on Embedded Computer Systems: Architectures, Modeling, and Simulation (SAMOS), Samos, Greece, 19–23 July 2015; pp. 197–202. [CrossRef]
45. Wang, L.; Lv, P.; Liu, L.; Han, J.; Leung, H.; Wang, X.; Yin, S.; Wei, S.; Mak, T. A Lifetime Reliability-Constrained Runtime Mapping for Throughput Optimization in Many-Core Systems. *IEEE Trans. Comput. Aided Des. Integr. Circuits Syst.* **2019**, *38*, 1771–1784. [CrossRef]
46. Haghbayan, M.; Miele, A.; Rahmani, A.M.; Liljeberg, P.; Tenhunen, H. A lifetime-aware runtime mapping approach for many-core systems in the dark silicon era. In Proceedings of the Design, Automation Test in Europe Conference & Exhibition (DATE), Dresden, Germany, 14–18 March 2016; pp. 854–857.
47. Haghbayan, M.; Miele, A.; Rahmani, A.M.; Liljeberg, P.; Tenhunen, H. Performance/Reliability-Aware Resource Management for Many-Cores in Dark Silicon Era. *IEEE Trans. Comput.* **2017**, *66*, 1599–1612. [CrossRef]
48. Rathore, V.; Chaturvedi, V.; Singh, A.K.; Srikanthan, T.; Rohith, R.; Lam, S.; Shafique, M. HiMap: A hierarchical mapping approach for enhancing lifetime reliability of dark silicon manycore systems. In Proceedings of the Design, Automation Test in Europe Conference & Exhibition (DATE), Dresden, Germany, 19–23 March 2018; pp. 991–996. [CrossRef]
49. Raparti, V.Y.; Kapadia, N.; Pasricha, S. ARTEMIS: An Aging-Aware Runtime Application Mapping Framework for 3D NoC-Based Chip Multiprocessors. *IEEE Trans. Multi-Scale Comput. Syst.* **2017**, *3*, 72–85. [CrossRef]
50. Bauer, L.; Henkel, J.; Herkersdorf, A.; Kochte, M.A.; Kühn, J.M.; Rosenstiel, W.; Schweizer, T.; Wallentowitz, S.; Wenzel, V.; Wild, T.; et al. Adaptive multi-layer techniques for increased system dependability. *It-Inf. Technol.* **2015**, *57*, 149–158. [CrossRef]

51. Das, A.; Kumar, A.; Veeravalli, B. Reliability-driven task mapping for lifetime extension of networks-on-chip based multiprocessor systems. In Proceedings of the Design, Automation Test in Europe Conference & Exhibition (DATE), Grenoble, France, 18–22 March 2013; pp. 689–694. [CrossRef]

52. Das, A.; Kumar, A.; Veeravalli, B. Reliability and Energy-Aware Mapping and Scheduling of Multimedia Applications on Multiprocessor Systems. *IEEE Trans. Parallel Distrib. Syst.* **2016**, *27*, 869–884. [CrossRef]

53. Das, A.; Kumar, A.; Veeravalli, B. Communication and migration energy aware design space exploration for multicore systems with intermittent faults. In Proceedings of the Design, Automation Test in Europe Conference & Exhibition (DATE), Grenoble, France, 18–22 March 2013; pp. 1631–1636. [CrossRef]

54. Bolchini, C.; Carminati, M.; Miele, A.; Das, A.; Kumar, A.; Veeravalli, B. Run-time mapping for reliable many-cores based on energy/performance trade-offs. In Proceedings of the IEEE International Symposium on Defect and Fault Tolerance in VLSI and Nanotechnology Systems (DFTS), New York, NY, USA, 2–4 October 2013; pp. 58–64. [CrossRef]

55. Namazi, A.; Safari, S.; Mohammadi, S.; Abdollahi, M. SORT: Semi Online Reliable Task Mapping for Embedded Multi-Core Systems. *ACM Trans. Model. Perform. Eval. Comput. Syst.* **2019**, *4*. [CrossRef]

56. Kriebel, F.; Rehman, S.; Shafique, M.; Henkel, J. AgeOpt-RMT: Compiler-Driven Variation-Aware Aging Optimization for Redundant Multithreading. In Proceedings of the Design Automation Conference (DAC), DAC '16, Austin, TX, USA, 5–9 June 2016; Association for Computing Machinery: New York, NY, USA, 2016. [CrossRef]

57. Nahar, B.; Meyer, B.H. RotR: Rotational redundant task mapping for fail-operational MPSoCs. In Proceedings of the IEEE International Symposium on Defect and Fault Tolerance in VLSI and Nanotechnology Systems (DFTS), Amherst, MA, USA, 12–14 October 2015; pp. 21–28. [CrossRef]

58. Axer, P.; Sebastian, M.; Ernst, R. Reliability Analysis for MPSoCs with Mixed-Critical, Hard Real-Time Constraints. In Proceedings of the IEEE/ACM/IFIP International Conference on Hardware/Software Codesign and System Synthesis (CODES), Taipei Taiwan, 9–14 October 2011; Association for Computing Machinery: New York, NY, USA, 2011; pp. 149–158. [CrossRef]

59. Pathan, R.M. Real-time scheduling algorithm for safety-critical systems on faulty multicore environments. *Real Time Syst.* **2017**, *53*, 45–81. [CrossRef]

60. Safari, S.; Hessabi, S.; Ershadi, G. LESS-MICS: A Low Energy Standby-Sparing Scheme for Mixed-Criticality Systems. *IEEE Trans. Comput. Aided Des. Integr. Circuits Syst.* **2020**, *39*, 4601–4610. [CrossRef]

61. Saraswat, P.K.; Pop, P.; Madsen, J. Task Migration for Fault-Tolerance in Mixed-Criticality Embedded Systems. *SIGBED Rev.* **2009**, *6*. [CrossRef]

62. Liu, G.; Lu, Y.; Wang, S. An Efficient Fault Recovery Algorithm in Multiprocessor Mixed-Criticality Systems. In Proceedings of the IEEE International Conference on High Performance Computing and Communications (HPCC)/ IEEE International Conference on Embedded and Ubiquitous Computing (EUC), Zhangjiajie, China, 13–15 November 2013; pp. 2006–2013.

63. Ranjbar, B.; Nguyen, T.D.A.; Ejlali, A.; Kumar, A. Online Peak Power and Maximum Temperature Management in Multi-core Mixed-Criticality Embedded Systems. In Proceedings of the Euromicro Conference on Digital System Design (DSD), Chalkidiki, Greece, 28–30 August 2019; pp. 546–553.

64. Iacovelli, S.; Kirner, R.; Menon, C. ATMP: An Adaptive Tolerance-based Mixed-criticality Protocol for Multi-core Systems. In Proceedings of the International Symposium on Industrial Embedded Systems (SIES), Graz, Austria, 6–8 June 2018; pp. 1–9.

65. Al-bayati, Z.; Meyer, B.H.; Zeng, H. Fault-tolerant scheduling of multicore mixed-criticality systems under permanent failures. In Proceedings of the IEEE International Symposium on Defect and Fault Tolerance in VLSI and Nanotechnology Systems (DFT), Storrs, CT, USA, 19–20 September 2016; pp. 57–62.

66. O'Connor, P.P.; Kleyner, A. *Practical Reliability Engineering*, 5th ed.; Wiley Publishing: Hoboken, NJ, USA, 2012.

67. Esmaeilzadeh, H.; Blem, E.; Amant, R.S.; Sankaralingam, K.; Burger, D. Dark silicon and the end of multicore scaling. In Proceedings of the 2011 38th Annual International Symposium on Computer Architecture (ISCA), San Jose, CA, USA, 4–8 June 2011; Association for Computing Machinery: New York, NY, USA, 2011.

68. Buttazzo, G.C. *Hard Real-Time Computing Systems: Predictable Scheduling Algorithms and Applications*; Springer Science & Business Media: Berlin/Heidelberg, Germany, 2011; Volume 24.

69. Ranjbar, B.; Nguyen, T.D.A.; Ejlali, A.; Kumar, A. Power-Aware Run-Time Scheduler for Mixed-Criticality Systems on Multi-Core Platform. *IEEE Trans. Comput. Aided Des. Integr. Circuits Syst.* **2020**. [CrossRef]

70. Sahoo, S.S.; Veeravalli, B.; Kumar, A. Markov Chain-based Modeling and Analysis of Checkpointing with Rollback Recovery for Efficient DSE in Soft Real-time Systems. In Proceedings of the IEEE International Symposium on Defect and Fault Tolerance in VLSI and Nanotechnology Systems (DFT), Frascati, Italy, 19–21 October 2020; pp. 1–6. [CrossRef]

71. Manolache, S.; Eles, P.; Peng, Z. Task Mapping and Priority Assignment for Soft Real-Time Applications under Deadline Miss Ratio Constraints. *ACM Trans. Embed. Comput. Syst.* **2008**, *7*. [CrossRef]

72. Cheng, E.; Mirkhani, S.; Szafaryn, L.G.; Cher, C.Y.; Cho, H.; Skadron, K.; Stan, M.R.; Lilja, K.; Abraham, J.A.; Bose, P.; et al. CLEAR: Cross-Layer Exploration for Architecting Resilience - Combining Hardware and Software Techniques to Tolerate Soft Errors in Processor Cores. In Proceedings of the Design Automation Conference (DAC), Austin, TX, USA, 5–9 June 2016; Association for Computing Machinery: New York, NY, USA, 2016. [CrossRef]

73. Henkel, J.; Bauer, L.; Zhang, H.; Rehman, S.; Shafique, M. Multi-Layer Dependability: From Microarchitecture to Application Level. In Proceedings of the Design Automation Conference (DAC), San Francisco, CA, USA, 1–5 June 2014; Association for Computing Machinery: New York, NY, USA, 2014; pp. 1–6. [CrossRef]

74. Frantz, A.P.; Cassel, M.; Kastensmidt, F.L.; Cota, E.; Carro, L. Crosstalk- and SEU-Aware Networks on Chips. *IEEE Des. Test* **2007**, *24*, 340–350. [CrossRef]

75. Lehtonen, T.; Liljeberg, P.; Plosila, J. Online Reconfigurable Self-Timed Links for Fault Tolerant NoC. Available online: https://www.hindawi.com/journals/vlsi/2007/094676/ (accessed on 12 December 2020).

76. Vitkovskiy, A.; Soteriou, V.; Nicopoulos, C. A fine-grained link-level fault-tolerant mechanism for networks-on-chip. In Proceedings of the IEEE International Conference on Computer Design, Amsterdam, The Netherlands, 3–6 October 2010; pp. 447–454. [CrossRef]

77. Wells, P.M.; Chakraborty, K.; Sohi, G.S. Adapting to Intermittent Faults in Multicore Systems. *SIGOPS Oper. Syst. Rev.* **2008**, *42*, 255–264. [CrossRef]

78. Lehtonen, T.; Wolpert, D.; Liljeberg, P.; Plosila, J.; Ampadu, P. Self-Adaptive System for Addressing Permanent Errors in On-Chip Interconnects. *IEEE Trans. Very Large Scale Integr. Syst.* **2010**, *18*, 527–540. [CrossRef]

79. Yu, Q.; Ampadu, P. Transient and Permanent Error Co-management Method for Reliable Networks-on-Chip. In Proceedings of the ACM/IEEE International Symposium on Networks-on-Chip, Grenoble, France, 3–6 May 2010; pp. 145–154. [CrossRef]

80. Rehman, S.; Chen, K.; Kriebel, F.; Toma, A.; Shafique, M.; Chen, J.; Henkel, J. Cross-Layer Software Dependability on Unreliable Hardware. *IEEE Trans. Comput.* **2016**, *65*, 80–94. [CrossRef]

81. Weichslgartner, A.; Wildermann, S.; Gangadharan, D.; Glaß, M.; Teich, J. A Design-Time/Run-Time Application Mapping Methodology for Predictable Execution Time in MPSoCs. *ACM Trans. Embed. Comput. Syst.* **2018**, *17*. [CrossRef]

82. Pourmohseni, B.; Wildermann, S.; Glaß, M.; Teich, J. Hard Real-Time Application Mapping Reconfiguration for NoC-Based Many-Core Systems. *Real Time Syst.* **2019**, *55*, 433–469. [CrossRef]

83. Safari, S.; Ansari, M.; Ershadi, G.; Hessabi, S. On the Scheduling of Energy-Aware Fault-Tolerant Mixed-Criticality Multicore Systems with Service Guarantee Exploration. *IEEE Trans. Parallel Distrib. Syst.* **2019**, *30*, 2338–2354. [CrossRef]

84. Rambo, E.A.; Ernst, R. Replica-Aware Co-Scheduling for Mixed-Criticality. In *Proceedings of the Euromicro Conference on Real-Time Systems (ECRTS)*; Leibniz International Proceedings in Informatics (LIPIcs); Bertogna, M., Ed.; Schloss Dagstuhl–Leibniz-Zentrum fuer Informatik: Dagstuhl, Germany, 2017; Volume 76, pp. 20:1–20:20. [CrossRef]

85. Bolchini, C.; Miele, A. Reliability-Driven System-Level Synthesis for Mixed-Critical Embedded Systems. *IEEE Trans. Comput.* **2013**, *62*, 2489–2502. [CrossRef]

86. Kang, S.; Yang, H.; Kim, S.; Bacivarov, I.; Ha, S.; Thiele, L. Reliability-aware mapping optimization of multi-core systems with mixed-criticality. In Proceedings of the Design, Automation Test in Europe Conference & Exhibition (DATE), Dresden, Germany, 24–28 March 2014; pp. 1–4.

87. Kang, S.h.; Yang, H.; Kim, S.; Bacivarov, I.; Ha, S.; Thiele, L. Static Mapping of Mixed-Critical Applications for Fault-Tolerant MPSoCs. In Proceedings of the Design Automation Conference (DAC), San Francisco, CA, USA, 1–5 June 2014; Association for Computing Machinery: New York, NY, USA, 2014; pp. 1–6. [CrossRef]

88. Choi, J.; Yang, H.; Ha, S. Optimization of Fault-Tolerant Mixed-Criticality Multi-Core Systems with Enhanced WCRT Analysis. *ACM Trans. Des. Autom. Electron. Syst.* **2018**, *24*. [CrossRef]

89. Jiang, W.; Hu, H.; Zhan, J.; Jiang, K. Work-in-Progress: Design of Security-Critical Distributed Real-Time Applications with Fault-Tolerant Constraint. In Proceedings of the International Conference on Embedded Software (EMSOFT), Torino, Italy, 30 September–5 October 2018; pp. 1–2.

90. Zeng, L.; Huang, P.; Thiele, L. Towards the Design of Fault-Tolerant Mixed-Criticality Systems on Multicores. In Proceedings of the International Conference on Compilers, Architectures and Synthesis for Embedded Systems (CASES), Pittsburgh, PA, USA, 2–7 October 2016; Association for Computing Machinery: New York, NY, USA, 2016. [CrossRef]

91. Caplan, J.; Al-bayati, Z.; Zeng, H.; Meyer, B.H. Mapping and Scheduling Mixed-Criticality Systems with On-Demand Redundancy. *IEEE Trans. Comput.* **2018**, *67*, 582–588. [CrossRef]

92. Saraswat, P.K.; Pop, P.; Madsen, J. Task Mapping and Bandwidth Reservation for Mixed Hard/Soft Fault-Tolerant Embedded Systems. In Proceedings of the IEEE Real-Time and Embedded Technology and Applications Symposium (RTAS), Stockholm, Sweden, 12–15 April 2010; pp. 89–98.

93. Bagheri, M.; Jervan, G. Fault-Tolerant Scheduling of Mixed-Critical Applications on Multi-processor Platforms. In Proceedings of the IEEE International Conference on Embedded and Ubiquitous Computing (EUC), Milano, Italy, 25–29 August 2014; pp. 25–32.

94. Kajmakovic, A.; Diwold, K.; Kajtazovic, N.; Zupanc, R.; Macher, G. Flexible Soft Error Mitigation Strategy for Memories in Mixed-Critical Systems. In Proceedings of the IEEE International Symposium on Software Reliability Engineering Workshops (ISSREW), Berlin, Germany, 28–31 October 2019; pp. 440–445.

95. Liu, Y.; Xie, G.; Tang, Y.; Li, R. Improving Real-Time Performance Under Reliability Requirement Assurance in Automotive Electronic Systems. *IEEE Access* **2019**, *7*, 140875–140888. [CrossRef]

96. Thekkilakattil, A.; Dobrin, R.; Punnekkat, S. Mixed criticality scheduling in fault-tolerant distributed real-time systems. In Proceedings of the International Conference on Embedded Systems (ICES), Coimbatore, India, 3–5 July 2014; pp. 92–97.

97. Koc, H.; Karanam, V.K.; Sonnier, M. Latency Constrained Task Mapping to Improve Reliability of High Critical Tasks in Mixed Criticality Systems. In Proceedings of the IEEE Annual Information Technology, Electronics and Mobile Communication Conference (IEMCON), British Columbia, VA, Canada, 17–19 October 2019; pp. 0320–0324.

98. Rambo, E.A.; Donyanavard, B.; Seo, M.; Maurer, F.; Kadeed, T.M.; De Melo, C.B.; Maity, B.; Surhonne, A.; Herkersdorf, A.; Kurdahi, F.; et al. The Self-Aware Information Processing Factory Paradigm for Mixed-Critical Multiprocessing. *IEEE Trans. Emerg. Top. Comput.* **2020**. [CrossRef]

99. LogiCORE IP. Soft Error Mitigation Controller v3. 4. Available online: https://www.xilinx.com/support/answers/54733.html (accessed on 12 December 2020).

100. Santos, R.; Venkataraman, S.; Kumar, A. Scrubbing Mechanism for Heterogeneous Applications in Reconfigurable Devices. *ACM Trans. Des. Autom. Electron. Syst.* **2017**, *22*. [CrossRef]

101. Koch, D.; Haubelt, C.; Teich, J. Efficient Hardware Checkpointing: Concepts, Overhead Analysis, and Implementation. In Proceedings of the International Symposium on Field Programmable Gate Arrays (FPGA), Monterey, CA, USA, 18–20 February 2007.

102. Lee, G.; Agiakatsikas, D.; Wu, T.; Cetin, E.; Diessel, O. TLegUp: A TMR code generation tool for SRAM-based FPGA applications using HLS. In Proceedings of the International Symposium on Field-Programmable Custom Computing Machines (FCCM), Napa, CA, USA, 30 April–2 May 2017; pp. 129–132.

103. Srinivasan, S.; Krishnan, R.; Mangalagiri, P.; Xie, Y.; Narayanan, V.; Irwin, M.J.; Sarpatwari, K. Toward Increasing FPGA Lifetime. *IEEE Trans. Dependable Secur. Comput.* **2008**, *5*, 115–127. [CrossRef]

104. Stott, E.; Cheung, P.Y.K. Improving FPGA Reliability with Wear-Levelling. In Proceedings of the International Conference on Field Programmable Logic and Applications (FPL), Chania, Crete, Greece, 5–7 September 2011; pp. 323–328. [CrossRef]

105. Angermeier, J.; Ziener, D.; Glaß, M.; Teich, J. Stress-Aware Module Placement on Reconfigurable Devices. In Proceedings of the International Conference on Field Programmable Logic and Applications (FPL), Chania, Crete, Greece, 5–7 September 2011; pp. 277–281. [CrossRef]

106. Zhang, H.; Bauer, L.; Kochte, M.A.; Schneider, E.; Braun, C.; Imhof, M.E.; Wunderlich, H.; Henkel, J. Module diversification: Fault tolerance and aging mitigation for runtime reconfigurable architectures. In Proceedings of the IEEE International Test Conference (ITC), Anaheim, CA, USA, 6–13 September 2013; pp. 1–10. [CrossRef]

107. Zhang, H.; Kochte, M.A.; Schneider, E.; Bauer, L.; Wunderlich, H.; Henkel, J. STRAP: Stress-aware placement for aging mitigation in runtime reconfigurable architectures. In Proceedings of the IEEE/ACM International Conference on Computer-Aided Design (ICCAD), Austin, TX, USA, 2–6 November 2015; pp. 38–45. [CrossRef]

108. Ghaderi, Z.; Bozorgzadeh, E. Aging-aware high-level physical planning for reconfigurable systems. In Proceedings of the Asia and South Pacific Design Automation Conference (ASP-DAC), Macao, China, 25–28 January 2016; pp. 631–636. [CrossRef]

109. Dumitriu, V.; Kirischian, L.; Kirischian, V. Run-Time Recovery Mechanism for Transient and Permanent Hardware Faults Based on Distributed, Self-Organized Dynamic Partially Reconfigurable Systems. *IEEE Trans. Comput.* **2016**, *65*, 2835–2847. [CrossRef]

110. Sahoo, S.S.; Nguyen, T.D.A.; Veeravalli, B.; Kumar, A. Lifetime-aware design methodology for dynamic partially reconfigurable systems. In Proceedings of the Asia and South Pacific Design Automation Conference (ASP-DAC), Jeju Island , Korea, 22–25 January 2018; pp. 393–398.

111. Sahoo, S.; Nguyen, T.; Veeravalli, B.; Kumar, A. Multi-objective design space exploration for system partitioning of FPGA-based Dynamic Partially Reconfigurable Systems. *Integration* **2019**, *67*, 95–107. [CrossRef]

112. Santos, R.; Venkataraman, S.; Das, A.; Kumar, A. Criticality-aware scrubbing mechanism for SRAM-based FPGAs. In Proceedings of the International Conference on Field Programmable Logic and Applications (FPL), Munich, Germany, 2–4 September 2014; pp. 1–8.

113. Santos, R.; Venkataraman, S.; Kumar, A. Dynamically adaptive scrubbing mechanism for improved reliability in reconfigurable embedded systems. In Proceedings of the Design Automation Conference (DAC), San Francisco, CA, USA, 7–11 June 2015; pp. 1–6.

114. Ahmadian, H.; Nekouei, F.; Obermaisser, R. Fault recovery and adaptation in time-triggered Networks-on-Chips for mixed-criticality systems. In Proceedings of the International Symposium on Reconfigurable Communication-centric Systems-on-Chip (ReCoSoC), Madrid, Spain, 12–17 July 2017; pp. 1–8.

115. Guha, K.; Majumder, A.; Saha, D.; Chakrabarti, A. Reliability Driven Mixed Critical Tasks Processing on FPGAs Against Hardware Trojan Attacks. In Proceedings of the Euromicro Conference on Digital System Design (DSD), Prague, Czech Republic, 29–31 August 2018; pp. 537–544.

116. Mandal, S.; Sarkar, S.; Ming, W.M.; Chattopadhyay, A.; Chakrabarti, A. Criticality Aware Soft Error Mitigation in the Configuration Memory of SRAM Based FPGA. In Proceedings of the International Conference on VLSI Design and International Conference on Embedded Systems (VLSID), Delhi, India, 5–9 January 2019; pp. 257–262.

117. Heiner, J.; Sellers, B.; Wirthlin, M.; Kalb, J. FPGA partial reconfiguration via configuration scrubbing. In Proceedings of the International Conference on Field Programmable Logic and Applications (FPL), Prague, Czech Republic, 31 August–2 September 2009; pp. 99–104. [CrossRef]

118. Sahoo, S.S.; Veeravalli, B.; Kumar, A. Cross-layer fault-tolerant design of real-time systems. In Proceedings of the IEEE International Symposium on Defect and Fault Tolerance in VLSI and Nanotechnology Systems (DFT), Storrs, CT, USA, 19–20 September 2016; pp. 63–68. [CrossRef]

119. Carter, N.P.; Naeimi, H.; Gardner, D.S. Design techniques for cross-layer resilience. In Proceedings of the Design, Automation Test in Europe Conference & Exhibition (DATE), Dresden, Germany, 8–12 March 2010; pp. 1023–1028. [CrossRef]

120. Santini, T.; Rech, P.; Sartor, A.; Corrêa, U.B.; Carro, L.; Wagner, F.R. Evaluation of Failures Masking Across the Software Stack. Available online: http://www.median-project.eu/wp-content/uploads/12_II-1_median2015.pdf (accessed on 12 December 2020).
121. Sahoo, S.S.; Nguyen, T.D.A.; Veeravalli, B.; Kumar, A. QoS-Aware Cross-Layer Reliability-Integrated FPGA-Based Dynamic Partially Reconfigurable System Partitioning. In Proceedings of the International Conference on Field-Programmable Technology (FPT), Naha, Japan, 11–15 December 2018; pp. 230–233.
122. Götzinger, M.; Rahmani, A.M.; Pongratz, M.; Liljeberg, P.; Jantsch, A.; Tenhunen, H. The Role of Self-Awareness and Hierarchical Agents in Resource Management for Many-Core Systems. In Proceedings of the IEEE International Symposium on Embedded Multicore/Many-core Systems-on-Chip (MCSOC), Hanoi, Vietnam, 12–14 September 2018; pp. 53–60. [CrossRef]
123. Rambo, E.A.; Kadeed, T.; Ernst, R.; Seo, M.; Kurdahi, F.; Donyanavard, B.; de Melo, C.B.; Maity, B.; Moazzemi, K.; Stewart, K.; et al. The Information Processing Factory: A Paradigm for Life Cycle Management of Dependable Systems. In Proceedings of the IEEE/ACM/IFIP International Conference on Hardware/Software Codesign and System Synthesis (CODES), New York, NY, USA, 13–18 October 2019; pp. 1–10.
124. Burns, A.; Davis, R.I. A Survey of Research into Mixed Criticality Systems. *ACM Comput. Surv.* **2017**, *50*. [CrossRef]
125. Mittal, S. A Survey of Techniques for Approximate Computing. *ACM Comput. Surv.* **2016**, *48*. [CrossRef]
126. Das, A.; Shafik, R.A.; Merrett, G.V.; Al-Hashimi, B.M.; Kumar, A.; Veeravalli, B. Reinforcement Learning-Based Inter- and Intra-Application Thermal Optimization for Lifetime Improvement of Multicore Systems. In Proceedings of the Design Automation Conference (DAC), San Francisco, CA, USA, 1–5 June 2014; Association for Computing Machinery: New York, NY, USA, 2014; pp. 1–6. [CrossRef]
127. Zhang, J.; Sadiqbatcha, S.; Gao, Y.; O'Dea, M.; Yu, N.; Tan, S.X.D. HAT-DRL: Hotspot-Aware Task Mapping for Lifetime Improvement of Multicore System Using Deep Reinforcement Learning. In Proceedings of the ACM/IEEE Workshop on Machine Learning for CAD, Virtual Event, Iceland, 16–20 November 2020; Association for Computing Machinery: New York, NY, USA, 2020; pp. 77–82.

Journal of
Low Power Electronics and Applications

Article

Framework for Design Exploration and Performance Analysis of RF-NoC Manycore Architecture

Habiba Lahdhiri [1,*], Jordane Lorandel [1], Salvatore Monteleone [1], Emmanuelle Bourdel [1] and Maurizio Palesi [2]

[1] ETIS UMR 8051, CY Cergy Paris University, ENSEA, CNRS, F-95000 Cergy, France;
    jordane.lorandel@ensea.fr (J.L.); salvatore.monteleone@ensea.fr (S.M.); emmanuelle.bourdel@ensea.fr (E.B.)
[2] Department of Electrical, Electronic, and Computer Engineering, University of Catania, 95125 Catania, Italy;
    maurizio.palesi@dieei.unict.it
* Correspondence: habiba.lahdhiri@ensea.fr

Received: 17 September 2020; Accepted: 31 October 2020; Published: 3 November 2020

Abstract: The Network-on-chip (NoC) paradigm has been proposed as a promising solution to enable the handling of a high degree of integration in multi-/many-core architectures. Despite their advantages, wired NoC infrastructures are facing several performance issues regarding multi-hop long-distance communications. RF-NoC is an attractive solution offering high performance and multicast/broadcast capabilities. However, managing RF links is a critical aspect that relies on both application-dependent and architectural parameters. This paper proposes a design space exploration framework for OFDMA-based RF-NoC architecture, which takes advantage of both real application benchmarks simulated using Sniper and RF-NoC architecture modeled using Noxim. We adopted the proposed framework to finely configure a routing algorithm, working with real traffic, achieving up to 45% of delay reduction, compared to a wired NoC setup in similar conditions.

Keywords: RF; NoC; OFDMA; simulator; routing; reconfigurable

---

1. Introduction

The significant integration of a large number of cores into the same chip for creating multi-/many-core Systems-on-Chips (SoCs) created new challenges for designers. The Network-on-Chip (NoC) paradigm has been promoted as a viable solution to deal with multi-/many-core emerging trends. Despite its strengths, NoCs have significant performance limitations due to the high latency and power consumption resulting from long multi-hop wired links used to deliver the data, especially in long-range communications across the chip. Several interconnect technologies have been proposed based on photonic, 3D, and Radio-Frequency (RF) to overcome this issue. Hybrid architectures were also introduced, combining multiple interconnect technologies.

Photonic solutions provide a way to reach near speed-of-light communications across on-chip wires [1,2]. These approaches achieve very low latency, but they face the problem of the considerable area dedicated to the signal conditioning circuitry. In this case, the optical NoC is introduced to enable high-speed links and negligible power dissipation. However, signal noise and waveguide losses are not negligible.

3D-NoC is an interesting approach to address the problem of the interconnection scale. This architecture responds to future multi-/many-core architectures' requirements by exploiting short vertical links between adjacent layers to improve network performance [3,4] considerably. However, the advantages of this technology cannot neglect thermal problems as the number of layers increases.

Another approach based on radio-frequency waves is the RF-NoC interconnect. It provides flexible communication and single-hop long-range communication, aiming at reducing latency.

This technology is based on the transmission of electromagnetic waves through the chip, allowing high bandwidth communication and low delay. Two types of radio-frequency interconnects exist, the first one making use of antennas and leading to free space communication (wireless), the other one exploiting communication using a waveguide (wired RF). The latter is similar to wireless propagation in terms of CMOS compatibility, high throughput, low overall power consumption, and near-light speed signals. Solutions using antennas have greater flexibility, but they also increase consumption and suffer from less immunity to interference when compared to waveguide [5]. Besides, the waveguide provides a communication channel, perfectly known at the design phase. Moreover, design and sizing of RF elements for RF-NoC architectures based on waveguides have already been proposed in the literature, demonstrating the feasibility of the approach. The interested reader could find more details in [6–8].

To take advantage of all the benefits of these new technologies, an efficient multiple access technique is required to share the spectrum resources among the different elements wishing to communicate. Many multiple access techniques exist, such as Frequency Division Multiple Access (FDMA), Time-Division Multiple Access (TDMA), and Wavelength Division Multiple Access (WDMA). To achieve high spectral efficiency, a multi-carrier modulation approach, namely Orthogonal Frequency Division Multiplexing (OFDM), is used. Among the significant advantages provided by OFDMA, it achieves high spectral efficiency and allows a flexible resource allocation while being a robust multi-carrier modulation against inter-carrier interference.

Regarding the NoC architecture, many parameters have an impact on power and performance. This is the case of the traffic occurring inside the NoC, which depends mainly on the applications running on the system. Synthetic traffics (e.g., Transpose Matrix and Random) are good choices for a first study but may not reflect traffics generated from real applications and scenarios. The NoC topology is also a crucial parameter. It leverages the choice of routing and selection algorithms as well as micro-architectural NoC parameters. All of these parameters have to be jointly considered when evaluating the performance of such architectures.

In particular, when adding a second interconnection layer based on RF, it becomes very complicated for a designer to make decisive choices that will ultimately have a relevant impact on power and performance figures. For example, the resource allocation strategy of RF interconnects, as well as the NoC routing policy, have to be finely defined to efficiently balance the traffic over the wired NoC or RF links.

Given the number of design choices, a need has emerged for simulation tools capable of simulating these emerging architectures. There are two main categories of simulators: (1) application-level simulators to analyze the behavior of a given application running on a specified multi-core architecture, and (2) Cycle-accurate NoC simulators, which perform fine-grained simulation of the NoC architecture, leading to more accurate power and latency results.

This paper introduces a simulation framework based on Noxim and Sniper simulators, enabling design space exploration for RF-NoC OFDMA architectures while considering real application traffic. The use of RF-NoC architectures with OFDMA brings some interesting advantages since OFDMA can adjust the channel usage to serve single o multiple users (the processing elements) simultaneously. In this sense, OFDMA is a very good option for low bandwidth applications, also thanks to the better frequency reuse and low latency. At best of our knowledge, there are no other simulation frameworks that allow the evaluation of such architectures by finely tuning the routing algorithm parameters for OFDMA RF-NoCs such as the ones introduced in [9,10].

The remainder of this paper is structured as follows. A comparative study of NoC simulators is presented in Section 2. Then, the considered RF-NoC architecture based on OFDMA is presented in Section 4, and the proposed framework is detailed in Section 5. Simulations results are presented in Section 6.1. Finally, conclusions are drawn in Section 7.

2. NoC Simulators

Before proceeding with the hardware implementation or the emulation of a NoC design, the use of a simulation framework is almost mandatory. A good simulation framework, indeed, allows minimizing implementation costs through an early estimation of different figures of merit before the physical implementation of the system, helping in the process of making the right design decisions suitable for the considered scenario. Most NoC simulators are developed in C++ or SystemC, and some of them in Java. Simulators written in Java are usually high-level simulators. They offer better code portability but lead to less efficiency. Simulators can be classified depending on their accuracy (e.g., cycle-accurate, and discrete event-driven) or depending on their programming abstraction level (i.e., high/low). In the following, most adopted existing NoC simulators are introduced with a particular focus on Noxim.

Booksim is a cycle-accurate simulator written in C++ by Dally and Towles from the University of Stanford in the USA [11]. Booksim is the first version not intended for a specific on-chip environment but mostly a generic simulator. This version was extended to overcome limitations in order to include some advanced features and technologies for on-chip networks. Booksim2 provides a wide diversity of topologies such as mesh, torus, tree, and butterfly. It supports a variety of routing algorithms and several options to customize the micro-architecture of routers to simulate.

DARSIM is a cycle-level, parallel simulator from the Massachusetts Institute of Technology (MIT) [12]. It allows simulating both 2D and 3D mesh architectures. DARSIM provides a large advanced set of NoC parameters such as different virtual channel (VC) allocation and memory models. The simulator offers diverse routing algorithms due to its highly parameterized routing table-based, which provides two possibilities: running the simulation from application traces or synthetic patterns. One of the strengths of this simulator is the ability of the hardware configuration, such as bandwidth, pipeline depth, and geometry. Besides, it allows to split the tasks between cores equally and achieves cycle-accurate simulations.

HNOCS (Heterogeneous NoC Simulator) [13] is dedicated to heterogeneous NoC architectures and is based on OMNet++. OMNet++ provides C++ APIs to a wide range of services to describe in detail the network topology. Moreover, the basic elements for the network configuration (routing algorithms/topologies/VC), HNOCS simulator provides parallelism, various Quality-of-Service (QoS), different arbitrary technologies, and power estimation. It offers three different router types, asynchronous, synchronous, and synchronous virtual output queue and performance statistics such as throughput, VC acquisition, and transfer latency.

Nigram is a cycle-accurate and discrete event simulator developed in SystemC by the Malaviya National Institute of Technology India and the University of Southampton UK [14]. It provides various network configuration commands to simulate different NoC architectures such as routing algorithms (source, XY, odd-even, adaptive), topologies (Tree, Torus, Mesh, and Ring), two flow control techniques (deflection and wormhole). The simulation statistics include throughput and latency.

Noxim is developed by the group of computer architectures at the University of Catania [15]. It is a low level, open-source, and cycle-accurate simulator written in C++/SystemC. Noxim provides various configuration parameters such as packet and buffer sizes, packet injection rate, different routing algorithms (XY, Odd-Even, West-first, North-last), traffic distributions (Random, Transpose, Bit-reversal, Butterfly Shuffle, Table Based traffic, hotspot), structures, and topologies. In addition to the wired NoC simulation, Noxim also supports Wireless NoC (WiNoC) evaluation and provides power consumption, throughput, and latency as performance analysis. Access Noxim is an extended version that supports 3D NoC architecture and adaptive routing [16].

Orion 3.0 is a simulator dedicated to evaluating the power performance of the NoC. It provides component dynamic and leakage power models. Orion3.0 [17] overcomes the limitations of the Orion simulator by supporting power models estimated from actual post-placement and routing layout and area.

SunFloor-3D is the extended version of the SunFloor simulator. SunFloor is able to generate a system specification that allows designing NoC architectures from a set of defined input constraints (energy, area, and model). SunFloor-3D is dedicated to 3D-NoC architectures [18] and provides many advanced features such as the placement of components in the 3D layers. It enables the characterization of the core assignment and communication bandwidth.

3. System Simulators

In the following, the two most adopted system-level simulators, namely Gem5 and Sniper, are introduced with a particular focus on Sniper, which, together with Noxim, has been chosen to evaluate the considered RF-NoC architecture.

Gem5 is one of the more general simulators that come to the aid of computer architecture researchers. It is the result of the combination of two simulators GEMS [19] and M5 [20]. GEMS provides a flexible and detailed memory system and multiple cache protocols. GEMS simulator supports many commercial Instruction Set Architectures (ISAs) such as x86 (64 bits), MIPS, ARM, ALPHA, SPARK, and PowerPC and implicates Linux boot on ARM, ALPHA, and x86. Gem5 [21,22] also includes the best features of M5, especially the highly configurable environment to simulate various processor models. Specialized versions of Gem5 exist, for example Gem5-gpu [23] which is a simulator dedicated to heterogeneous CPU-GPU.

Sniper is a multi-core simulator based on the infrastructure of Graphite [24]. Sniper allows parallel, fast and accurate simulations and supports both homogeneous and heterogeneous multi-core architectures [25]. The principal simulator feature is the core model based on interval simulation. Sniper is considered as a high-speed simulator due to the interval simulation, which raises the abstraction level. It is useful for core and system-level studies that need details more than the typical one-IPC models. It includes SPLASH-2 (Barnes, Cholesky, FFT, FMM, Lu, Ocean, Radix, Radiosity, and Raytrace) and Parsec (Blackscholes, Bodytrack, Canneal, Dedup, Facesim, Ferret Fluidanimate, Freqmine, Raytrace, Streamcluster, Swaptions, VIPS, and x264) benchmarks in order to evaluate the NoC architecture. Sniper also provides SimAPI interfaces and Python to monitor and control its behavior at run time.

This section, together with the previous one provided a comparison of simulation tools to help decide on the suitable simulator regarding NoC designs and proposals starting from both NoC-/system-level available simulation tools. Each of these tools has its own peculiarities when adopted in a standalone or combined fashion. For example, Booksim2 provides a highly flexible simulation environment that allows fine-grained management of many elements, such as buffer size, virtual channels, and routing algorithms, and Gem5, coupled with Garnet2.0, offers the support to Full System (FS) simulations. Tables 1 and 2 summarize the different NoC and system simulators, respectively. These simulators, detailed above, are just a representative set of the existing possibilities taken into account in this research. By the way, no simulator found in literature includes all evaluation criteria at the same time. The proposed framework, detailed in Section 5, is based on two simulators: Noxim and Sniper. Noxim has been preferred to other NoC simulators since it already supports Wireless NoC architectures. Therefore, it offers an already established starting base and core elements such as Radio Hubs to simulate long-distance, single-hop communications. Also, Noxim comes with a tool, namely noxim explorer, that helps the user run batch simulations after defining the ranges of values for the simulator's parameters. Radio Hubs and Noxim Explorer have been extended to support the use case presented in the submitted manuscript. For what concerns Sniper, it has been chosen for its flexibility, the availability of its SimAPI to control the simulator's behavior at run-time, and the fact it allows tracing the traffic of real applications running on multi-core NoC-based architectures.

Table 1. NoC Simulators Comparison.

Simulator	Team	Language	Abstraction Level	Topologies	Benchmark Support	Heterogeneous	Ref
BookSim	University of Stanford	C++	High	Many	-	-	[11]
DARSIM	MIT	C++	High	Any	+	-	[12]
HNOCS	Technion Israel Institute of Technology	OMNET++	High	2D/3D Mesh	+	+	[13]
Nigram	University of Southampton	SystemC	Low	Any	-	-	[14]
Noxim	University of Catania	SystemC	Low	Many	-	-	[15]

Table 2. System Simulators Comparison.

Simulator	Team	Language	Abstraction Level	Topologies	Benchmark Support	Heterogeneous	Ref
Sniper	Ghent University	SystemC	High	Many	+	+	[25]
Gem5	AMD, ARM,HP, MIPS,Princeton, MIT, etc.	C++	High	Many	+	+	[21,22]
Gem5-GPU	AMD, ARM,HP, MIPS,Princeton, MIT, etc.	C++	High	Any	+	+	[23]

4. RF-NoC OFDMA Architecture

The communication between cores in a conventional NoC is ensured by wired links and multiple switches/routers. To overcome the latency and power consumption issue, we selected a hybrid topology based on RF links for single-hop long-distance communications. In this section, we present the considered RF-NoC OFDMA architecture introduced in [9]. In this work, RF-NoC based on waveguide is preferred over the more widespread WiNoC based on mm-wave antennas since wired RF transmission lines are considered a more suitable candidate for the implementation of high-speed EM propagation-based on-chip interconnects with consolidated CMOS technology.

4.1. Topology

The topology defines the physical layout and the connections between nodes in the network. It impacts network performance and cost since the topology constrains the minimum number of hops a packet must perform to reach its destination. There are two main classes of topologies: direct and indirect. In the case of direct topologies, each node of the NoC consists of both a Processing Element (PE) and a router. Therefore, nodes are able to both perform computation and manage the communication towards other nodes. These topologies are called direct because each node has a direct (point-to-point) link to a subset of other nodes in the network; a mesh, as the one shown in Figure 1, is a classic example of direct topology. In the case of indirect topologies, computation and communication (packet routing) features are managed in separated nodes and, in particular, each computation node (PE) is connected to a switching node (or router) that enables the communication with other computation nodes. A classic example of indirect topology is Multi-stage Interconnect Networks (MINs) Figure 1.

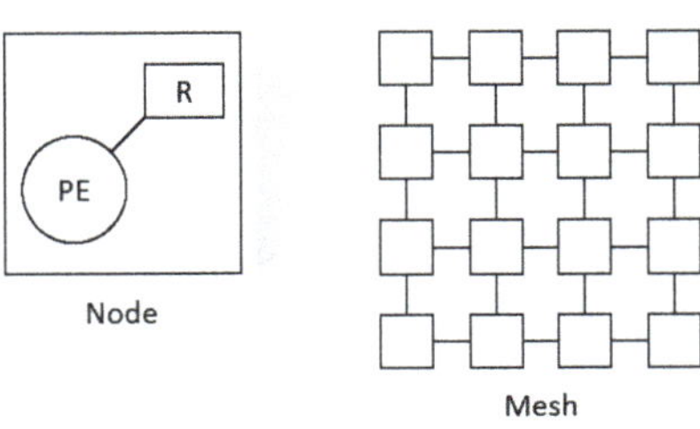

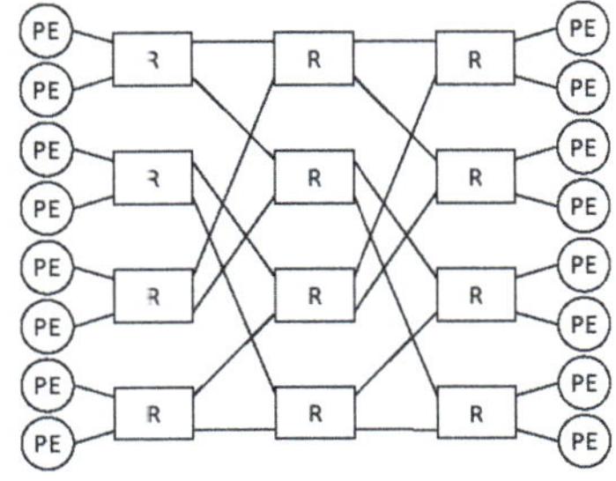

Figure 1. Representation of a 4 × 4 mesh and of an 8 nodes Multistage Interconnection Network as representative examples of direct and indirect topologies, respectively.

The considered RF-NoC architecture has a direct topology. It presents two levels of hierarchy covering the communication among cores (though a wired interconnect) and clusters (through an RF interconnect). Clusters are sets of cores providing another layer of hierarchy. In the considered architecture, 1024 cores are divided into 32 clusters, containing 32 cores. The communication within a cluster is handled through a 2D mesh wired NoC since the average path length is short compared to the global network. Figure 2 shows the wired links connecting cores within the cluster. Note that wired links between adjacent routers of separate clusters also exist. Moreover, each cluster contains a Radio Hub (RH) that attaches to it the four routers located in the cluster center. The RH is the component that leads to the second level of the hierarchy, i.e., the communications between clusters through the RF waveguide. Thus, each cluster features an RF-NoC Interface, located at its center, to access the waveguide. It is connected on one side to the four central routers, as illustrated in Figure 2, and on the other side to the RF waveguide.

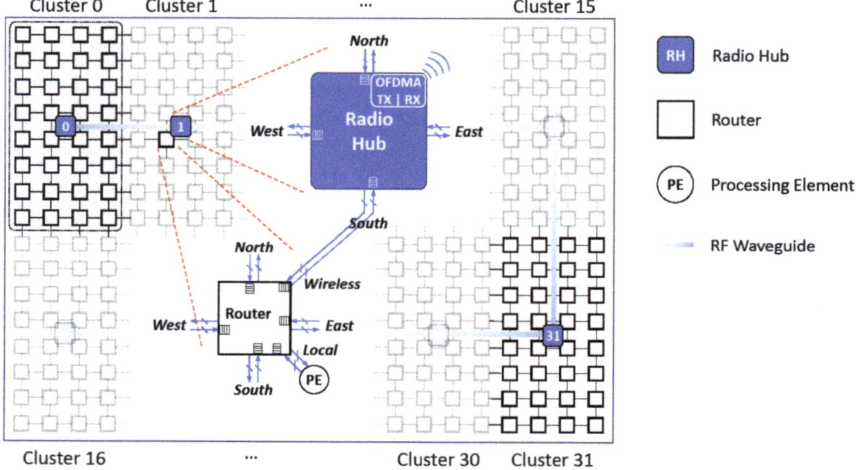

Figure 2. RF-NoC architecture.

4.2. OFDMA for RF Resource Allocation

As mentioned previously, Orthogonal Frequency-Division Multiple Access (OFDMA) is used. This approach allows for achieving high spectral efficiency by dividing the bandwidth into several orthogonal narrow sub-channels. The use of OFDMA allows simultaneous communications between multiple radio hubs using different frequency channels. In our configuration, a bandwidth B of 10 GHz is divided into 1024 sub-carriers. A frequency spacing of 9.76 MHz between each sub-carrier is thus obtained. The OFDMA symbol duration of T_s is computed as follows and is equal to 102.4 ns:

$$T_s = \frac{N_{sc}}{B} \tag{1}$$

with N_{sc} representing the number of sub-carriers.

The data rate R can be changed to transmit more or less information per OFDMA symbol, by modifying the modulation order M:

$$R = \frac{M \cdot N_{sc}}{T_s} \tag{2}$$

with M representing the number of bits per QAM symbol: this number is 2 for QPSK, 4 for 16-QAM, and so on.

Each cluster can transmit data through its RF interface using a group 32 contiguous sub-carriers but can receive the entire bandwidth, making possible multicast and broadcast communications between clusters. Based on the given configuration, the maximal binary throughput per channel can reach 625 Mbit/s when QSPK is chosen, while omitting the possible use of synchronization techniques. The overall theoretical throughput is 20Gbit/s for QSPK. The block diagram of the OFDMA transmitter and receiver is presented in Figure 3. Each cluster has its own OFDMA transceiver. However, to effectively exploit RF-NoC architectures, one of the main problems is the definition of an appropriate routing algorithm.

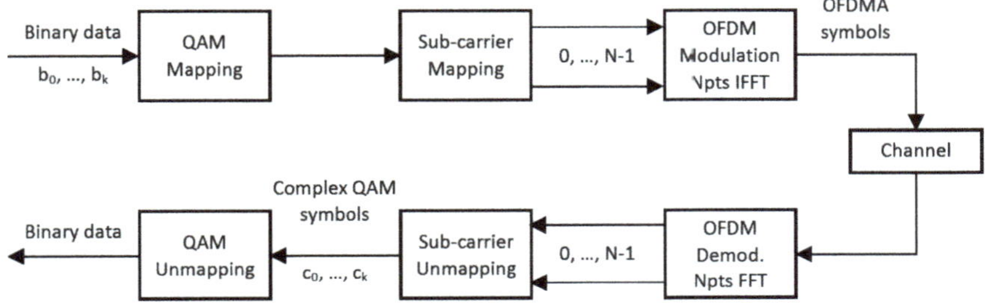

Figure 3. Block Diagram of the OFDMA transmitter and receiver paths.

4.3. Threshold-Based Routing Algorithm

We used a routing algorithm suitable for RF-NoC [9] that is based on the comparison of two distances. The first distance regards the wired path separation between the packet source and destination, and the second distance the wireless path using the RF link. If the wireless distance is greater than the wired distance, then the packets are transmitted using the wireless link, leading to a reduction of the network's average delay. A naive solution could be to take the minimum distance but this could rapidly create a bottleneck at the radio hub. Therefore, an adjusting threshold γ is defined to control the utilization of the RF link to avoid network congestion [9]. In the next section, we investigate the impact of the threshold value of the considered routing algorithm by comparing performance figures obtained using a proposed framework against synthetic traffics used into Noxim.

5. Proposed Framework

In this section, we detail the proposed framework, depicted in Figure 4, which is specific for the performance evaluation of RF-NoC architectures. This framework is based on the combination of two existing simulators: Sniper and Noxim. We selected Noxim as in its released version it already supports wireless communications thus its extension to RF-NoCs allows us to have in the same framework the availability of three different NoC architectures, namely, traditional wired NoC, heterogeneous wired WiNoC, and heterogeneous wired RF-NoC. In addition, to obtain the communication patterns generated by an application, it needs executing the application on a multi-core simulator and tracing all the communication flows induced during the execution of the application. To this end, we used Sniper.

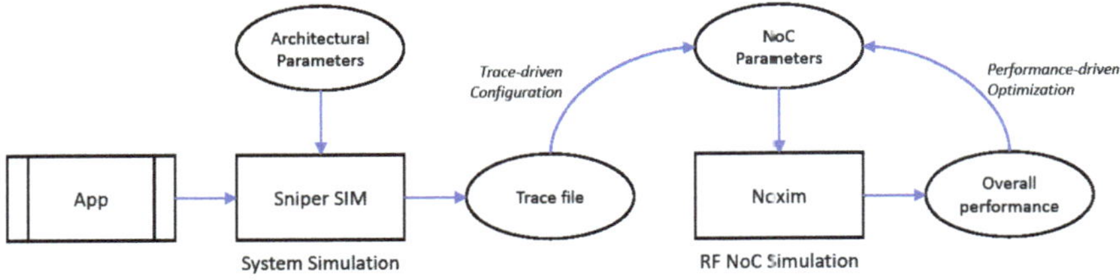

Figure 4. Flow of the proposed Framework.

5.1. Sniper NoC Configuration

The major exploited benefits offered by Sniper are the integrated benchmarks that enable fast tests with common tasks, the possibility to write custom test applications, and the full details of the interconnection network, core models, and cache. Sniper was thus used to model our NoC topology and to obtain communication traces from real application benchmarks. Sniper includes several folders that provide built-in tools and configuration files for the simulation parameters. The main folders used in our framework are: config which contains the NoC configuration file that describes the network to be simulated. Figure 5 gives an example of NoC configuration specifying the number of cores, memory cache levels, network topology, cores concentration, and bandwidth in bit per cycle; Benchmarks, which contains various benchmarks such as those from SPLASH-2, PARSEC, and SPEC CPU®2006 sets. test in which a set of applications to evaluate the network such are collected. The initial set of applications come from Sniper (e.g., FFT) but it is possible to add the custom applications to test; Finally tools folder includes python-coded tools to analyze simulation results.

After providing the NoC configuration file with all network parameters and the chosen application, Sniper produces a set of output files containing (i) general information related to the simulation (sim.info), (ii) the final configuration of the simulated architecture (sim.cfg), (iii), the results of the simulation in the form of a table (sim.out), and (iv) other statistics related to the execution (sim.stats).

```
[general]
total_cores = 1024       # Total number of cores in the simulation

[perf_model/cache]
levels = 2               # Number of cache level

[network/emesh_hop_by_hop]
link_bandwidth = 64   # Per-link, per-direction bandwidth in bits/cycle (In bits/cycle)
hop_latency = 2       # Per-hop latency in core cycles (In cycles)
concentration = 1     # Number of cores per network stop / network interface (must be >= last-level-cache/shared_cores)
dimensions = 2        # Dimensions (1 for line/ring, 2 for 2-D mesh/torus)
wrap_around = false   # Use wrap-around links (false for line/mesh, true for ring/torus)
size = "32:32"        # ":"-separated list of size for each dimension, default = auto
```

Figure 5. Sniper NoC configuration.

5.2. Trace File

Results provided by Sniper are not directly exploitable by Noxim. Thus, we developed a python script to format the output results provided by Sniper for Noxim properly. In more detail, Sniper provides a tool called *SIFT* that allows for trace recording. We extract the communication statistics from this trace file and generate the appropriate traffic-based routing table for Noxim.

In more details, firstly, the total number of exchanged packets between cores per link and the number of cycles are extracted. Secondly, the packet injection rates per link are computed, to finally generate the corresponding routing table.

To calculate the packet injection rate for each source/destination pair, we use a tool provided by Sniper, namely *dumpstats*, which provides simulation statistics. After the statistics have been stored in a file, the information regarding the timing and size of each communication is parsed by a python script introduced to extrapolate the number of packets P exchanged per link as well as the number of cycles C for each core to get the actual PIR (Packet Injection Rate) using the following equation:

$$PIR = \frac{P}{C} \quad (packet/cycle/node) \tag{3}$$

From these statistics, the total number of communication occurring inside the RF-NoC is easily derived. The python script, proposed for PIR's evaluation, is then able to generate the traffic table. This traffic table is a text file in which each line represents the communication between a source and a destination and their associated PIR. Figure 6 shows an example of a few lines of the generated traffic table. This format of the traffic table is supported by the Noxim simulator. From the user point of view, all the previous steps are automatically done by the framework.

```
Source   Destination  PIR
0        1020         1e-06
1        84           7e-06
12       341          1e-06
20       537          1e-06
```

Figure 6. Traffic table format.

5.3. RF-NoC Simulation under Noxim

The choice of Noxim simulator is based on its capability of supporting WiNoC topologies. It includes a fundamental component of wireless interconnects which is the radio-hub. The radio-hub allows single-hop links between faraway nodes in order to avoid multiple wired hops. It provides also the channel component which abstracts a flit transmission using a given wireless frequency. Noxim makes use of the Transaction Level Model (TLM) to simulate wireless communications. It provides also an energy model that includes both wireless and wired energy consumption. However, OFDMA is not supported natively. This will be detailed after. To perform simulations using Noxim, a YAML configuration file that contains all NoC parameters has to be filled. This file is divided into four parts: in (i) *Topology and Structure* are defined all necessary details of the components for the considered NoC architecture, such as the number of cores, router buffer size, radio-hub configuration (attached nodes, buffer size, access technique), and channel data rate (bit/s). Then (ii) the *Workload* part contains various data traffic models (uniform, butterfly, transpose, hot-spot), the packet injection rate, and the packet size. The parameters, such as the routing algorithm, channel access technique, and the choice between wired/wireless communication, are defined in part (iii) *Dynamic behavior*. Finally, the *Simulation* section collects parameters regarding the simulation setup itself, such as the number of cycles, warm-up time, reset time, and the level for statistics details. In addition to the traffic models provided by Noxim, it gives the possibility to simulate a real application by mapping its communication graph into custom table-based traffic. This table-based traffic allows defining the source/destination pairs with the packet injection rate, its statistical distribution, and traffic volume to be injected. Thanks to this feature, we can easily use in the proposed framework the generated traffic table detailed above, which is supported by Noxim.

After defining Noxim inputs in accordance with Sniper configuration, the simulator provides a set of performance statistics at the end of each simulation in order to evaluate the simulated architecture. In particular, they are: *received packets*, that reports the total number of packets effectively delivered at their destinations; the *average communication delay*, calculated as the difference between the clock cycles in which the packet is generated and consumed by the destination, respectively; the *network throughput*, defined as the ratio between the total received flits and the simulation duration in clock cycles. Finally, *energy consumption* summarizes the energy consumption of links, routers, radio-hubs, and network interfaces. Starting from the existing features, Noxim was extended to support RF-NoC OFDMA architecture. Noxim implements the token-ring technique to access the radio channels, and only one radio hub can transmit information on the wireless link at a time. As a consequence, we extended Noxim to support OFDMA and concurrent accesses to wireless channels. In addition, the threshold-based routing algorithm in Section 4.3 was also integrated into the simulator. Noxim provides a tool called *Noxim explorer* which is dedicated to the design space exploration. It allows for the execution of a set of simulations with different configuration parameters. We extended *Noxim explorer* to perform various simulations with different threshold values and consequently study the impact of threshold to the topology and the traffic distribution.

Regarding input Noxim parameters, some of them are directly defined according to Sniper configuration file, such as the number of cores, the topology, link bandwidth, etc. However, the user could still define other NoC architectural parameters as well as RF-related parameters e.g., number of sub-carriers, total frequency bandwidth, etc.

6. Performance Evaluation and Experimental Results

The traffic distribution strongly affects the performance of the network. In this part, we compare the results of synthetic traffic natively included in Noxim and the results obtained using the proposed framework, which integrates traffic generated from a real application. This comparison aims to validate the accuracy of the framework and draw conclusions about the choice of the threshold value for different application scenarios and different topologies. Moreover, this framework aims at showing the interest of automatically support real application traffic during design space exploration of OFDMA-based RF-NoC.

6.1. Synthetic Traffic Results

We define three different application scenarios according to the amount of long-distance communications, namely scenarios 1, 2, and 3, with their respective percentage of long-distance communications 75%, 50%, and 25%. This approach allows to classify the results according to the traffic pattern. In addition, to have a fair comparison between synthetic and real traffics, the total number of communications inside the RF-NoC remains approximately the same for all the experiments. This lets us study the impact of the threshold value most efficiently. From the following application specifications and the NoC topology, we generated a table-based traffic, depending on the network size, the number of hops to discriminate between short and long-range communications, and the packet injection rate, using a custom python script. Note that, for the considered topology, we define a communication as "long-distance communication" when the distance from the source to destination is greater than 8 wired hops, and then it is not necessary to exploit the RF link between adjacent clusters. The generated table-based traffic is used as input for Noxim.

Figure 7 reports threshold values for 32×32, 16×32 and 16×16 RF-NoC architectures under the three different traffic scenarios, with their respective delay reductions. We choose three different packet injection rate values to study the evolution of the threshold and the delay reduction value under different traffic loads. In these results, the total number of communications is 1×10^5, which is similar for the real application traffic generated from Sniper and the synthetic ones, and all communications have the same packet injection rate. This total number of communications remains constant to have a consistent comparison. The reported threshold for different PIR values and different topologies in the following results refers to the appropriate threshold to reach the maximum network delay reduction.

Starting from the 32×32 architecture results, we notice that for a PIR of 5×10^{-6}, the threshold remains constant (5 hops) regardless of the scenario, and the delay reduction is about 53% for the first scenario with a slight decrease for other applications. Then, the threshold value increases for PIR equal to 5×10^{-5}, with values of 25, 20, and 10 hops for the scenarios with 75%, 50%, and 25% of long-range communications, and we reach a significant delay reduction. Finally, for PIR equal to 5×10^{-4}, we observe that the threshold and the delay reduction are decreasing because the network enters in the saturation zone and is no longer able to manage the traffic load. We notice that the latency reduction decreases with the percentage of long-distance communications; for example, for a PIR value of 5×10^{-6}, we achieve 53% of delay reduction in the first scenario and 33% in the third application scenario.

For the 16×32 architecture, it can be seen that we have the same trend with a slight degradation of the threshold value and we reach a delay reduction of 79% in the case of 75% long-distance communications for a PIR equal to 5×10^{-5}. For this same PIR value, we notice that this topology allows achieving better latency reduction compared to 32×32 topology under the first traffic scenario.

From the results of 16×16 architecture, we can see that the threshold decreases with the network size Figure 7c. This value is between 10 and 5 hops for this topology for different application scenarios and PIR values. We can also note that the delay reduction is less significant for this architecture, particularly for low PIR values, i.e., 5×10^{-6} and regardless of the type of traffic.

The presented results show that the threshold value is impacted by the application scenario, which refers to the percentage of long-distance communications in an application, and the traffic load, represented by the PIR value. These results also give an idea about the percentage of latency reduction that could be achieved compared to a wired NoC, which helps to decide on the use of the RF link and topology choice.

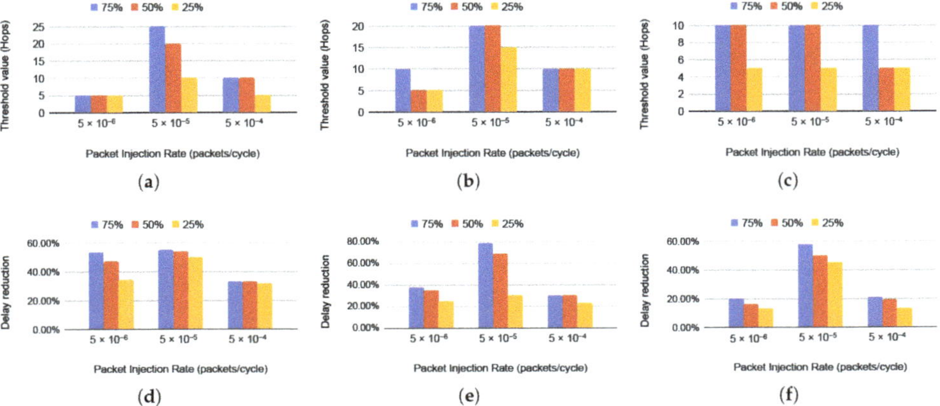

Figure 7. Threshold values for 32×32 (**a**), 16×32 (**b**), and 16×16 (**c**) architectures, with respective delay reductions (**d–f**).

6.2. Design Space Exploration

In this section, we present an example of a design space exploration using the proposed framework. Following the steps shown in Figure 4, we chose Splash2-FFT as benchmark provided by Sniper to evaluate three different RF-NoC topologies 32×32, 16×32 and 16×16. Table 3 includes simulation parameters. Then, we generate the corresponding traffic table for Noxim to get performance statistics. Finally, the following results illustrated in Figure 8 were obtained.

Table 3. Sniper NoC configuration.

Parameter	Value
Cores number	1024/512/256
Memory cache levels	2
Memory model	emesh hop by hop
Core model	Nehalem
System model	magic
Hop latency (cycles)	2
Core concentration per tile	1
Link bandwidth (bits/cycle)	64

Figure 8a shows the average delay of both wired NoC and RF-NoC for different threshold values. We notice a high latency when the threshold is between zero and five hops, which reflect the overuse of the RF link. The RF utilization is about 85%, with a threshold value of zero. It means that most of the packets are routed towards the radio hubs, which leads to network congestion. However, the latency gets reduced in a significant way when the threshold increases until it reaches the value of 25 hops. This threshold value leads to a delay reduction of 45%, as shown in the graph depicted in Figure 8b, which confirms the importance of the choice of the threshold to attend a maximum delay reduction. Once the threshold is greater than 25 hops, the average delay increases again.

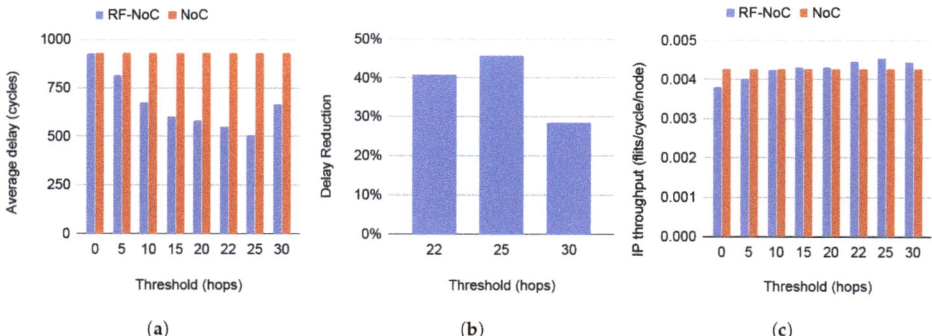

Figure 8. Average latency (**a**), delay reduction (**b**), and network throughput (**c**) in the considered 32 × 32 architecture under Splash2-FFT benchmark.

The IP throughput metric was also evaluated for all architectures depending on the threshold value as illustrated by Figure 8c. We notice that the RF-NoC provides the same throughput as the conventional NoC expected for low threshold values, which is due to network congestion.

The same steps were applied for 16 × 32 and 16 × 16 topologies and Figure 9 shows the threshold value evolution and the maximum reached delay reduction. We notice that the threshold value and the delay reduction increases along with the network size for the considered application. From Figure 9 we can conclude that for Splash2-FFT, in a 32 × 32 topology, and 25 hops as threshold form the best combination leading for the highest network latency reduction, as illustrated in Figure 8b showing the importance of the threshold selection.

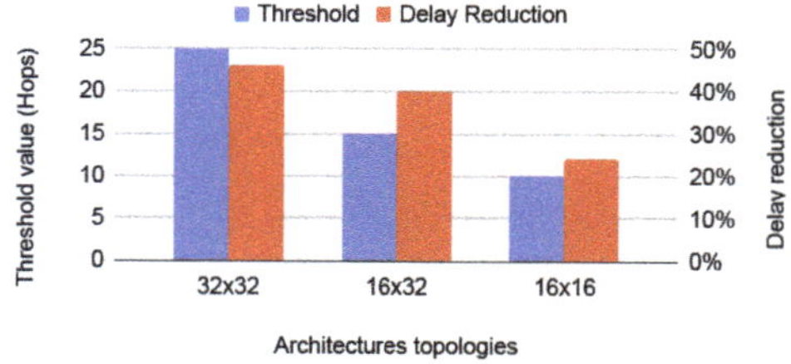

Figure 9. Threshold and maximum delay reduction percentage of 32 × 32, 16 × 32 and 16 × 16 architectures.

6.3. Results Comparison

After presenting the results obtained with the synthetic traffic and the results provided by the proposed framework, we compare these two outcomes. For this, we consider the example of the Splash2-FFT application for the 32 × 32 NoC architecture. The first step is the characterization of the application reported in Table 4 in order to identify the closest scenario and PIR value. Then, we compare these values with results reported in Figure 7a,d. The considered application has 75% of long-range communications and the nearest PIR value is 5×10^{-5}. Note that there is a small difference between the total number of communications inside the RF-NoC between the synthetic traffic and the real one (17,560 communications) that has to be taken into account, that's why we pass to the PIR value of 5×10^{-5}. The suitable threshold value is 25 hops, which is proved by the proposed framework in Figure 8a. For delay reduction, we reach almost the same percentage (about 50%) compared to the conventional NoC. If we apply the same steps for 16 × 32 and 16 × 16 topologies, we found almost the same threshold value, but there is a difference in the percentage of the delay reduction. The synthetic traffic reports a higher percentage of latency reduction in Figure 7e,f compared to results obtained from the use of the framework in Figure 9 which is due to the difference in the total number of communications that is not negligible in these topologies.

Table 4. Application characterization.

Benchmark	Splash2-FFT
Topology	32×32
Packet size (flit)	8
Flit size (bit)	64
Average PIR (packets/cycle)	8×10^{-6}
% of long range communication entries	80%
% of long range communication	60%
Total number of nodes communication (see Section 5.2)	117,560

6.4. Simulation Time

An important feature of a design exploration framework is the simulation time. Figure 10 represents the simulation times when simulating Splash2-FFT benchmark for the 3 considered topologies i.e., 32×32, 16×32 and 16×16 using the framework. All simulations were done on a DELL Latitude 5580 computer, with Intel core i7 processor, 16 Gb RAM, running ubuntu 16.04 LTS. The simulation time of Sniper depends on the benchmark, whereas the simulation time of Noxim depends on the number of clock cycles we want to simulate. In this example, we set the simulation time to 10k clock cycles in Noxim. That is, actual communication flows simulated by Sniper are replaced with statistical communication flows in Noxim. According to Figure 10, we observe that the simulation times increase along with the benchmark complexity. Thus, the fraction of simulation time of Sniper dominates the total simulation time. Even if we change Noxim's input parameters (threshold value, RF bandwidth, modulation order, etc.) the results may remain roughly the same.

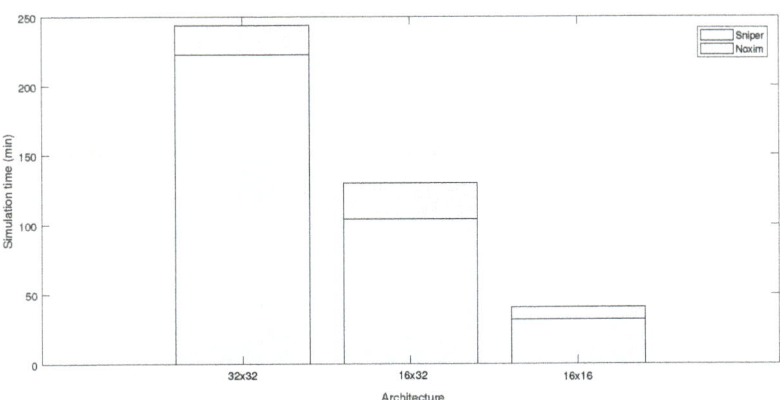

Figure 10. Simulation times for 3 architectures obtained using the framework.

7. Conclusions

In this paper, we presented a design space exploration framework for RF-NoC architectures. This framework is based on the joint use of the Sniper simulator, allowing to take advantage of real application benchmarks, and an extended version of Noxim, which supports OFDMA and integrates a suitable routing algorithm. We compared the results obtained from a real application using the proposed framework with those coming from an equivalent synthetic traffic. We demonstrated that the framework provides an efficient way to consider application-dependent and RF-NoC related

J. Low Power Electron. Appl. **2020**, 10, 37

parameters to achieve the best delay reduction. In this study, a delay reduction of respectively 45%, 30% and 25% were obtained by the RF-NoC for the considered real application, with the appropriate threshold value. As future works, further investigations with different types of benchmarks e.g., stream and RF-NoC configuration, will be done.

Author Contributions: Conceptualization, H.L. and J.L.; funding acquisition, E.B. and M.P.; investigation, H.L., J.L., S.M., E.B. and M.P.; methodology, H.L., J.L. and S.M.; project administration, E.B. and M.P.; software, H.L. and S.M.; supervision, J.L., E.B. and M.P.; validation, H.L. and J.L.; visualization, H.L.; writing—original draft, H.L.; writing—review & editing, J.L., S.M., E.B. and M.P. All authors have read and agreed to the published version of the manuscript.

Funding: This work has been supported by the following institutions/grants: (i) the Italian Ministry of Economic Development (MISE) within the research program "UE-PON Imprese e Competitività 2014-2020 Contratto di sviluppo M9 (CDS 000448)"—CUP: C32F18000100008; (ii) the CY Advanced Studies Institute at the CY Cergy Paris Université (formerly Université de Cergy-Pontoise) under the Paris Seine Initiative for Excellence ("Investissements d'Avenir" ANR-16-IDEX-0008); (iii) the Department of Electrical, Electronic, and Computer Engineering (DIEEI) at University of Catania within the research program "Piano per la Ricerca 2016/2018".

Conflicts of Interest: The authors declare no conflict of interest. The funders had no role in the design of the study; in the collection, analyses, or interpretation of data; in the writing of the manuscript, or in the decision to publish the results.

References

1. Mo, K.H.; Ye, Y.; Wu, X.; Zhang, W.; Liu, W.; Xu, J. A hierarchical hybrid optical-electronic network-on-chip. In Proceedings of the 2010 IEEE Computer Society Annual Symposium on VLSI, Lixouri, Greece, 5–7 July 2010; pp. 327–332.

2. Sharma, K.; Sehgal, V.K. Modern architecture for photonic networks-on-chip. *J. Supercomput.* **2020**, 1–21. [CrossRef]

3. Ye, T.T.; Micheli, G.D.; Benini, L. Analysis of power consumption on switch fabrics in network routers. In Proceedings of the 39th Annual Design Automation Conference, New Orleans, LA, USA, 10–14 June 2002; pp. 524–529.

4. Manna, K.; Mathew, J. A Constructive Heuristic for Designing a 3D NoC-Based Multi-Core Systems. In *Design and Test Strategies for 2D/3D Integration for NoC-Based Multicore Architectures*; Springer: Berlin/Heidelberg, Germany, 2020; pp. 53–63.

5. Karkar, A.; Mak, T.; Tong, K.F.; Yakovlev, A. A survey of emerging interconnects for on-chip efficient multicast and broadcast in many-cores. *IEEE Circuits Syst. Mag.* **2016**, *16*, 58–72. [CrossRef]

6. Hamieh, M.; Ariaudo, M.; Quintanel, S.; Louët, Y. Sizing of the physical layer of a rf intra-chip communications. In Proceedings of the 2014 21st IEEE International Conference on Electronics, Circuits and Systems (ICECS), Marseille, France, 7–10 December 2014; pp. 163–166.

7. Hamieh, M.; Quintanel, S.; Ariaudo, M.; Louet, Y. A new interconnect method for radio frequency intra-chip communications using transistors-based distributed access. *Microw. Opt. Technol. Lett.* **2018**, *61*, [CrossRef]

8. Brière, A. Modélisation Système D'une Architecture D'interconnexion RF Reconfigurable pour les Many-Cœurs. Ph.D. Thesis, Université Pierre et Marie Curie—Paris VI, Paris, France, 2017.

9. Lahdhiri, H.; Lorandel, J.; Bourdel, E. Threshold-based routing algorithm for RF-NoC OFDMA architecture. In Proceedings of the 2019 IEEE 14th International Symposium on Reconfigurable Communication-centric Systems-on-Chip (ReCoSoC), York, UK, 1–3 July 2019; pp. 105–112.

10. Romera, T.; Brière, A.; Denoulet, J. Dynamically Reconfigurable RF-NoC with Distance-Aware Routing Algorithm. In Proceedings of the 2019 14th International Symposium on Reconfigurable Communication-Centric Systems-on-Chip (ReCoSoC), York, UK, 1–3 July 2019; pp. 98–104.

11. Jiang, N.; Becker, D.U.; Michelogiannakis, G.; Balfour, J.; Towles, B.; Shaw, D.E.; Kim, J.; Dally, W.J. A detailed and flexible cycle-accurate network-on-chip simulator. In Proceedings of the 2013 IEEE International Symposium on Performance Analysis of Systems and Software (ISPASS), Austin, TX, USA, 21–23 April 2013; pp. 86–96.

12. Lis, M.; Shim, K.S.; Cho, M.H.; Ren, P.; Khan, O.; Devadas, S. DARSIM: A parallel cycle-level NoC simulator. In Proceedings of the Sixth Annual Workshop on Modeling, Benchmarking and Simulation (MoBS), Saint Malo, France, 20 June 2010.

13. Ben-Itzhak, Y.; Zahavi, E.; Cidon, I.; Kolodny, A. HNOCS: Modular open-source simulator for heterogeneous NoCs. In Proceedings of the 2012 IEEE International Conference on Embedded Computer Systems (SAMOS), Samos, Greece, 18–20 July 2012; pp. 51–57.

14. Jain, L.; Al-Hashimi, B.; Gaur, M.; Laxmi, V.; Narayanan, A. NIRGAM: A simulator for NoC interconnect routing and application modeling. In Proceedings of the 2007 IEEE Design, Automation and Test in Europe conference, Nice, France, 13–16 March 2007; pp. 16–20.

15. Catania, V.; Mineo, A.; Monteleone, S.; Palesi, M.; Patti, D. Cycle-accurate network on chip simulation with noxim. *ACM Trans. Model. Comput. Simul. (TOMACS)* **2016**, *27*, 1–25. [CrossRef]

16. Jheng, K.Y.; Chao, C.H.; Wang, H.Y.; Wu, A.Y. Traffic-thermal mutual-coupling co-simulation platform for three-dimensional network-on-chip. In Proceedings of the 2010 IEEE International Symposium on VLSI Design, Automation and Test, Hsin Chu, Taiwan, 26–29 April 2010; pp. 135–138.

17. Kahng, A.B.; Lin, B.; Nath, S. ORION3. 0: A comprehensive NoC router estimation tool. *IEEE Embed. Syst. Lett.* **2015**, *7*, 41–45. [CrossRef]

18. Seiculescu, C.; Murali, S.; Benini, L.; De Micheli, G. SunFloor 3D: A tool for networks on chip topology synthesis for 3-D systems on chips. *IEEE Trans. Comput. Aided Des. Integr. Circuits Syst.* **2010**, *29*, 1987–2000. [CrossRef]

19. Martin, M.M.; Sorin, D.J.; Beckmann, B.M.; Marty, M.R.; Xu, M.; Alameldeen, A.R.; Moore, K.E.; Hill, M.D.; Wood, D.A. Multifacet's general execution-driven multiprocessor simulator (GEMS) toolset. *ACM SIGARCH Comput. Archit. News* **2005**, *33*, 92–99. [CrossRef]

20. Binkert, N.L.; Dreslinski, R.G.; Hsu, L.R.; Lim, K.T.; Saidi, A.G.; Reinhardt, S.K. The M5 simulator: Modeling networked systems. *IEEE Micro* **2006**, *26*, 52–60. [CrossRef]

21. Binkert, N.; Beckmann, B.; Black, G.; Reinhardt, S.K.; Saidi, A.; Basu, A.; Hestness, J.; Hower, D.R.; Krishna, T.; Sardashti, S.; et al. The gem5 simulator. *ACM SIGARCH Comput. Archit. News* **2011**, *39*, 1–7. [CrossRef]

22. Lowe-Power, J.; Ahmad, A.M.; Akram, A.; Alian, M.; Amslinger, R.; Andreozzi, M.; Armejach, A.; Asmussen, N.; Beckmann, B.; Bharadwaj, S.; et al. The Gem5 Simulator: Version 20.0+. *arXiv* **2020**, arXiv:2007.03152.

23. Power, J.; Hestness, J.; Orr, M.S.; Hill, M.D.; Wood, D.A. gem5-gpu: A heterogeneous cpu-gpu simulator. *IEEE Comput. Archit. Lett.* **2014**, *14*, 34–36. [CrossRef]

24. Miller, J.E.; Kasture, H.; Kurian, G.; Gruenwald, C.; Beckmann, N.; Celio, C.; Eastep, J.; Agarwal, A. Graphite: A distributed parallel simulator for multicores. In Proceedings of the HPCA-16 2010 IEEE The Sixteenth International Symposium on High-Performance Computer Architecture, Bangalore, India, 9–14 January 2010; pp. 1–12.

25. Carlson, T.E.; Heirman, W.; Eeckhout, L. Sniper: Exploring the level of abstraction for scalable and accurate parallel multi-core simulation. In Proceedings of the 2011 International Conference for High Performance Computing, Networking, Storage and Analysis, Seattle, WA, USA, 11 November 2011; pp. 1–12.

Publisher's Note: MDPI stays neutral with regard to jurisdictional claims in published maps and institutional affiliations.

Journal of
*Low Power Electronics
and Applications*

Article

PkMin: Peak Power Minimization for Multi-Threaded Many-Core Applications

Arka Maity [1,*], **Anuj Pathania** [2] and **Tulika Mitra** [1]

[1] School of Computing, National University of Singapore, Singapore 117417, Singapore; tulika@comp.nus.edu.sg

[2] Parallel Computing Systems, University of Amsterdam, 1090 GH Amsterdam, The Netherlands; a.pathania@uva.nl

[*] Correspondence: amaity@comp.nus.edu.sg

Received: 16 July 2020; Accepted: 24 September 2020; Published: 30 September 2020

Abstract: Multiple multi-threaded tasks constitute a modern many-core application. An accompanying generic Directed Acyclic Graph (DAG) represents the execution precedence relationship between the tasks. The application comes with a hard deadline and high peak power consumption. Parallel execution of multiple tasks on multiple cores results in a quicker execution, but higher peak power. Peak power single-handedly determines the involved cooling costs in many-cores, while its violations could induce performance-crippling execution uncertainties. Less task parallelization, on the other hand, results in lower peak power, but a more prolonged deadline violating execution. The problem of peak power minimization in many-cores is to determine task-to-core mapping configuration in the spatio-temporal domain that minimizes the peak power consumption of an application, but ensures application still meets the deadline. All previous works on peak power minimization for many-core applications (with or without DAG) assume only single-threaded tasks. We are the first to propose a framework, called *PkMin*, which minimizes the peak power of many-core applications with DAG that have multi-threaded tasks. *PkMin* leverages the inherent convexity in the execution characteristics of multi-threaded tasks to find a configuration that satisfies the deadline, as well as minimizes peak power. Evaluation on hundreds of applications shows *PkMin* on average results in 49.2% lower peak power than a similar state-of-the-art framework.

Keywords: peak-power management; many-core; directed acyclic task graphs

---

1. Introduction

A many-core application is made up of tens of tasks. All of the tasks are inherently multi-threaded. An accompanying generic Directed Acyclic Graph (DAG) models the execution dependency between the tasks and therefore determines the precedence order for execution [1]. The application must complete execution within a given hard deadline. One way to meet the deadline is to execute as many tasks as possible in parallel. Executing them with the maximum parallelization permitted under the DAG with all available cores results in a short execution time, but also results in high peak power.

The highest power consumption observed during the task's execution defines its peak power. Rated peak power consumption predominantly determines the cost (and weight) of cooling infrastructure that accompanies the many-core. Higher peak power also results in higher on-chip temperatures, which leads to reliability issues [2–4]. Higher temperatures can also trigger performance crippling thermal-throttling, which makes execution unpredictable [5]. We can minimize peak power by executing all of the tasks sequentially on a single core, but such an execution will violate the deadline.

The problem of peak power minimization [6,7] in many-cores is to determine a spatio-temporal task-to-core mapping (configuration) that still meets the application deadline, but minimizes the peak power. We make use of two well-known observations in the execution of multi-threaded tasks in order to efficiently solve the problem. First, Figure 1a shows that the execution time of tasks in a many-core application is discretely convex with increasing core allocation [8–11]. Second, Figure 1b shows that the peak power of tasks is discretely linear with increasing core allocation. These two observations open up the possibility of employing convex optimization in order to solve this problem.

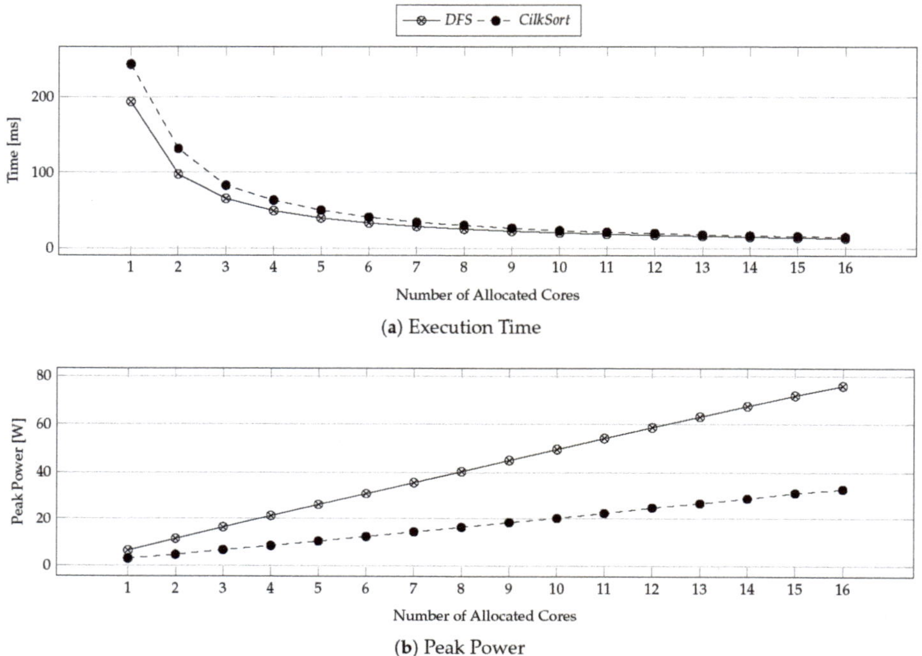

(**a**) Execution Time

(**b**) Peak Power

Figure 1. Execution characteristics of different tasks with different core allocations.

Most works on peak power minimization for many-core applications only assume single-threaded tasks. Therefore, they have no hope for the exploitation of execution characteristics that are shown in Figure 1, which are specific to multi-threaded tasks. In this work, we propose a framework called *PkMin* that minimizes the peak power of multi-threaded many-core applications with DAG under deadline constraint by exploiting the observations made in Figure 1. We evaluate *PkMin* against a similar state-of-the-art framework [12] while using hundreds of applications for a thorough evaluation. Empirical evaluations show that *PkMin* results in on average 48% lower peak power than the state-of-the-art.

2. Motivational Example

Figure 2 shows a motivational example of the problem of peak power minimization in many-cores. We use Sniper [13] multicore x86 simulator to execute the binaries that are associated with the task. Sniper directly reports the execution time values (in clock cycles) for the binaries and also generates traces that is then used by downstream tools, like McPAT [14], to estimate the power consumed by various per core components like L1 (I/D) caches, instruction fetch units, L2 caches, etc., as well uncore components, like memory controllers and DRAM subsystem. This methodology avoids the insertion of costly instrumentation hooks within the program. We account for the power consumption only on the "active" cores, i.e., those that are directly reponsible for execution of the application. Figure 2a gives a DAG for an application composed of four tasks that we need to execute on many-core within 425 ms. Tasks *A* and *B* are composed of *DFS*. Tasks *C* and *D* are composed of *CilkSort*.

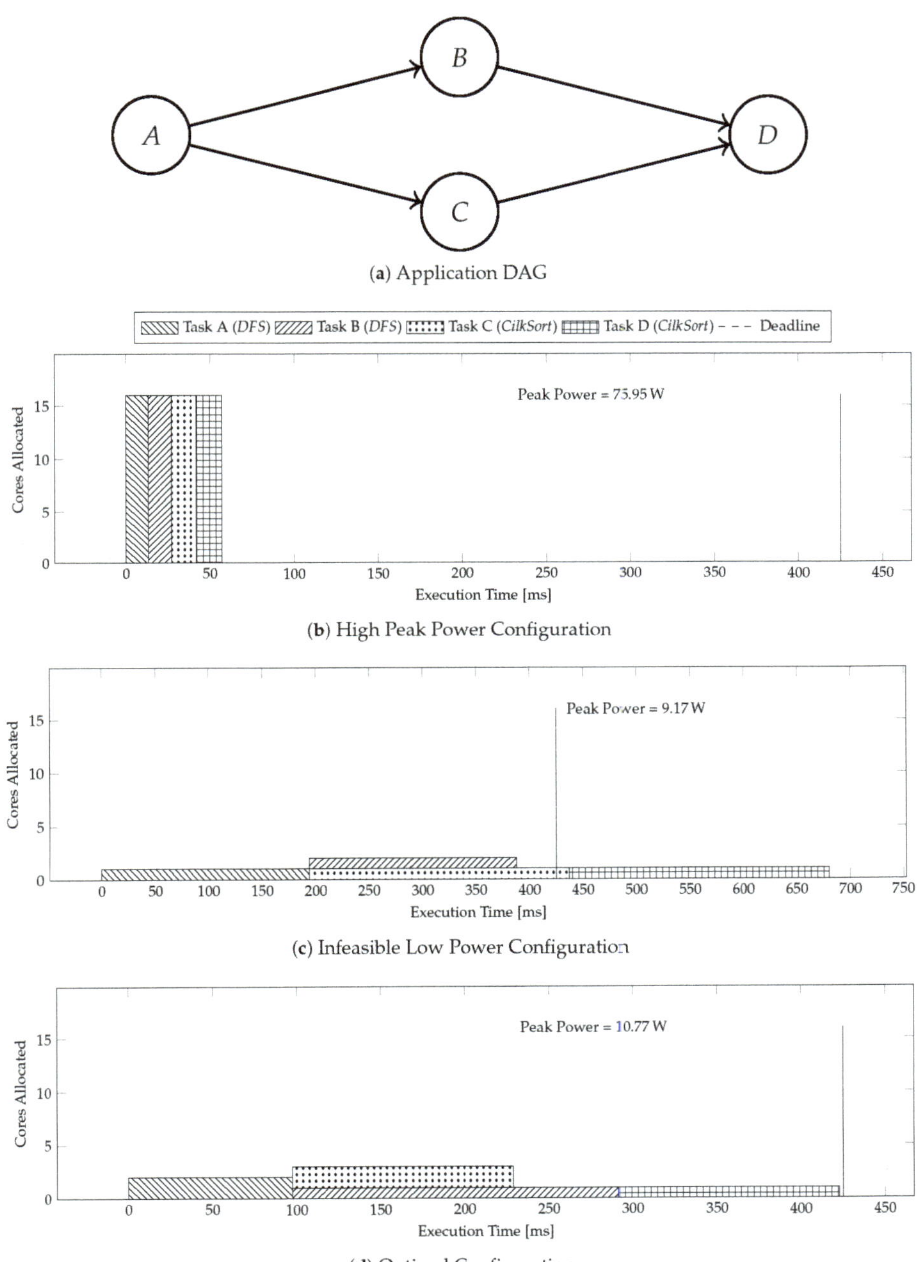

(**a**) Application DAG

(**b**) High Peak Power Configuration

(**c**) Infeasible Low Power Configuration

(**d**) Optimal Configuration

Figure 2. Motivational example for peak power minimization of multi-threaded many-core applications with Directed Acyclic Graph (DAG) under a deadline constraints.

Figure 2b shows the execution time of the benchmark in a configuration that executes the tasks serially with all available cores. The execution in Figure 2b meets the deadline, but leads to a high peak power of 75.95 W. Figure 2c shows the execution time of the application in a configuration that parallelizes the execution, but does not use multi-threading in tasks. Execution in Figure 2c leads to low peak power, but it violates the deadline. Figure 2d shows the execution time of the application in an optimal configuration generated by *PkMin* that allocates just enough cores to tasks, such that both precedence and deadline constraints are met, but with a minimal peak power of 10.77 W. Figure 2d results in 85.82% lower and only 17.45% higher peak power than the configuration in Figure 2b,c, respectively.

3. Related Work

Peak power minimization in the context of multi-/many-core scheduling is an active subject of research [15]. The problem is important in both embedded [16–18] as well as super-computing domain [19,20]. The authors in [21] propose an algorithm to minimize peak power for an application with a task graph without a deadline. The authors of [22] were the first to study the problem of peak power minimization in the context of task graph-based many-core applications with a deadline. They propose an optimal algorithm that schedules a set of independent tasks from the application on a many-core to minimize application's peak power while meeting its deadline. The authors of [23] work on the same problem, but proposed an alternative light-weight heuristic algorithm. Most recently, the authors in [24] proposed an algorithm to minimize peak power for many-core applications with DAG under reliability constraints. All of the works that target the peak power minimization problem directly assume the tasks with DAG within the many-core application to be single-threaded and thereby individually schedulable only on a single core. On the contrary, we focus on tasks that are all individually multi-threaded, wherein the level of multi-threading in each of them is an independently configurable design knob for solving the peak power minimization problem.

4. Convex Optimization Sub-Routine for Solving Serialized DAG

This section provides the details of the convex optimization sub-routine that is at the heart of *PkMin*. This sub-routine solves the problem of peak power minimization for a many-core application with a serial DAG optimally in the continuous domain and near-optimally in the discrete domain, as the problem is NP-Hard in the discrete domain. *PkMin* uses this sub-routine to perform peak power minimization of any generic DAG in Section 5.

System Model: we need to execute an application with $M \in \mathbb{N}$ multi-threaded tasks indexed while using i on a many-core with $N \in \mathbb{N}$ cores. Tasks execute serially in an order ordained by the serialized DAG. N_i is the maximum number of cores that can be allocated to the task i. Each task i is executed with $C_i \in \mathbb{R}_+$ number of cores with the domain constraint $1 \leq C_i \leq N_i$. The number of cores C_i allocated to the task i in practice needs to be discrete. However, we, at first, assume it to be a non-negative real value greater than equal to 1 and less than equal to N_i for tractability.

Based on the observations made in Figure 1a, we assume execution time $\tau_i : \mathbb{R}_+ \to \mathbb{R}_+$ of task i with C_i cores allocated to be a univariate convex function of the number of allocated cores. In the domain $[1, N_i] \in \mathbb{R}_+$ allocated cores, the execution time $\tau_i(C_i)$ is a monotonically decreasing convex function of the number of allocated cores C_i. Based on observations made in Figure 1b, we assume the power $\rho_i : \mathbb{R}_+ \to \mathbb{R}_+$ of a task i with C_i cores allocated to be a univariate linear function of the number of allocated cores.

Execution Model: the execution time of the application in totality $\tau : \mathbb{R}_+^M \to \mathbb{R}_+$ with configuration (core allocation vector) $\vec{C} = \langle C_1, C_2, ..., C_M \rangle \in \mathbb{R}_+^M$ is the sum of the execution time of the individual tasks.

$$\tau(\vec{C}) = \sum_{i=1}^{M} \tau_i(C_i) \tag{1}$$

Because the execution time of each task is individually convex and the sum of convex functions is a convex function, then the application's execution time $\tau(\vec{C})$ is a multivariate convex function of configuration $\vec{C}$. Figure 3a shows the non-negative convex execution time surface of a two-task application.

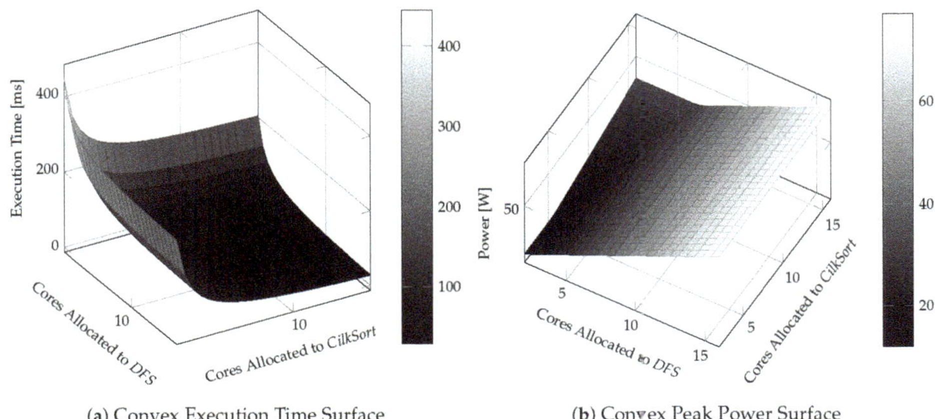

(a) Convex Execution Time Surface (b) Convex Peak Power Surface

Figure 3. Characteristics for a two-task application (with a serialized DAG) with different core allocations for each task in the continuous domain.

Peak Power Model: peak power of the application $\hat{\rho} : \mathbb{R}_+^M \to \mathbb{R}_+$ with configuration $\vec{C}$ is given by task with the maximum power amongst all of the tasks.

$$\hat{\rho}(\vec{C}) = \max_{i=1}^{M} \rho_i(C_i) \tag{2}$$

The application's peak power $\hat{\rho}(\vec{C})$ is a multivariate convex function of configuration $\vec{C}$, as the peak power of each task is individually linear and the max of linear functions is a convex function. Figure 3b shows the non-negative convex peak power surface of a two-task application.

Deadline Model: the peak power function $\hat{\rho}(\vec{C})$ attains its lowest value when all of the tasks execute with bare minimum cores ($\forall_i\ C_i = n_i$), but this is only permitted when there is no constraint on the execution time. The application's hard deadline of $\hat{\tau} \in \mathbb{R}_+$ put a constraint $\tau(\vec{C}) \leq \hat{\tau}$ on its execution time. The deadline $\hat{\tau}$ divides the domain for minimization of peak power function $\hat{\rho}(\vec{C})$ into feasible and infeasible regions.

Given a deadline, we are required to minimize the peak power over the feasible region $F \subset \mathbb{R}_+^M$. We now prove the feasible region F to be a convex set. Let $\vec{C}_x, \vec{C}_y \in F$ be two feasible configurations. By feasibility definition

$$\tau(\vec{C}_x) \leq \hat{\tau} \text{ and } \tau(\vec{C}_y) \leq \hat{\tau} \tag{3}$$

Because $\tau(\vec{C})$ is a convex function,

$$\tau(\lambda\vec{C}_x + (1-\lambda)\vec{C}_y) \leq \lambda\tau(\vec{C}_x) + (1-\lambda)\tau(\vec{C}_y)$$

Using Equation (3), we obtain

$$\begin{aligned}
\tau(\lambda\vec{C}_x + (1-\lambda)\vec{C}_y) &\leq \lambda\hat{\tau} + (1-\lambda)\hat{\tau} \\
&\leq \hat{\tau}
\end{aligned}$$

Therefore, F is a convex set, since, for any $\vec{C}_x, \vec{C}_y \in F$ and any $\lambda \in [0,1]$, we have $\tau(\lambda\vec{C}_x + (1-\lambda)\vec{C}_y) \in F$. Figure 4 shows the feasible region F for an application with two tasks with a given hard deadline.

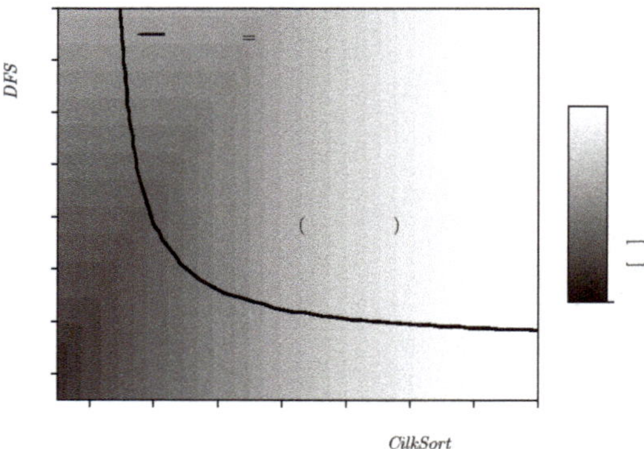

Figure 4. Feasible region for a two-task application (with a serialized DAG) with different number of cores allocated to each task given a hard deadline.

Solution: our problem reduces to minimizing a convex peak power function $\hat{\rho}(\vec{C})$ over the feasible convex set F as its domain. Formally, it can also be summarized, as follows

$$
\begin{aligned}
&\text{minimize} \quad && \hat{\rho}(\vec{C}) \\
&\text{subject to} \quad && \tau(\vec{C}) \le \hat{t} \\
& && 1 \le C_i \le N_i \; \forall C_i \in \mathbb{R}_+
\end{aligned}
\tag{4}
$$

We solve the above convex optimization problem in the continuous domain using *NLOpt* [25] tool that internally solves the problem whlileusing the Method of Moving Asymptotes (MMA) [26] algorithm. The problem-solving time is insignificant, even with an extremely large number of tasks.

Solution Discretization: when we modify the constraint in Equation (4) to force cores allocated to the task to be integers i.e., $\forall C_i \in \mathbb{Z}_+$ in the domain $[1, N_i] \in \mathbb{Z}_+$ instead of real-numbers, the problem becomes an NP-Hard Convex Mixed Integer Non-Linear Programming (CMINLP) problem [27]. Still, we can expect local optima in the discrete domain to be close to the discrete global optimum, because the optimization is tractable using convex programming in its relaxed continuous domain, unlike an arbitrary optimization problem [28].

We first solve Equation (4) to obtain the optimal real-valued configuration in the continuous domain. We then round up all of the individual real-valued core allocations to the nearest integer. The rounding up/down decision is tricky, because there are $2^{\text{task graph size}}$ possibilities, all of which cannot be exhaustively tested for optimality. Two choices are immediately obvious, i.e., to round-down all the allocations or round-up all of them. Because the execution time is non-increasing with increasing core allocation for any task (Figure 1a), rounding down can make the allocation infeasible while rounding up will preserve the feasibility, although it not guaranteed to be optimal in the discrete domain. For example, if the optimal configuration in a continuous domain for a three task application is found to be $\vec{C} = \langle 4.1, 3.4, 5.9 \rangle$, then we round up to discrete configuration $\vec{C} = \langle 5, 4, 6 \rangle$.

5. Peak Power Minimization with PkMin

The problem of peak power minimization for many-core applications with precedence and deadline constraints is inherently a multi-dimensional bin-packing problem, which is well-known to be NP-Hard [29]. We introduce a framework, called *PkMin*, which solves the problem near-optimally by exploiting the observations in Figure 1. At the heart of *PkMin* is a convex optimization sub-routine that can solve the problem optimally for a serialized DAG in a continuous domain. Section 4 provides the details of the sub-routine. The problem is NP-Hard, even with a serialized DAG, in the discrete domain [27]. Therefore, the sub-routine extrapolates the real-valued solution to the discrete domain.

In general, an application DAG allows for the possibility of multiple tasks to be run in parallel. For a given deadline, parallel execution allows for tasks to individually stretch out further along in the time domain more than a serialized execution. Therefore, we can execute tasks with a fewer number of cores being allocated to them individually and still meet the deadline. Tasks execute with a lower peak power (Figure 1b) with a smaller number of cores. However, parallel execution is not guaranteed to lower the peak power of the application, because the peak power of tasks that execute in parallel adds up.

Figure 5 shows the functioning of *PkMin* with the help of a flowchart. *PkMin* begins by serializing the DAG for an application by applying a topological sorting procedure [30]. It then passes the DAG to a convex optimization sub-routine (Section 4) that computes an allocation for the serialized DAG. *PkMin* then enumerates all of the task pairs that can be executed in parallel by computing the transitive closure [31] of the DAG that exposes the pairwise independent tasks. A pair of independent tasks is then "stacked" together to form a single unified task.

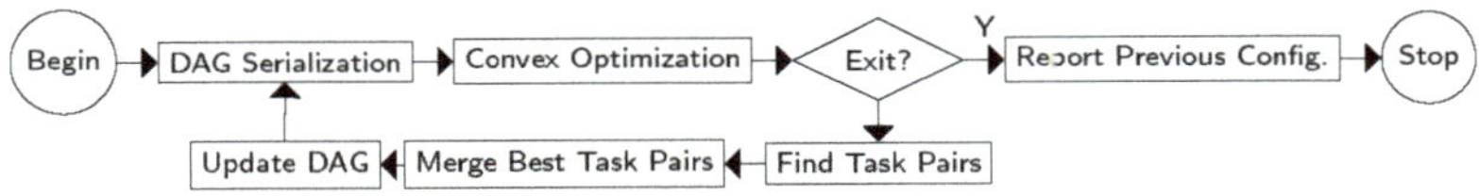

Figure 5. Execution Flow for *PkMin*.

All of the sub-tasks in the unified task always execute with the same number of cores. The execution time and peak power of the unified task is the max and sum of the execution times and peak powers of the sub-tasks, respectively. The sum operator preserves the linearity of power characteristics, while the maximum operator preserves the convexity of execution time characteristics. The new unified task has characteristics that are similar to its constituent tasks. If there are multiple task pairs to choose from, *PkMin* chooses the task pair that gives the greatest reduction in execution time on unification.

The unified task replaces its sub-tasks in the original DAG. *PkMin* then serializes the modified DAG and passes it to the convex optimization sub-routine again in order to obtain a new feasible configuration. It repeats the process of DAG modification, followed by convex optimization iteratively until an exit condition is encountered in one of the following ways.

1. The new configuration yields a higher peak power than the previous configuration, i.e., a local minimum is reached.
2. There are no more candidate task pairs that can be parallelized.

PkMin reports the configuration from the previous iteration as the final solution configuration.

Working Example

This section explains the functioning of *PkMin* with the help of a working example. Figure 6 visualizes the steps that were taken by *PkMin* to solve the motivational example shown in Figure 2. *PkMin* begins with the original DAG that is shown in Figure 2a. It then serializes the DAG, as shown in Figure 6a. It then runs the convex optimization module from Section 4 in order to obtain the core allocation for the serialized DAG, as shown in Figure 6b. It then tries to stack Task B and Task C together by combining them into a new Task $B||C$ and create a new DAG, as shown in Figure 6c.

Figure 6d gives the characteristics of unified Task $B||C$, which inherits the convexity properties of the parent tasks under equal core distribution. Because the DAG in Figure 6c is already serialized, then *PkMin* can directly operate on it. *PkMin* runs convex optimization sub-routine again on the unified DAG to obtain a new core allocation as shown in Figure 6e. However, allocation in Figure 6e has worse peak power than the allocation in Figure 6b, and, therefore, the algorithm terminates with allocation in Figure 6b as the reported solution.

When compared to the worst-case peak power in Figure 2b, the solution reported by *PkMin* has 85.15% lower peak power. The solution reported by *PkMin* has only 4.27% higher peak power when compared to the optimal solution in Figure 2d.

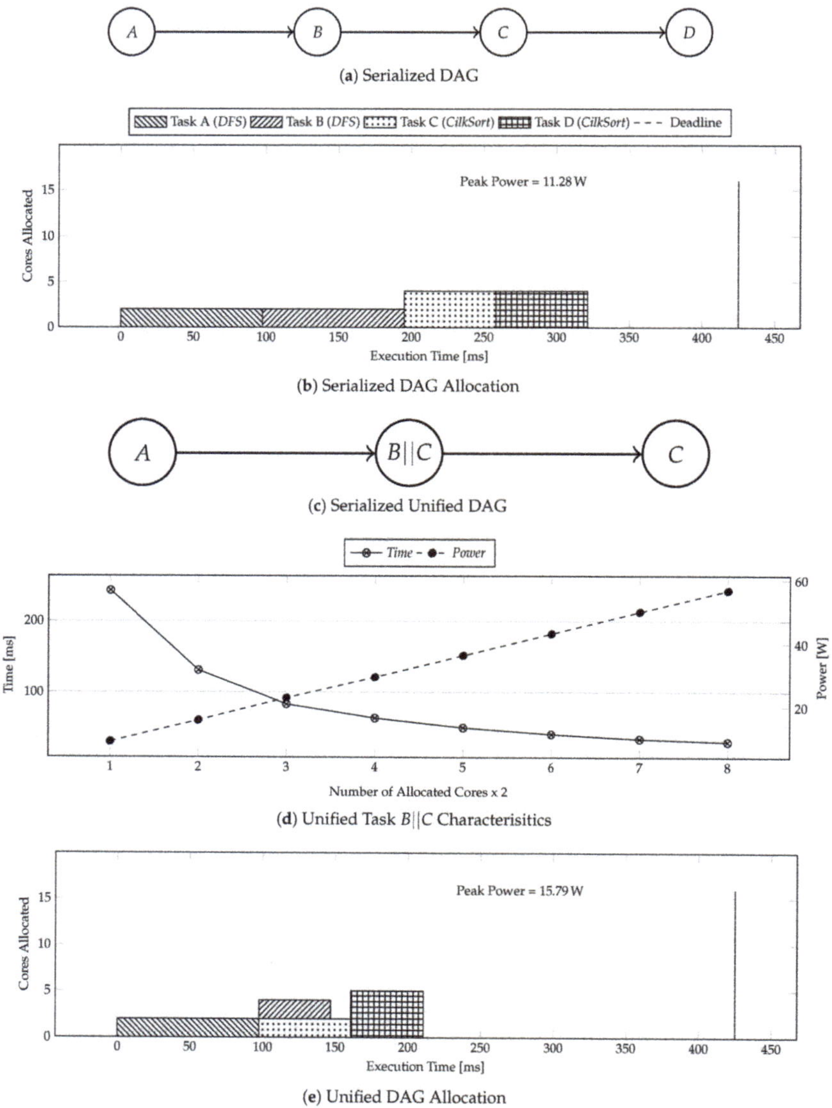

Figure 6. Working example for peak power minimization of the motivational example that is shown in Figure 2 using *PkMin*.

6. Experimental Evaluation

Experimental Setup: we use Sniper simulator [32] to simulate the execution of multi-threaded many-core applications. The simulated multi-core is composed of eight tiles—with two cores each—arranged in a 4 × 2 grid connected while using a Network on Chip (NoC) with hop latency of four cycles and link bandwidth of 256 bits. Two cores within the tile share a 1 MB L2 cache. Cores implement *Intel x86* Instruction Set Architecture (ISA) and run at a frequency of 4 GHz with each core holding a 32 KB private L1 data and instruction caches. Many-core's power consumption is provided by the integrated *McPat* [14] assuming a 22 nm technology node fabrication.

Application Task Graphs: we use a set of five benchmarks—*CilkSort, DFS, Fibonacci, Pi,* and *Queens*—from *Lace* benchmark suite [33] to create our tasks. In order to generate random DAGs of size N, we first sample with replacement N tasks from the benchmark set, thereafter with a probability p, we add an edge between select pair of nodes, such that the acyclic property of the resulting directed graph is preserved. The setup allows for us to thoroughly evaluate *PkMin* with an arbitrarily large number of tasks while simultaneously generating a large number of randomized applications for a given number of tasks.

Application Deadline: setting up arbitrarily short deadlines will render application execution infeasible. In order to set up a feasible deadline, we first note the minimum execution time that is achievable by all of the benchmarks, as, for example, illustrated in Figure 1a for DFS and CilkSort. Let B be the benchmark execution time that is worst among all of the benchmarks considered. We then set the deadline to $B \cdot N$ for an application task graph with N tasks. This ensures the existence of a feasible solution. This is also a fairly tight deadline, as all of the tasks are forced to execute with maximum available cores, if they choose to execute one after the other in a serial fashion. If the application deadline is relaxed further, then other execution configurations with much lower cores (and hence peak power) may become feasible.

Baseline: we are unaware of any work that also solves the problem of peak power minimization for multi-threaded many-cores applications with DAG under deadline constraints. The authors of [12] propose a framework, called *D&C*, which uses a divide and conquer algorithm to minimize execution time for multi-threaded many-core applications with DAG under a peak power constraint. Therefore, *D&C* solves dual of the problem solved by *PkMin*. We modify *D&C* to *DCPace* that solves the same problem as *PkMin* by replacing the constraint from peak power to deadline and replacing the objective function from minimizing executing time to minimizing peak power. Modification keeps the underlying algorithm's ethos intact. DCPace thus acts as a suitable baseline for *PkMin*.

DCPace begins by allocating cores to tasks, to run them in their most energy-efficient configuration, i.e., the one with minimum energy consumed. It then generates an intermediate schedule by scheduling the tasks at the earliest possible time permitted under precedence constraints. It then identifies the midpoint of the schedule along the time axis. All of the tasks that are actively executing at the midpoint must be independent. *DCPace* divides the task into three bins beg, mid, and end. All three bins are assigned a sub-deadline that is equal to the third of the original deadline. All independent tasks in mid are greedily scheduled, such that the bin's peak power is minimized under the available core and sub-deadline constraints. This is done using a strip packing heuristic, like Next-Fit Decreasing Height (NFDH) [34]. In a strip-packing problem, a collection of rectangles of different height and width are to be packed on a rectangular bin, with a fixed width and unbounded height, such that no two rectangles overlap. This problem is NP-complete. NFDH begins by sorting the rectangles in decreasing order of their heights. After that, it packs the rectangle in a left-justified fashion until the next rectangle can no longer fit in the remaining space to the right. A new level is defined as the packing restarts from the remaining rectangles. *DCPace* continues to divide recursively mid and end using their midpoint. Recursion breaks when a bin becomes a singleton with only one task.

Power and Energy Consumption Analyses: Figure 7 illustrates the working of DCPace and PkMin algorithms. In this experiment, we use tasks graph with 100 tasks and set the deadline to 1700 million clock cycles or 425 ms. DCPace chooses the most energy-efficient core allocation to execute each task, which is not changed thereafter. Given the task execution time and task peak power characteristics, as shown in Figure 1, the minimum energy allocation can only occur either when all of the cores are allocated, or a minimum number of core is allocated to each task. Figure 7b, shows the variation in the total cores allocated as the application execution proceeds in time. Both DCPace and PkMin show considerable variations in the total-cores allocated, although the former exclusively varies between the maximum and the minimum possible allocations. Because the goal of PkMin is to reduce peak power exclusively, its allocations amongst tasks under PkMin are such that any two different non-overlapping tasks have almost similar peak power consumption when compared to DCPace. PkMin exploits the convexity properties of the task characteristics in order to achieve this "equivalent power" allocations. The power trace of application execution in Figure 7a under PkMin has almost no peaks and troughs as compared to DCPace.

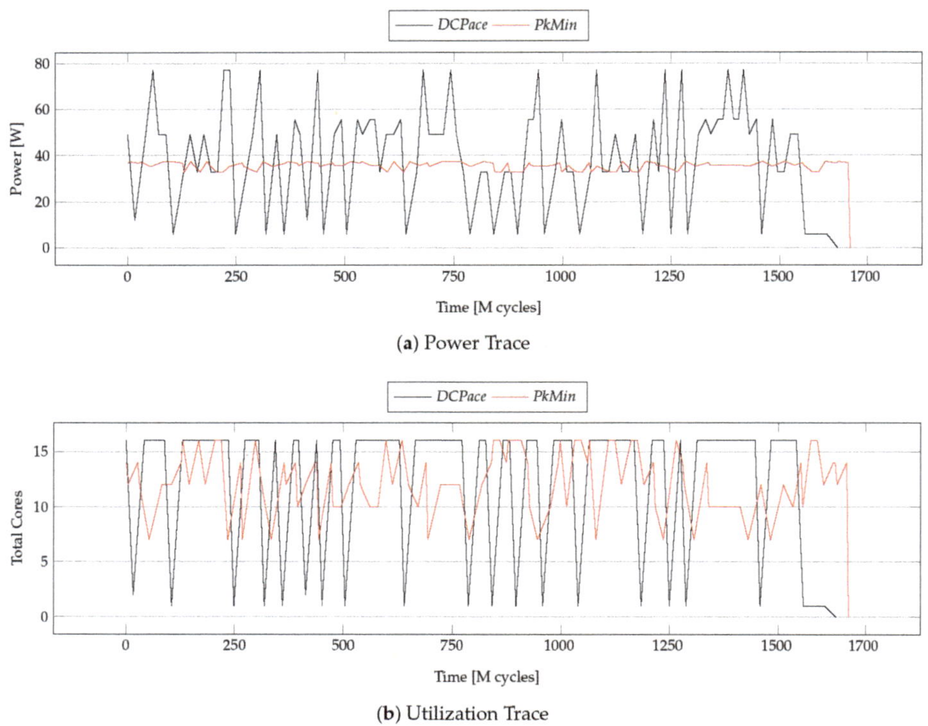

(**a**) Power Trace

(**b**) Utilization Trace

Figure 7. Power consumption and core utilization trace.

Performance Evaluation: we evaluate the efficacy of *PkMin* in minimizing peak power for applications with an increasing number of tasks. We also evaluate the same applications using *DCPace* to put the performance of *PkMin* in context.

First, we show the peak power savings for application task graph of sizes varying from 10 to 100. We set the deadline to $17 \cdot N$ million clock cycles, where the best possible execution time of each task is 17 million clock cycles and N is the number of tasks in the application. Figure 8a shows that *PkMin* has, on average, 48% lower peak power when compared to the *DCPace*. The energy consumption of PkMin is, however, around 0.8% higher than the DCPace, as illustrated in Figure 8b.

Figure 9 orthogonally shows the efficacy of *PkMin* in minimizing the peak power of hundred random applications, each with 100 tasks. The deadline is similarly set to 1700 million clock cycles. *PkMin* results in lower peak power than *DCPace*, with only 1% additional energy overhead.

Figure 10 shows the efficacy of *PkMin* in minimizing peak power in a random application (with 100 tasks) as its deadline is relaxed. In this case, the improvement of peak power for *PkMin* over *DCPace* comes at the cost of worsening energy consumption. However, a significant reduction in the peak power of approximately 88% is possible with less than 10% increase in energy consumed.

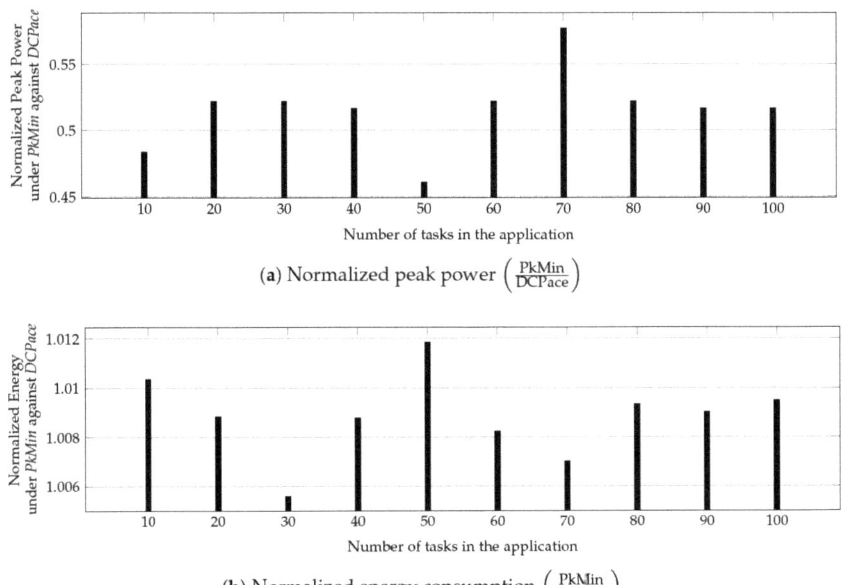

(a) Normalized peak power $\left(\frac{\text{PkMin}}{\text{DCPace}}\right)$

(b) Normalized energy consumption $\left(\frac{\text{PkMin}}{\text{DCPace}}\right)$

Figure 8. Application performance under *PkMin* normalized with respect to *DCPace*. Application size varies from 10 tasks to 100 tasks.

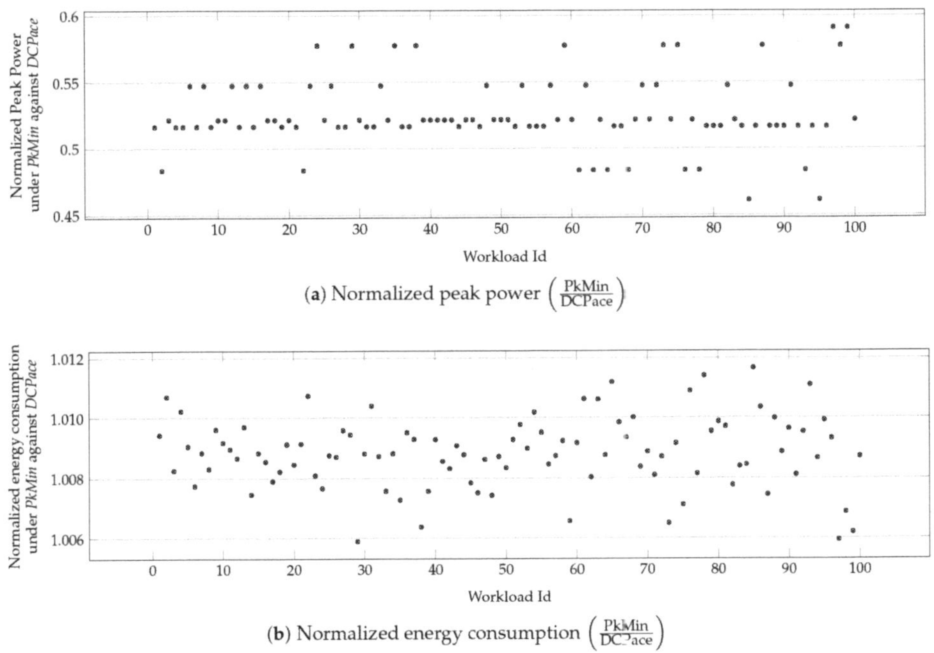

(a) Normalized peak power $\left(\frac{\text{PkMin}}{\text{DCPace}}\right)$

(b) Normalized energy consumption $\left(\frac{\text{PkMin}}{\text{DCPace}}\right)$

Figure 9. Application performance under *PkMin* with 100 task applications normalized against their performance under *DCPace*.

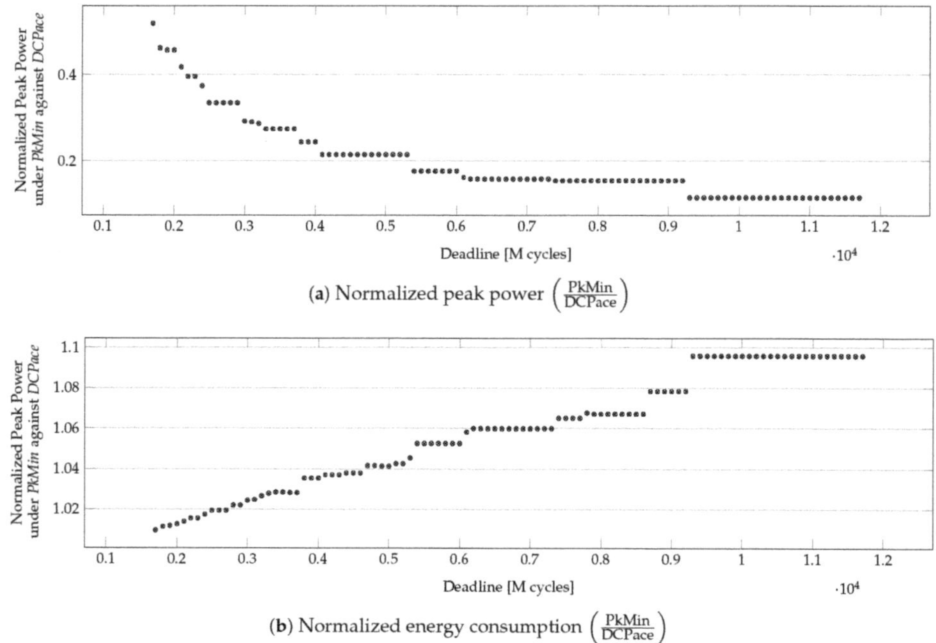

(**a**) Normalized peak power $\left(\frac{\text{PkMin}}{\text{DCPace}}\right)$

(**b**) Normalized energy consumption $\left(\frac{\text{PkMin}}{\text{DCPace}}\right)$

Figure 10. Peak power under *PkMin* for a 100-task application with different deadlines that are normalized against its peak power under *DCPace*.

Scalability: *PkMin* uses *NLOpt* internally, which has a low polynomial-time computational complexity. It invokes *NLOpt* at the max number of tasks $|M|$ times, keeping the computational complexity still polynomial. *PkMin* also uses topological sort and transitive closure graph algorithms that also have a worst-case polynomial computational complexity of $O(|M|)$ and $O(|M|^2)$, respectively. This low polynomial-time computational complexity makes *PkMin* highly scalable. Figure 11 shows the increase in worst-case problem-solving time that is required under *PkMin* with an increase in the number of tasks in applications. For a 100-task application, *PkMin* requires 1.3 s to compute the near-optimal configuration.

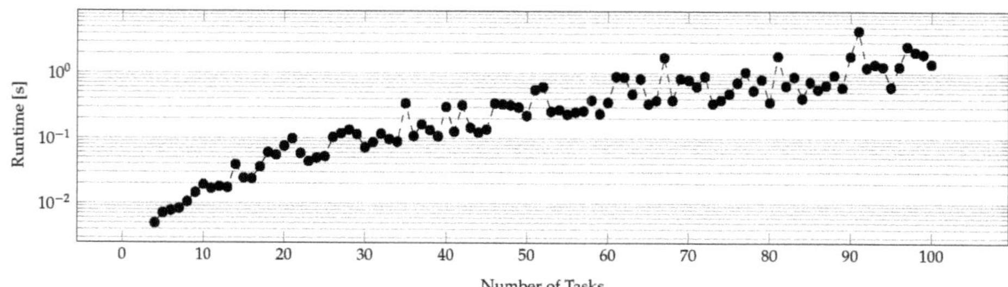

Figure 11. Runtime of *PkMin* for applications with different number of tasks.

7. Conclusions

We introduced a framework, called *PkMin*, in this work that solves the problem of peak power minimization for a multi-thread many-core application with DAG under a deadline. *PkMin* exploits the execution characteristics of multi-threaded tasks in many-core applications to optimally solve the problem for a serialized DAG in the continuous domain while using convex optimization. It then uses the convex optimization sub-routine to solve the problem near-optimally for any generic DAG.

Empirical evaluations on hundreds of applications show configurations obtained under *PkMin* have, on average, up to 48% lower peak power than similar state-of-the-art with less than 1% additional total energy. The peak power savings can be further increased to 88% with less than 10% energy overheads whenever the deadline is relaxed. *PkMin* has polynomial-time computation complexity with negligible problem-solving overheads, which makes it suitable for use at both run-time and design-time.

Author Contributions: Conceptualization, A.M., A.P. and T.M.; methodology, A.M., A.P. and T.M.; software, A.M.; validation, A.M.; formal analysis, A.P. and A.M.; resources, T.M.; writing–original draft preparation, A.P. and A.M.; writing–review and editing, A.M., A.P. and T.M.; supervision, T.M.; project administration, T.M.; funding acquisition, T.M. All authors have read and agreed to the published version of the manuscript.

Funding: This work was supported by the National Research Foundation, Prime Minister's Office, Singapore under its IndustryIHL Partnership Grant NRF2015-IIP003.

Conflicts of Interest: The authors declare no conflict of interest.

References

1. Singh, A.K.; Jigang, W.; Kumar, A.; Srikanthan, T. Run-time Mapping of Multiple Communicating Tasks on MPSoC Platforms. *Procedia Comput. Sci.* **2010**, *1*, 1019–1026. [CrossRef]
2. Kriebel, F.; Shafique, M.; Rehman, S.; Henkel, J.; Garg, S. Variability and Reliability Awareness in the Age of Dark Silicon. *IEEE Des. Test* **2015**, *33*, 59–67. [CrossRef]
3. Salehi, M.; Shafique, M.; Kriebel, F.; Rehman, S.; Tavana, M.K.; Ejlali, A.; Henkel, J. dsReliM: Power-constrained Reliability Management in Dark-Silicon Many-Core Chips under Process Variations. In Proceedings of the 2015 International Conference on Hardware/Software Codesign and System Synthesis (CODES+ISSS), Amsterdam, The Netherlands, 4–9 October 2015.
4. Ma, Y.; Chantem, T.; Dick, R.P.; Hu, X.S. Improving System-Level Lifetime Reliability of Multicore Soft Real-Time Systems. *IEEE Trans. Very Large Scale Integr. Syst.* **2017**, *25*, 1895–1905. [CrossRef]
5. Pagani, S.; Bauer, L.; Chen, Q.; Glocker, E.; Hannig, F.; Herkersdorf, A.; Khdr, H.; Pathania, A.; Schlichtmann, U.; Schmitt-Landsiedel, D.; et al. Dark Silicon Management: An Integrated and Coordinated Cross-Layer Approach. *Inf. Technol.* **2016**, *58*. [CrossRef]
6. Pathania, A.; Khdr, H.; Shafique, M.; Mitra, T.; Henkel, J. QoS-Aware Stochastic Power Management for Many-Cores. In Proceedings of the 2018 55th ACM/ESDA/IEEE Design Automation Conference (DAC), San Francisco, CA, USA, 24–28 June 2018.
7. Pathania, A.; Khdr, H.; Shafique, M.; Mitra, T.; Henkel, J. Scalable Probabilistic Power Budgeting for Many-Cores. In Proceedings of the Design, Automation & Test in Europe Conference & Exhibition (DATE), Lausanne, Switzerland, 27–31 March 2017.
8. Pathania, A.; Venkataramani, V.; Shafique, M.; Mitra, T.; Henkel, J. Distributed Scheduling for Many-Cores Using Cooperative Game Theory. In Proceedings of the 2016 53nd ACM/EDAC/IEEE Design Automation Conference (DAC), Austin, TX, USA, 5–9 June 2016.
9. Pathania, A.; Venkataramani, V.; Shafique, M.; Mitra, T.; Henkel, J. Distributed Fair Scheduling for Many-Cores. In Proceedings of the 2016 Design, Automation & Test in Europe Conference & Exhibition (DATE), Dresden, Germany, 14–18 March 2016.
10. Pathania, A.; Venkatramani, V.; Shafique, M.; Mitra, T.; Henkel, J. Optimal Greedy Algorithm for Many-Core Scheduling. *IEEE Trans. Comput.-Aided Des. Integr. Circuits Syst.* **2017**, *36*, 1054–1058. [CrossRef]
11. Venkataramani, V.; Pathania, A.; Shafique, M.; Mitra, T.; Henkel, J. Scalable Dynamic Task Scheduling on Adaptive Many-Core. In Proceedings of the 2018 IEEE 12th International Symposium on Embedded Multicore/Many-Core Systems-on-Chip (MCSoC), Hanoi, Vietnam, 12–14 September 2018.
12. Demirci, G.; Marincic, I.; Hoffmann, H. A Divide and Conquer Algorithm for DAG Scheduling Under Power Constraints. In Proceedings of the International Conference for High Performance Computing, Networking, Storage, and Analysis, Dallas, TX, USA, 11–16 November 2018.

13. Carlson, T.E.; Heirman, W.; Eeckhout, L. Sniper: Exploring the Level of Abstraction for Scalable and Accurate Parallel Multi-Core Simulation. In Proceedings of the International Conference for High Performance Computing, Networking, Storage and Analysis (SC), Seatle, WA, USA, 12–18 November 2011.

14. Li, S.; Ahn, J.H.; Strong, R.D.; Brockman, J.B.; Tullsen, D.M.; Jouppi, N.P. McPAT: An Integrated Power, Area, and Timing Modeling Framework for Multicore and Manycore Architectures. In Proceedings of the 2009 42nd Annual IEEE/ACM International Symposium on Microarchitecture (MICRO), New York, NY, USA, 12–16 December 2009.

15. Singh, A.K.; Shafique, M.; Kumar, A.; Henkel, J. Mapping on multi/many-core systems: Survey of current and emerging trends. In Proceedings of the 2013 50th ACM/EDAC/IEEE Design Automation Conference (DAC), Austin, TX, USA, 29 May–7 June 2013.

16. Rapp, M.; Pathania, A.; Henkel, J. Pareto-optimal power-and cache-aware task mapping for many-cores with distributed shared last-level cache. In Proceedings of the International Symposium on Low Power Electronics and Design (ISLPED), Seattle, WA, USA, 23–25 July 2018.

17. Rapp, M.; Pathania, A.; Mitra, T.; Henkel, J. Prediction-Based Task Migration on S-NUCA Many-Cores. In Proceedings of the Design, Automation & Test in Europe Conference & Exhibition (DATE), Florence, Italy, 25–29 March 2019.

18. Rapp, M.; Sagi, M.; Pathania, A.; Herkersdorf, A.; Henkel, J. Power-and Cache-Aware Task Mapping with Dynamic Power Budgeting for Many-Cores. *IEEE Trans. Comput.* **2019**, *69*, 1–13. [CrossRef]

19. Bartolini, A.; Borghesi, A.; Libri, A.; Beneventi, F.; Gregori, D.; Tinti, S.; Gianfreda, C.; Altoè, P. The DAVIDE Big-Data-Powered Fine-Grain Power and Performance Monitoring Support. In Proceedings of the 15th ACM International Conference on Computing Frontiers, Ischia, Italy, 8–10 May 2018.

20. Oleynik, Y.; Gerndt, M.; Schuchart, J.; Kjeldsberg, P.G.; Nagel, W.E. Run-Time Exploitation of Application Dynamism for Energy-Efficient Exascale Computing (READEX). In Proceedings of the 2015 IEEE 18th International Conference on Computational Science and Engineering, Porto, Portugal, 21–23 October 2015.

21. Lee, B.; Kim, J.; Jeung, Y.; Chong, J. Peak Power Reduction Methodology for Multi-Core Systems. In Proceedings of the International SoC Design Conference (ISOCC), Seoul, Korea, 22–23 November 2010.

22. Lee, J.; Yun, B.; Shin, K.G. Reducing Peak Power Consumption in Multi-Core Systems without Violating Real-Time Constraints. *IEEE Trans. Parallel Distrib. Syst.* **2014**, *25*, 1024–1033

23. Munawar, W.; Khdr, H.; Pagani, S.; Shafique, M.; Chen, J.J.; Henkel, J. Peak Power Management for Scheduling Real-Time Tasks on Heterogeneous Many-Core Systems. In Proceedings of the 20th IEEE International Conference on Parallel and Distributed Systems (ICPADS), Hsinchu, Taiwan, 16–19 December 2014.

24. Ansari, M.; Yeganeh-Khaksar, A.; Safari, S.; Ejlali, A. Peak-Power-Aware Energy Management for Periodic Real-Time Applications. *IEEE Trans. Comput. Aided Des. Integr. Circuits Syst.* **2019**, *39*, 779–788. [CrossRef]

25. Johnson, S.G. The NLopt Nonlinear-Optimization Package. Available online: https://github.com/stevengj/nlopt (accessed on 25 September 2020).

26. Svanberg, K. A Class of Globally Convergent Optimization Methods Based on Conservative Convex Separable Approximations. *SIAM J. Optim.* **2002**, *12*, 555–573. [CrossRef]

27. Bonami, P.; Kilinç, M.; Linderoth, J. Algorithms and Software for Convex Mixed Integer Nonlinear Programs. In *Mixed Integer Nonlinear Programming*; Springer: Berlin/Heidelberg, Germany, 2012.

28. Moriguchi, S.; Tsuchimura, N. Discrete L-Convex Function Minimization Based on Continuous Relaxation. *Pac. J. Optim.* **2009**, *5*, 227–236.

29. Chekuri, C.; Khanna, S. On Multidimensional Packing Problems. *J. Comput.* **2004**, *33*, 837–851. [CrossRef]

30. Pearce, D.J.; Kelly, P.H. A Dynamic Topological Sort Algorithm for Directed Acyclic Graphs. *J. Exp. Algorithmics* **2007**. [CrossRef]

31. Ioannidis, Y.E.; Ramakrishnan, R. Efficient Transitive Closure Algorithms. In Proceedings of the 1988 VLDB Conference: 14th International Conference on Very Large Data Bases, Los Angeles, CA, USA, 29 August–1 September 1988.

32. Pathania, A.; Henkel, J. HotSniper: Sniper-based toolchain for many-core thermal simulations in open systems. *IEEE Embed. Syst. Lett.* **2018**, *11*, 54–57. [CrossRef]
33. Van Dijk, T.; van de Pol, J.C. Lace: Non-Blocking Split Deque for Work-Stealing. In *European Conference on Parallel Processing (Euro-Par)*; Springer: Berlin/Heidelberg, Germany, 2014.
34. Coffman, E.G., Jr.; Garey, M.R.; Johnson, D.S.; Tarjan, R.E. Performance Bounds for Level-Oriented Two-Dimensional Packing Algorithms. *SIAM J. Comput.* **1980**, *9*, 808–826, doi:10.1137/0209062. [CrossRef]

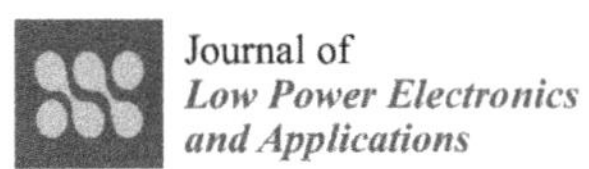

Journal of
*Low Power Electronics
and Applications*

Article

Intra- and Inter-Server Smart Task Scheduling for Profit and Energy Optimization of HPC Data Centers

Sayed Ashraf Mamun [1,*,†], Alexander Gilday [2,†], Amit Kumar Singh [3], Amlan Ganguly [1], Geoff V. Merrett [4], Xiaohang Wang [5] and Bashir M. Al-Hashimi [2]

[1] Department of Computer Engineering, Rochester Institute of Technology, New York, NY 14623, USA; axgeec@rit.edu
[2] Department of Electronics and Computer Science (ECS), University of Southampton, Southampton SO171BJ, UK; adg1n17@soton.ac.uk (A.G.); bmah@ecs.soton.ac.uk (B.M.A.-H.)
[3] School of Computer Science and Electronic Engineering (CSEE), University of Essex, Colchester CO43SQ, UK; a.k.singh@essex.ac.uk
[4] Department of Electronic and Software Systems, University of Southampton, Southampton SO171BJ, UK; gvm@ecs.soton.ac.uk
[5] School of Software Engineering, South China University of Technology, Guangzhou 510006, China; baikeina@163.com
* Correspondence: sam7753@rit.edu
† These authors contributed equally to this work.

Received: 13 August 2020; Accepted: 29 September 2020; Published: 14 October 2020

Abstract: Servers in a data center are underutilized due to over-provisioning, which contributes heavily toward the high-power consumption of the data centers. Recent research in optimizing the energy consumption of High Performance Computing (HPC) data centers mostly focuses on consolidation of Virtual Machines (VMs) and using dynamic voltage and frequency scaling (DVFS). These approaches are inherently hardware-based, are frequently unique to individual systems, and often use simulation due to lack of access to HPC data centers. Other approaches require profiling information on the jobs in the HPC system to be available before run-time. In this paper, we propose a reinforcement learning based approach, which jointly optimizes profit and energy in the allocation of jobs to available resources, without the need for such prior information. The approach is implemented in a software scheduler used to allocate real applications from the Princeton Application Repository for Shared-Memory Computers (PARSEC) benchmark suite to a number of hardware nodes realized with Odroid-XU3 boards. Experiments show that the proposed approach increases the profit earned by 40% while simultaneously reducing energy consumption by 20% when compared to a heuristic-based approach. We also present a network-aware server consolidation algorithm called Bandwidth-Constrained Consolidation (BCC), for HPC data centers which can address the under-utilization problem of the servers. Our experiments show that the BCC consolidation technique can reduce the power consumption of a data center by up-to 37%.

Keywords: high performance computing; data centers; resource allocation; profit; energy consumption; machine learning; reinforcement learning; server consolidation

---

1. Introduction

High Performance Computing (HPC) data centers typically contain a large number of computing nodes each consisting of multiple processing cores. The size and performance of these systems continue to increase which causes concerns to be raised over higher energy requirements [1–4]. Research estimates that data centers worldwide account for 1.1–1.5% of global electricity use [5]. It is therefore important to take measure that reduce this energy consumption.

Different techniques have been proposed in the literature to improve the energy efficiency of the data center. Dynamic Power Management (DPM) and Dynamic Voltage Frequency Scaling (DVFS) are popular techniques to reduce the power consumption of under-utilized resources. Consolidation of virtual machines (VMs) running in different servers into fewer servers to enable aggressive DPM or DVFS has become a major focus area in the research community [6–13].

However, server-consolidation with the sole objective of power reduction can impact the performance of the HPC data center negatively if the network is incapable of supporting the resultant aggregated traffic patterns or hotspot scenarios [14]. Therefore, careful attention to the impact of consolidation on network performance is necessary.

HPC data centers maintain queues of jobs which arrive periodically and must schedule these jobs to be executed in order to produce a profit. It is common to assign values to jobs which imply their level of importance compared to other jobs: higher value equates to higher importance [15]. Value is typically assigned based on expected profit earned from the completion of the job. In HPC systems, the scheduling of jobs is influenced by their value; typically a resource management system will attempt to maximize its profits by allocating its limited resources to the highest-value jobs in the queue. This is especially true when jobs arrive at a rate higher than the rate at which the system can process and execute them. Typically, the value obtained by completion of a job is time-dependent which reflects the necessity to schedule as early as possible.

Existing approaches which optimize resource management use fast heuristics to very quickly find practical allocations for the dynamically arriving jobs. The use of heuristics over more complicated algorithms reduces overhead in the delay of allocations and also lowering the resource requirement of the resource management system itself. Research has also been conducted which considers profiling results from design-time testing to improve both the run-time computational complexity and the quality of the heuristics [16,17]. While the results of these researches show significant improvement over other approaches, the technique is only applicable in select situations due to the required accurate information and assumption that there is little deviation in resources required by the jobs in the system. The challenge to overcome these disadvantages is to design an algorithm that accounts for this variation and builds up a full history of information about jobs in real-time instead of requiring the information to exist already.

In an orthogonal direction, novel data center network (DCN) technologies, leveraging emerging interconnection paradigms such as millimeter-wave (mmWave) interconnects have been proposed to reduce the power consumption of the networking equipment [18,19]. Wireless data center architectures have been proposed where Top-of-Rack (ToR) switches are interconnected with mmWave links while the intra-rack communication is achieved through traditional Ethernet [20–23]. Alternatively, server-centric wireless DCNs where direct wireless links are used for server-to-server communication have also been designed [24,25]. These wireless data center architectures can be considered as viable alternate for traditional wired architecture for HPC computing for reducing even more power consumption. Furthermore, designing adequate server consolidation techniques can result in further power saving.

Contribution: This paper attempts to address this challenge by using ideas from reinforcement learning techniques and designing a novel server consolidation technique. The resource management system performs allocations that optimize both profit (value) and energy using a combination of light-weight heuristics and historic run-time results. The system considers the scheduling problem as a Multi-Armed Bandit (MAB) model where each individual job allocation is a possible action. The approach uses a novel algorithm, inspired by the Upper-Confidence Bound technique [26], collects profiling information at run-time (exploration) to optimize future allocations of jobs (exploitation). We also propose a network-aware approach to server-consolidation called Bandwidth Constrained Consolidation (BCC) and study its impact on a data center. While consolidating tasks can reduce the power consumption of data centers, due to the arrival of new tasks and completion of existing tasks, the consolidated utilization profile of the servers may change adversely affecting the power consumption over time. Hence, the BCC consolidation algorithm should be repeated periodically. Hence we also propose a method to find the optimal inter-consolidation time for a data center and derive a mathematical formulation to estimate the optimal inter-consolidation time. This will enable optimally scheduling consolidation in a data center without the need for extensive simulations and measurements to achieve the optimality.

Paper Organization: Section 2 presents work related to this paper. Section 3 introduces the problem, including the definition of a job and its value, the MAB model, and the model of HPC systems. The novel, reinforcement learning-based approach is introduced in Section 3.4. In Section 4 we introduce the network aware server consolidation technique, traffic model, our proposed algorithm, and mathematical model to estimate optimal inter-consolidation time. Experimental results are presented in Sections 5 and 6 concludes the paper.

2. Related Work

It is proven that the use of market-inspired resource allocation heuristics provides promising results in the common situation that HPC systems are overloaded with more jobs than they can handle [27]. These heuristics use some implementation of a value for jobs, both fixed [28] and changing over time [15]. Most of them choose the highest value job first, which might consume too many resources, leaving limited resources for jobs arriving in the future. Therefore, resources required for each job should be optimized.

Other heuristics exist such as value density (value divided by resource requirement) which addresses the issue of the highest value job consuming too many resources [29–31]. This heuristic instead will prefer jobs which are small and have reasonably high value over very large jobs with high value. However, this heuristic, and others like it, do not consider energy consumption.

A number of reinforcement learning techniques were compared for scheduling tasks on large-scale distributed systems [32]. In this comparison, energy efficiency was considered by attempting to maximize CPU utilization. This intuitively increases energy efficiency by reducing the wasted energy of having CPUs powered on but in an idle state. Similar reinforcement learning techniques are explored for data centers [33]. However, these researches only considered the fulfillment of the service level agreement (SLA) which provides a fixed value when jobs are completed before a specified deadline. This research instead considers the common case where the value of a job changes gradually as a function of time, known as a value-curve. In addition, the energy efficiency optimization does not focus on the reduction of energy consumption directly.

A report identified that the use of profiling results in jointly optimizing the value and energy of a job [17]. Further research also expanded the optimizations to monitoring and adapting the allocations during tasks' execution by migrating to a different set of resources [34]. Both approaches require profiling of jobs at design-time, and the latter also requires the ability for jobs to pause execution and resume on a different set of resources. In contrast, the bandit-based technique in this paper attempts to similarly predict the value and energy of jobs without relying on the assumptions that prior information is obtainable and migration of jobs is possible.

In a complementary direction, under-utilization of the servers in a data center has always been observed mainly due to the over-provisioning for the peak demand hours [35]. In [6], various formulations of the cost-aware application placement problem for servers were first introduced without considering network performance. Similarly, in [7], a system was proposed that optimizes power consumption, performance benefits, and transient costs incurred by server consolidation. In [8] an efficient power-aware resource scheduling strategy was proposed that reduces data center power consumption based on live VM migration. A framework for VM migration and placement was proposed in [9] considering both the network topology and network traffic demands to minimize energy consumption while satisfying as many network-demands as possible. In [10], energy-aware VM placement was proposed where application dependencies were considered to reduce network energy consumption. In [11], a network-aware VM consolidation scheme was proposed for solving combined

VM consolidation problems to conserve the energy of the data center. In [12], a heuristic to control VM migration based on prioritizing VMs with steady capacity was proposed. In addition to server consolidation, an opportunistic approach to reduce power consumption is proposed in [36]. From all of these studies, it is clear that if an adequate consolidation algorithm can be designed, a significant amount of power reduction can be possible for the HPC data centers.

Orthogonally, various designs have been proposed to address DCN design issues such as energy consumption, cabling complexity, scalability, and over-subscription. One popular topology used today in data center networks is a fat-tree topology. To address oversubscription and other issues in wired DCNs many alternative DCN architectures have been proposed such as BCube, DCell, DOS, VL2, and Helios [35,37]. However, these innovations still rely on copper or optical cables and do not mitigate the challenges due to high power consumption, design, and maintenance of a DCN with physical links. To alleviate the issues of DCNs with power-hungry switching fabrics and bundles of cables wireless data centers with mm-wave inter-rack links are envisioned in [18,20,21]. Most of the recent works on wireless data centers propose interconnecting entire racks of servers as units with 60 GHz wireless links primarily in order to utilize the commodity Ethernet switching between servers inside individual racks [18]. Phased antenna arrays or directional horn antennas are used to establish wireless links between ToRs in the entire data center [21–23]. Line-of-Sight (LoS) communication paths are necessary between the antennas for reliable communication in a wireless data center [21]. In [25] a novel wireless DCN architecture, based on 60 GHz wireless links between the individual servers of namely, S2S-WiDCN was proposed which drastically reduces the power consumption of the network portion of the data center while sustaining comparable performance. Hence, adopting an adequate wireless architecture for the HPC environment can result in significant power saving.

3. System and Problem Definition for Scheduling Problem

Figure 1 shows a simplified model of a typical HPC data center. Users submit their jobs to the data center which stores them in some data structure such as a queue until they can be allocated. Attempted allocations usually occur upon a change in the system, such as new jobs arriving or current allocations finishing which frees resources.

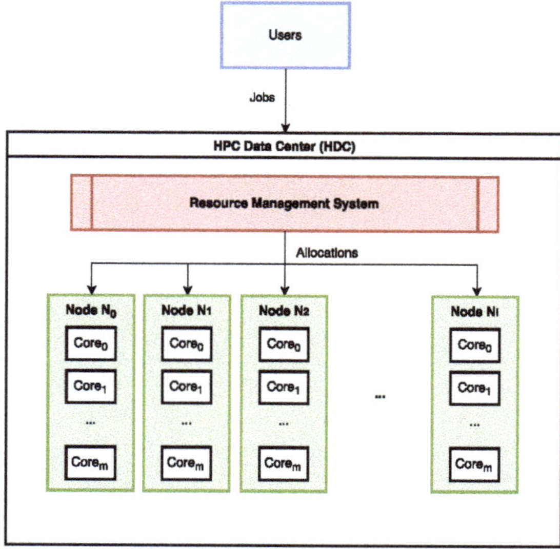

Figure 1. The model of the system targeted by this paper. A High Performance Computing (HPC) data center containing multiple nodes with many-core CPUs.

3.1. HPC System

The HPC Data Center (HDC) consists of a resource management system (RMS) connected to a number of different nodes $(N_0, N_1, ..., N_i)$ each containing a set of processing cores $(Core_0, Core_1, ..., Core_m)$. Each node represents a physical server in the data center being considered. The processing cores are homogeneous and communicate with each other via an interconnect. The RMS operates on its own set of resources and assigns arrived jobs to a set of cores within a single node. A single job is considered to use resources of a single node to avoid communication overhead between nodes. Further, to avoid migration overhead, it is assumed that jobs cannot be paused or migrated to a different set of resources during execution.

The HPC system is created in hardware as three Odroid-XU3 boards connected in a local area network (LAN). The CPU in these boards, Samsung Exynos5422, is powered by Arm® big.LITTLE™ architecture: four Cortex®-A15 cores at 2.0 GHz and four Cortex®-A7 cores at 1.4 GHz. As our model requires homogeneous cores in the nodes, only the A15 cores are used for job allocation. This conveniently allows the proposed RMS to execute solely on the A7 cores of one board instead of requiring separate hardware for its own set of resources. This represents a many HPC system with three nodes and many HPC systems are realized in the same way, where one node (server) or a set of cores act as the manager and other nodes (servers) are used to execute jobs after allocation [34]. Without the loss of generality and having better and more hardware availability, a large HPC system can be realized.

3.2. Jobs and Value Curves

Each job j is modeled as an tuple $J = (T; A)$, where T is the arrival time of the job and A is the application to be executed in order to complete the job. Each job j also has its own value curve function VC_j which converts a completion time of its execution to the value of the job to its user. These functions are typically monotonically-decreasing until reaching zero at a certain threshold of time, as shown in Figure 2. It is assumed that the value curves are pre-designed for each job and accurately reflect the economic importance to the user as the economic model is out of the scope of this paper.

The PARSEC benchmark suite was used as the set of applications to queue in the system. This is because the benchmark applications were designed to have a range of multi-threading and other resource requirements. The focus on emerging workloads means the jobs are representative of potential future workloads in all situations including but not exclusive to HPC systems. A number of applications from the suite were selected and value-curves were designed for each application based on testing the execution time across different numbers of cores. In particular, we considered PARSEC applications listed in Table 1. The table also represents the value-curve for each application in terms of execution time and the value achieved if job is executed by that time. The jobs are generated by selecting a random application from this list and assigning an arrival time to it. Short and long periods of no jobs arriving are created to realize periods of no users (such as nights and weekends); these allow the scheduler to "catch-up" on remaining jobs in the queue.

Table 1. Value at different execution times for PARSEC benchmark applications.

Application	Execution Time (s)				Value (Currency)			
blackscholes	4	6	9	12	100	80	50	0
bodytrack	6	8	11	14	100	80	50	0
dedup	5	7	9	12	100	80	50	0
facesim	8	15	18	22	100	80	50	0
ferret	22	26	29	31	100	80	50	0
fluidanimate	7	9	11	14	100	80	50	0
freqmine	9	15	17	20	100	80	50	0
streamcluster	22	28	35	42	100	80	50	0
vips	9	11	13	16	100	80	50	0

The value curves create a natural "soft deadline" which implies an ideal time for a job to finish, but also that violation of the deadline does not mean the job completion was irrelevant [38]. Instead, the value of the job is reduced depending on the extra time the user has had to wait for completion [15,39]. Major violations of the deadline will create no value for the user and therefore the energy consumed by the computation was wasted.

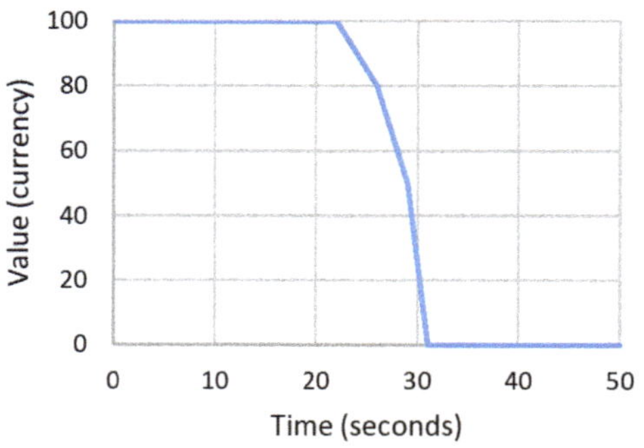

Figure 2. An example value curve of a job.

3.3. Problem Definition

The problem addressed by the paper is the allocation of jobs to a finite set of resources such that value obtained from completion of the jobs is maximized while energy consumption is minimized simultaneously. The problem is defined as follows:

- **Input:** Job queue $(j_1, ..., j_n)$, Value curve for each job VC_j, Nodes within the HPC data center $(N_0, ..., N_i)$.
- **Constraints:** Restricted available cores on the nodes in HDC.
- **Objective:** Jointly optimize overall value Val_{total} and energy consumption E_{total}, by maximizing the quotient Val_{total} / E_{total}.

The RMS needs to make very fast decisions on which node to allocate a job and which cores within that node should do the computation. It is assumed that any one job requires a minimum of one core and individual cores are never shared between multiple jobs.

3.4. Proposed Approach Based on Reinforcement Learning

This section describes the proposed approach. It first outlines the Adapted Multi-Armed Bandit (AMAB) model and discusses the assumptions and constraints of using such a model. Finally, the algorithm is described and explained in detail.

3.4.1. Adapted Multi-Armed Bandit Model

The MAB is a common model in reinforcement learning [40]. The model is of a game played in rounds from time $t = 1, ..., T$. At each round, the player must select a single action from a known set of actions $a_t \in A$. The environment then generates a reward $r_t \sim R^{a_t}$ where R^{a_t} is an unknown probability distribution. The goal of the game is to maximize the cumulative reward $\sum_{t=1}^{T} r_t$ by selecting actions which are likely to give high rewards. Due to the unknown nature of the reward distributions, players are required to first "explore" the possible actions to figure out the distributions before they can "exploit" the actions with the highest average probability.

For this paper, the problem is modeled as an adaptation of the MAB (AMAB) model. The possible actions are the different possible allocations of single jobs currently in the RMS queue, i.e., for each job there exists an action for each possible number of cores available for it to be allocated to.

There may be multiple jobs of the same type in the queue which will create duplicate allocation possibilities, however, they will likely have different schedule delays. The rounds are the allocation stages, which occur at every change in the job queue or available resources. Therefore, the time step between each round varies significantly. The rewards are $Val_j / Energy_j$ for each job after computation is complete.

The MAB model does not perfectly fit the described problem due to a number of assumptions: rounds are at discrete time steps, every action is available at every round, exactly one action is selected per round, and rewards are received instantly after selecting an action. As such, the problem model differs in the following ways: only a subset of all possible actions are available per round, multiple or no actions may be selected per round, and rewards at the start of some round in the future after selecting an action. To fit the model, this paper considers the use of the Upper Confidence Bound (UCB) algorithm [41], which is described in the next subsection.

3.4.2. Upper Confidence Bound Algorithm

The premise of the UCB algorithm is to model the uncertainty of information gathered through experimentation, allowing for exploitation to occur naturally as uncertainty is reduced. The UCB algorithm records the average rewards received for each action $\hat{r}_i$ alongside the number of times that action was chosen C_i. It uses this count of previous rewards to calculate the uncertainty of the recorded average. This uncertainty and the average are combined to give a largest possible estimate for the actual mean of the reward distribution.

$$\hat{\mu}_i = \hat{r}_i + \sqrt{\frac{2\log(\frac{1}{\delta})}{C_i}} \tag{1}$$

where δ is a confidence value which is usually chosen or search for via parameter optimization. If the value is very small then the result is optimistic of a higher possible mean, while a high value implies less optimism. Possible values for δ are explored in experimentation. The algorithm simply chooses the action with the maximum $\hat{\mu}_i$ as it is predicted to be the action with the best average rewards. As actions are repeated, the uncertainty represented by the second part of Equation (1) decreases due to the increased C_i. However, the equation relies on $C_i \neq 0$. Due to this requirement, the algorithm must first attempt each action at least once before estimating the means. Usually, implementations of UCB will spend the first k rounds, where $k = |A|$, selecting each action in turn. This gives an initial estimate for the average reward, though with a high uncertainty.

The final selection rule for UCB is as follows.

$$a_t = \begin{cases} \arg\max_i \hat{\mu}_i, & \text{if } t > k; \\ t, & \text{otherwise}. \end{cases} \tag{2}$$

However, UCB makes various assumptions which do not fit our adapted model, such as that the rewards are in the interval $[0, 1]$. The next section describes these assumptions and the required adaptations to be compatible with our HPC system.

3.4.3. Proposed Algorithm for Confidence-Based Approach

For the purpose of this paper, the algorithm created will be referred to as the Confidence-Based Approach (CBA) from its UCB inspiration. The CBA algorithm features a similar selection rule to UCB, however, it is modified to accommodate the new assumptions. The first modification is the selection: as every action is not available on every round, it cannot guarantee exploration of action t for $t \leq k$. Instead, it always attempts to find the maximum possible mean from the available actions. When an action i with $C_i = 0$ is encountered, the algorithm overrides the search for the maximum possible mean and instead allocates according to the job and number of cores of that action. This ensures full exploration of all possible actions that are encountered by the scheduler as soon as they are encountered.

The rewards from each action i are not guaranteed to be in the interval $[0, 1]$ either, as they are dependent on the user-specified value curve and the energy consumed by the system. This means that Equation (1) is not an accurate representation of the highest possible mean of an action. However, knowing the typical interval for the rewards can be used to scale the uncertainty part of the equation up to somewhat compensate for the difference. This can be done relatively easily using the specifications for the processing cores to obtain expected power and adding constraints to the maximum of the user specified value-curves. For our jobs, described in Section 3.2, typical reward values (>75%) were in the range $[0, 5]$ so we use a scaling multiplier of 5.

The reward, value divided by energy, is not strictly dependent on the job and number of allocated cores. Instead, it is derived from the sum of the computation time of the job, which is directly dependent on the allocation, and also the delay in scheduling of the job (time spent in the queue waiting). Due to this, the algorithm does not record the average reward from allocations but instead records the average computation time $\hat{t}_i$ and energy $\hat{e}_i$. During the scheduling process, it combines this average computation time with the current schedule delay of the job to estimate the expected average value for the current moment in time. The average energy consumed is not affected by schedule delay so is used directly to calculate the expected reward.

The final scheduler is shown in Algorithm 1. The loop on line 1 of Algorithm 1 is the initialization of the data required to estimate the average rewards. The full list of possible allocations is the list of every combination of the type of job and number of cores, i.e., for each job there is an entry for every number of cores between its maximum number of cores and 1. Lines 7–31 show the allocation "rounds" in the bandit algorithm. First, all finished jobs are accounted for with appropriate updates of resources and historical data, then newly arrived jobs are added to the queue, and finally, the algorithm attempts to allocate jobs to newly freed resources if possible. The selection rule is on line 24 and is nearly identical to the rule described in Equation (2). Possible values for δ are explored during experimentation.

The Odroid boards used for the experiment support monitoring the energy consumption of the quad-core Cortex®-A15. The power sensor can only detect the energy consumed by the entire processor and not the individual cores. This means that it is impossible to get an accurate recording of energy use for a single job using less than four cores. Instead, the energy recorded over the duration of a job is simply an estimate and will vary based on other jobs running simultaneously on the nodes. Therefore, most jobs will have their energy consumption over-estimated and the reward $Val_j / Energy_j$ will be underestimated. Experimentation for different levels of confidence using the δ parameter can give insight into how the variation in energy consumption estimates affects the learning algorithm. $Energy_{total}$ for the full system can be still be recorded accurately for comparison.

Algorithm 1: CBA Resource Allocation.

    Input: Incoming Jobs, HPC Data Center HDC.
    Output: Resource Allocation for Incoming Jobs.

1 **for** *i in possible single allocations* **do**
2 **end**
3 $\hat{t}_i \Leftarrow 0$;
4 $\hat{e}_i \Leftarrow 0$;
5 $C_i \Leftarrow 0$;
6 **while** *1* **do**
7   **if** *any running_job(s) have finished or job(s) arrive* **then**
8   Update data center resources;
9   Update $\hat{t}_i$, $\hat{e}_i$ and C_i for all jobs finished;
10   Update *jobQueue*;
11   **while** *Any possible allocations* **do**
12   *selectedAllocation* $\Leftarrow null$;
13   *maxU* $\Leftarrow 0$;
14   **for** *j in jobQueue* **do**
15   **for** *c = maxAvailableCores to 1* **do**
16                $i \Leftarrow allocation(j,c)$;
17   **if** $C_i = 0$ **then**
18   *selectedAllocation* $\Leftarrow i$;
19   **go to** 31;
20   **end**
21                $curDelay \Leftarrow curTime - j.arrivalTime$;
22                $\hat{v}_i \Leftarrow j.getValueAtTime(curDelay + \hat{t}_i)$;
23                $\hat{r}_i \Leftarrow \frac{\hat{v}_i}{\hat{e}_i}$;
24                $u \Leftarrow \hat{r}_i + 5\sqrt{\dfrac{2\log\left(\frac{1}{\delta}\right)}{C_i}}$;
25   **if** $u > maxU$ **then**
26   *selectedAllocation* $\Leftarrow i$;
27   *maxU* $\Leftarrow u$;
28   **end**
29   **end**
30   **end**
31   Allocate job according to *selectedAllocation*;
32   Update data center resources;
33   **end**
34   **end**
35 **end**

---

4. System and Problem Definition for Network Aware Server Consolidation

To augment the efficient scheduling algorithm discussed above, we propose a server consolidation algorithm which will ensure optimal resource utilization under the performance constraints of the data center network.

4.1. Network Aware Server Consolidation

Server consolidation is a process where VMs running in one server are relocated to one or more different servers. However, as discussed earlier we propose a network-aware consolidation approach which takes into account the traffic interaction between the VMs running on the servers. In order for the VM migration approach to be network or traffic-aware, we first need to understand the nature of traffic interaction over the data center network.

4.2. Traffic Pattern Model

The traffic pattern in a data center network can be modeled in terms of multiple parameters such as flow arrival rates, flow injection rates, flow sizes, flow completion time and proportion of inter-rack and intra-rack flows [42]. In [25] a novel wireless DCN architecture, based on 60 GHz wireless links between the individual servers of namely, S2S-WiDCN was proposed which drastically reduces the power consumption of the network portion of the data center while sustaining comparable performance. The proposed network aware server consolidation can be adopted for S2S-WiDCN or conventional wired fat-tree data center networks. In the S2S-WiDCN, there are six separate directional antenna arrays in the vertical plane of the server, and another one array on the top of the server. Therefore, seven simultaneous links from a server can co-exist at the same time. We represent the number of possible simultaneous links per server as θ. Let F be a vector whose elements are the number of existing flows along each sector determined from the number of flows existing in each server based on their destinations and the routing protocol. Let f denote the traffic flow rate. It is to be noted, that the flow rate f, has a Gaussian distribution [42,43]. Therefore, to support 99.86% (one-sided z-distribution) of the flow rates, the required channel throughput should be

$$r = f_{\mu+3\sigma}F, \tag{3}$$

where elements of the vector r, are the required channel throughput in each of the sectors and $f_{\mu+3\sigma}$ is the value of the flow rate which is three standard deviations higher than the mean. For the S2S-WiDCN, to accommodate multiple channel access, a single 60 GHz IEEE802.11ad link is subdivided into n_{OFDM} number of separate OFDM channels. Therefore, the bandwidth of each OFDM channel is given by,

$$B_{\text{W}} = B_{\text{60GHZ}} / n_{\text{OFDM}}, \tag{4}$$

where B_{60GHZ} is the bandwidth of the single physical channel. For a wired network, B_{W} would be equal to the bandwidth of the connected wire link. Therefore, to reduce the adverse effect of server consolidation on network performance, the following inequality must be satisfied for all wireless links or sectors from each server in the S2S-WiDCN,

$$r_x < B_{\text{W}} \ \forall \, x \tag{5}$$

where r_x is an element of r. If the inequality in (5) cannot be satisfied due to high flow rates, consolidation will result in worsening of data center network performance as discussed in the results.

Moreover, it has been observed from the measurement of a variety of data centers in [43], a large proportion of the server-to-server traffic flows, up to 80%, are intra-rack, meaning between servers in the same rack. Only a small remaining proportion of about 20% is inter-rack, or between servers in different racks. Therefore, to reduce the effective load on the network while consolidation, VMs that communicate more often should be migrated into the same physical server. Hence, in addition to reducing server underutilization, co-location of highly communicating VMs is also a desirable goal as it will reduce both power consumption and network traffic. This way, in our consolidation algorithm, we considered both the inequality of (5) and the proportion of inter and intra-rack traffic to make it network-aware.

4.3. The Network-Aware Consolidation Algorithm

The primary goal for consolidation is to reduce the total power consumption by reducing the number of active servers as well as network utilization. The underlying assumption is that the computational requirement for every VM running in the HPC data center and the injection rates of every flow from each VM is known and readily available. In a single server, multiple VMs can run at a single instance. However, during the server consolidation, we considered all the VMs running as a single entity, meaning if migration is possible, all the VMs running on the server would be migrated to the new physical server for consolidation. The migrations happen in *online* mode following live migration [10]. While this will reduce the granularity of the consolidation, it is a more scalable approach suitable for large data centers with thousands of servers. Moreover, the task-level granularity for a network-aware consolidation requires the knowledge of traffic flow per task, which is difficult to model, predict, or access in large data centers. Data center traffic rates are modeled usually among entire servers [42] limiting us to design consolidation algorithms at a server-level granularity.

We assume that every server has the same computational capacity and VMs running on a server utilizes a variable percent, collectively which can be represented by u. Let us assume that the maximum permissible utilization, without any significant degradation in performance or violation of legal contracts of any server, is D_u. D_u is a manufacturer specified parameter and can vary from model to model. The pseudo-code for implementing the BCC is shown in Algorithm 2. At first, all the servers running in the data center are divided into smaller clusters, such that servers within a cluster have a large number of flows exchanged among themselves, whereas servers in different clusters have a much smaller number of flows exchanged among them. Such a clustering places highly communicating servers in the same cluster. This intra-cluster consolidation reduces the communication among these highly communicating servers. This clustering is a Graph Partitioning Problem, which is to partition graph vertices into disjoint groups with minimum edge cut cost. The Kernighan–Lin algorithm [44] is adopted for the graph-partitioning tasks in our work. Here we treat servers as vertices and the number of flows going outside of the server as edge costs. After the partitioning, all the servers in each cluster are sorted according to their utilization u. The outer loop (line 5 in Algorithm 2) in the proposed algorithm chooses the candidate to migrate in the ascending order of utilization starting with the least utilized one. The inner loop (line 6 in Algorithm 2) chooses the destination to migrate in the descending order of utilization starting with the most utilized one. If the sum of the utilization of the candidates to migrate and the potential destination is less than D_u and each element of the vector sum of their required injection rates is less than the channel throughput per OFDM channel, the candidate is migrated to the destination. This flow rate related condition for migration is informed by our traffic model related constraint in (5). After a successful migration, the inner loop is broken out of, to choose the next server in the outer loop for potential migration. If either of the two conditions fails, the inner loop continues till the list of servers for the potential destination is exhausted. For each completion of the inner loop, the outer loop progresses to the next candidate for migration.

Algorithm 2: Algorithm for Bandwidth-Constrained Consolidation (BCC).

Input: Set of servers, $S = \{s_1, s_2, ..., s_N\}$; Server utilizations, $u = [\, u_1\ u_2,\ ...., u_N]^T \in \mathbb{R}^n_+$; Flow injection rate
  $R = [r_1, r_2, ..., r_N] \in \mathbb{R}^{\theta \times N}_+$; Communication cost $\Psi = \{\, \psi(i,j)\} \in \mathbb{R}^{N \times N}_+$
Output: Utilization profile u after BCC Consolidation

```
 1: Clustering:
 2: {S₁, S₂, ...., Sₖ} ∈ S ← Graph-partition(S, Ψ)                      ▷ Kernighan-Lin
 3: for m = 1 to k do
 4:    Sₘ ← sort (Sₘ)                               ▷ ascending order of server utilization u
 5:    for i = 1 to size (Sₘ) − 1 do                          ▷ loop for migrating server
 6:       for j = size (Sₘ) to i + 1 do                       ▷ loop for destination server
 7:          ũ = uᵢ + uⱼ
 8:          r̃ = rᵢ + rⱼ
 9:          if (ũ < Dᵤ and ∀ x ∈ {1, 2, ..., θ} r̃ₓ < B_W) then
10:             Migrate (sᵢ, sⱼ)                       ▷ sᵢ is the i-th entry of the sorted Sₘ
11:             break                                  ▷ Break from loop in line 6
12:          end if
13:       end for
14:    end for
15: end for
```

The necessary steps of the migration function (*Migrate*) used in the pseudo-code of BCC Algorithm 2 are shown in Algorithm 3. The function migrates server a to server b. At first, all the VMs running in migrating server a is migrated in the destination server b. So the utilization of the destination server increases which is the summation of the utilization of both the servers. Vector r_b is updated based on all the flows running on server b post migration. After successful migration, server a is put into the PowerNap state [45] having zero utilization. In the PowerNap state, most of the components of the server are powered down except the network interface card (NIC), the wireless transceivers, and a small portion of the CPU to get the signal for waking up when required.

Algorithm 3: Migration Function.

Input: Source and destination server for migration
Output: Updated utilization profile u after single migration
1: **function** MIGRATE(server a, server b)
2:      $b \leftarrow$ VMs running on a
3:      $u_b = u_a + u_b$
4:      $r_b = r_a + r_b$
5:      $a \rightarrow PowerNap$ state
6:      $u_a = 0$
7:      $r_a = 0$
8: **end function**

4.4. Complexity Analysis

Optimizing the performance of the scheduler in the data center has been a major research focus area for the last few years [46]. The calculation for the consolidation algorithm operations take place on the scheduler. The complexity of server consolidation over the entire data center to provide the optimal solution using exhaustive search method is $O(N^N)$ where N is the total number of servers in the entire data center. This is because N set of VMs can be potential candidates for migration to N servers in N ways. Therefore, each of N set of VMs has N options for potential migrations and for each of the N such scenarios the other sets of VMs also have all N options to create each possible migration scenario. However, this complexity is too high even for moderately large data centers. Therefore, we compare our proposed BCC consolidation algorithm with the Clustered Exhaustive Search (CES) algorithm, which finds the optimal migration within each cluster using the exhaustive search. We have adopted the Kernighan–Lin algorithm to do the clustering in the beginning of the BCC consolidation. If all the servers are equiprobable to have links between themselves, the computational complexity becomes $O(N^2 \log N)$ [44]. If the average number of servers in a cluster is n and if the number of the clusters formed is m, the computational complexity of the CES algorithm after clustering is $O(mn^n)$ and is an np-hard problem. With the clustering, the complexity of the CES algorithm is $O(N^2 \log N + n^n m)$. On the contrary, for the BCC algorithm, inside each cluster, the servers are sorted according to their utilization having a complexity of $O(n \log n)$ using merge sort [47]. Next, the two loops for finding the source and destination of the migration has the complexity of $O(n^2)$ in the worst case. So the overall complexity of BCC for all m clusters becomes $O(N^2 \log N + m^n \log n + mn^2)$. Therefore, BCC has a much lower complexity compared to the overall exhaustive search algorithm. As the clustering is similar in both CES and BCC, the difference in their complexity comes from the mechanism of determining candidates for migration. The complexity of BCC after clustering is $O(mn \log n + mn^2) \approx O(mn^2)$ which, is lower than that of the CES after clustering.

4.5. Optimizing the Inter-Consolidation Time

Due to the arrival of new tasks and the completion of existing tasks, the consolidated utilization profile of the servers may change over time. Therefore, the power consumption of the HPC data center may be adversely affected over time. Hence, the BCC consolidation algorithm should be repeated periodically. Repeating the consolidation too often might not reduce the total power consumption enough to justify the additional network traffic introduced as a result of the consolidation. On the contrary, delaying the consolidation can adversely affect the potential opportunity to save power. Hence, to determine the optimal time interval between two consecutive consolidations, an appropriate cost function, to capture the trade-off between power savings and network traffic is required. We define the expected value of the time-dependent cost function $C(t)$ for inter consolidation time interval as

$$C(t) = \kappa E[A(t)] - E[B(t)], \tag{6}$$

where $A(t)$ is migration cost related to the network traffic which represents the total traffic movement for the consolidation operation, $B(t)$ is the total power saving due to the consolidation, t represents the time interval between two consecutive consolidation operation, and κ is a scaling constant which captures the relative significance of network traffic and power savings. $E[\cdot]$ represents the expected value and is necessary as random task arrivals and completion make $A(t)$ and $B(t)$ random processes. At the optimal inter-consolidation time interval of t^*, the cost $C(t)$ should have the minimum value, that is,

$$t^* = \underset{t \in \mathbb{R}}{\operatorname{argmin}} \, C(t) \tag{7}$$

Figure 3 represents the timeline for the consolidation operation which shows two consecutive consolidation operations. u denotes the utilization profile of all the servers in the HPC data center, where $u = [u_1, u_2, ..., u_N]^T \in \mathbb{R}_+^N$ if N is the total number of servers in the data center. At time t_0, when the utilization profile is u_0, first consolidation operation takes place, and immediately after the consolidation, at time t_1, the utilization profile of the servers becomes u_1. After t seconds at t_2 the utilization profile of the servers becomes u_2 and a second consolidation is carried out which is completed at t_3 with a final utilization profile of u_3. Hence, it can be written that, $u_1 = \Gamma(u_0)$ and $u_3 = \Gamma(u_2)$, where Γ represents the consolidation operation. Furthermore, it holds that,

$$u_2 = u_1 + \delta t, \tag{8}$$

where $[\delta]_i$ is the task increase rate. $A(t)$ is directly related to the amount of traffic transferred through the network for the migration. If the average size of traffic per migration is v, then $A(t)$ can be represented by,

$$A(t) = v\left(\|u_1 + \delta t\|_0 - \|\Gamma(u_1 + \delta t)\|_0\right), \tag{9}$$

where $\|\cdot\|_0$ represents the ℓ_0-norm and returns the number of non-zero entries of its vector argument that is the total number of active servers. Therefore the difference between the ℓ_0-norms capture the total number of VMs migrating as a result of the consolidation.

Let, η_0, η_1, $\eta_2(t)$, and $\eta_3(t)$ represent the number of idle servers at time t_0, t_1, t_2, and t_3, respectively, where

$$\begin{aligned}
\eta_0 &= N - \|u_0\|_0, \\
\eta_1 &= N - \|\Gamma(u_0)\|_0, \\
\eta_2(t) &= N - \|u_1 + \delta t\|_0, \\
\eta_3(t) &= N - \|\Gamma(u_1 + \delta t)\|_0,
\end{aligned} \tag{10}$$

and N is the total number of servers in the data center. Hence, from (9) and (10), the expected value of $A(t)$ can be expressed as

$$E[A(t)] = v\left(\eta_3(t) - \eta_2(t)\right). \tag{11}$$

If the aggregate load running across the data center remains approximately constant between two consolidations, the expected number of idle servers after any consolidation operation will be similar, i.e., $\eta_1 \approx \eta_3(t)$ and does not depend on t. This assumption especially is valid when the granularity of the tasks are small compared to the capacity of an individual server. Hence, the expected value of $A(t)$ can be written as

$$E[A(t)] = v\left(\eta_1 - \eta_2(t)\right). \tag{12}$$

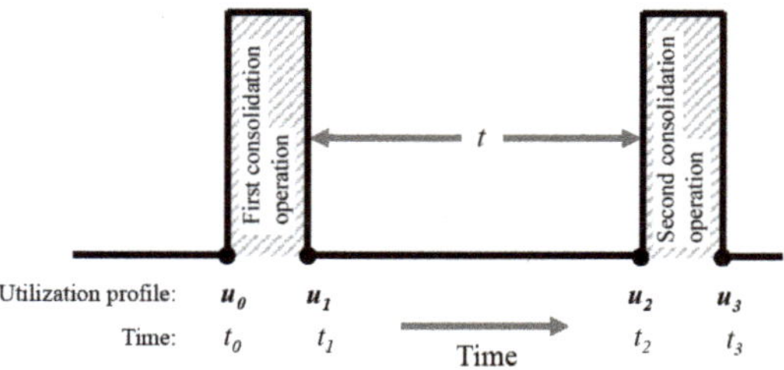

Figure 3. Timeline of consolidation operations.

On the other hand, the expected value of $B(t)$ can be estimated as

$$E[B(t)] = P_{idle}\eta_2(t) + P \cdot \|u_1\|_1 + (N - \eta_2(t))P_0 - P_{idle}\eta_1 - P \cdot \|u_2\|_1 - (N - \eta_1)P_0. \tag{13}$$

Here, P_{idle} is the power consumption per server in the PowerNap mode and P_0 represents the power consumption per server just after waking up from the PowerNap mode. P is the slope of the linear regime of the power profile of the server as discussed in Section 5.2.2. $\|\cdot\|_1$ represents the ℓ_1-norm and returns the sum of the utilization of all the active servers.

As the aggregate load across the data center is approximately constant over time, the total utilization of all active servers is approximately constant. Moreover, as the power consumption of the *active* servers is almost a linear function, it can be estimated that, $\|u_1\|_1 \approx \|u_2\|_1$ Hence, Equation (13) can be rewritten as

$$\begin{aligned} E[B] &= P_{idle}\big(\eta_2(t) - \eta_1\big) - P_0\big(\eta_2(t) - \eta_1\big) \\ &= \big(\eta_2(t) - \eta_1\big)\big(P_{idle} - P_0\big) \\ &= \big(\eta_2(t) - \eta_1\big)K \end{aligned} \tag{14}$$

Here $K = P_{idle} - P_0$ is a constant with respect to t. Combining Equations (6), (12) and (14), the estimated cost of the consolidation after time interval t can be found to be

$$\begin{aligned} C(t) &= \kappa v\big(\eta_1 - \eta_2(t)\big) - K\big(\eta_2(t) - \eta_1\big) \\ &= \big(\eta_1 - \eta_2(t)\big)(\kappa v + K) \\ &= \big(\eta_1 - \eta_2(t)\big)K' \end{aligned} \tag{15}$$

where $K' = (\kappa v + K)$ is a constant with respect to t. Thus to estimate t that minimizes the cost, we have to find the t that minimizes $\eta_2(t)$, though $\eta_2(t)$ is not known. Below we present a model for approximate $\eta_2(t)$.

To approximate $\eta_2(t)$, we consider that the servers follow the model of $M/M/1$ queuing processes [48], where λ and μ represent the new task arrival rate and task finishing rate per server, respectively. Hence, if a server initially has a utilization i, then t seconds later it will have utilization k, with the probability,

$$p_k^{(i)}(t) = e^{-(\lambda+\mu)t}\left[\rho^{\frac{k-i}{2}}I_{k-i}(at) + \rho^{\frac{k-i-1}{2}}I_{k+i+1}(at) + (1-\rho)\rho^k \sum_{j=k+i+2}^{\infty} \rho^{-j/2}I_j(at)\right], \tag{16}$$

where $\rho = \frac{\lambda}{\mu}$, $a = 2\sqrt{\lambda\mu}$ and $I_k = \sum_{m=0}^{\infty} \frac{(-1)^m}{m!\Gamma(m+k+1)}\left(\frac{x}{2}\right)^{2m+k}$ represents the modified Bessel function of the first kind of k-th order [48]. This model is valid only for $\lambda < \mu$, which is essentially true for sustenance of data centers of interest.

The probability that a node becomes idle at time t can be found from (16) by replacing k with zero. Hence the probability of a node becoming idle can be expressed as

$$p_0^{(i)}(t) = e^{-(\lambda+\mu)t}\left[\rho^{-\frac{i}{2}}I_{-i}(at) + \rho^{\frac{-i-1}{2}}I_{i+1}(at) + (1-\rho)\sum_{j=i+2}^{\infty}\rho^{-j/2}I_j(at)\right]. \tag{17}$$

Hence, the expected number of the idle nodes at $t_2 = t_1 + t$ can be expressed as,

$$\eta_2^{\text{model}}(t) = \sum_{l=0}^{N}\sum_{\mathcal{J}\subseteq[N]}\prod_{n\in\mathcal{J}}p_0^{i(n)}(t)\prod_{m\notin\mathcal{J}}\left(1 - v_0^{i(m)}(t)\right) \tag{18}$$

where $[N] = \{1,2,3,....,N\}$ and $\mathcal{J}$ is all the possible realizations of l idle nodes. In view of (18) and (15), the inter consolidation cost can be approximated as

$$C^{\text{model}}(t) = \left(\eta_1 - \eta_2^{\text{model}}(t)\right)K'$$
$$= \eta_1 K' - K'\sum_{l=0}^{N}\sum_{\mathcal{J}\subseteq[N]}\prod_{n\in\mathcal{J}}p_0^{i(n)}(t)\prod_{m\notin\mathcal{J}}\left(1 - p_0^{i(m)}(t)\right). \tag{19}$$

$C^{\text{model}}(t)$ in (19), can be calculated for any t, since $p_0^i(t)$ is known for any t. Thus optimal inter-consolidation time can be approximated by

$$t^*_{\text{model}} = \underset{t\in\mathcal{G}}{\text{argmin}}\, C^{\text{model}}(t) = \underset{t\in\mathcal{G}}{\text{argmax}}\, \eta_2^{\text{model}}(t), \tag{20}$$

where $\mathcal{G}$ is a finite-length fixed-step grid in $\mathbb{R}$. From this mathematical model, the optimal inter-consolidation time can be estimated without the need for thousands of simulations involving different random utilization profiles of servers. The accuracy of the mathematical model is verified with a Monte-Carlo simulation in Section 5.2.4.

5. Experimental Results

In this section, we discuss and evaluate the performance and effectiveness of both CBA and BCC algorithms. At first we discuss the results related to CBA, followed by the results related to BCC. Then we discuss the effect of combining CBA and BCC algorithms.

5.1. Experimental Results for CBA Algorithm

Initial experimentation was required to find a good confidence value δ. These experiments used a high arrival rate of jobs, though it is possible to repeat these experiments for other arrival rates if desired. A few values were tested in the range $[0.01, 0.95]$, however, the search for the optimal value was purposely shallow. This was due to the possibility that the search is not possible for systems implementing a similar approach in the future; instead, the experiments were designed to find a rough approximation which was more generalized (i.e., how optimistic should similar algorithms be for best results). The results of the parameter optimization are shown in Figure 4. The results of using different δ values were compared to a baseline of $\delta = 1$ which implied complete certainty in the recorded job averages as $\log(1) = 0$, so $\hat{\mu}_i = \hat{r}_i$. The graph shows that a high confidence level (0.95) was the most successful which implies there was little variation in the data and, therefore, the true mean was not plausibly significantly higher than previous readings. However, experiments using $\delta \leq 0.75$ performed worse than $\delta = 1$ which shows that it was better to be slightly overconfident in previous results than to be very optimistic in higher possible mean reward. Optimizing the parameter further is considered a secondary objective to the problem addressed by this paper, as such further experimentation in this area is recommended for systems with different assumptions which may affect reward distributions.

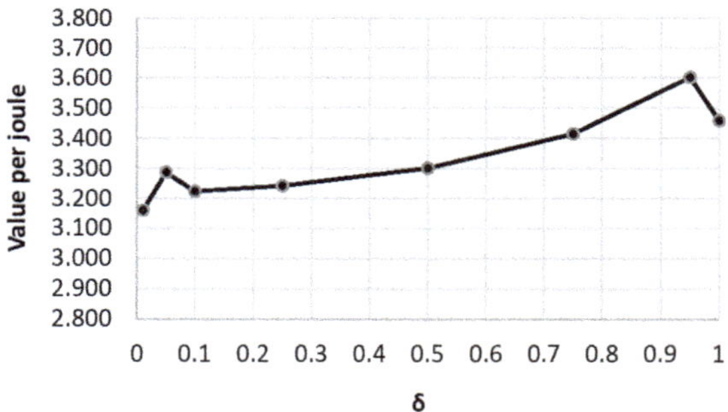

Figure 4. Value/energy for different confidence values.

To evaluate the scheduling algorithm in different situations, the arrival rate of jobs was adjusted to realize larger periods of time with high/low activity compared to the small periods as described in Section 3.2. The small periods of time realized as nights and weekends were kept the same length for consistency; the scheduler must handle a higher number of jobs in the same amount of time. In total, 50 days and nights were simulated of which 14 days were weekends. In a single day, the number of jobs which arrive depends on the arrival rate: 20 jobs arrived at the high rate, 12.5 jobs (on average) arrived at the medium rate, and eight jobs arrived at the low rate. In these experiments, we considered the $Val_{total}/Energy_{total}$ recordings of the previous experiments and also the percentage of jobs which provided no value to the customer: either they finished computation after computation or are rejected due to $\hat{r}_i = 0$.

5.1.1. Experimental Baselines

As discussed in Section 2, the approaches which considered the joint optimization of value and energy required prior information collected at design-time of jobs [17,34]. Similar approaches considered indirect energy optimization by minimizing CPU idle time [32] which was not a solution to the problem addressed by this paper. We compared results of experiments using the CBA approach to a simple heuristic approach. As no prior information was available, the heuristic used was the value obtained by the job if completion was instantaneous, i.e., heuristic was a value optimizing approach [15,28]. This approach, further referred to as the Heuristic-Based Approach (HBA), was implemented by replacing lines 22–24 with $u \Leftarrow j \cdot getValueAtTime(curDelay)$.

5.1.2. Profit and Energy Consumption Results at Varied Arrival Rates

Figure 5 displays the total value earned (profit) and the energy consumed by both HBA and CBA at the different arrival rates. At the high arrival rate, jobs arrived very frequently while at the low arrival rate the jobs arrived much less frequently. An interesting initial observation of the figure is that the value earned by HBA decreased as the arrival rate increased while the value earned by CBA increased instead. This was likely due to the poor estimation of value by the HBA algorithm: high schedule delays of jobs which arrived earlier but were delayed in preference of other jobs may still have lain at the peak of the value curves, earning less value when the computation time resulted in a value much further down the curve than the heuristic can predict. For the CBA algorithm, the increased arrival rate of jobs increased the total value as it had more options for allocation. It also tended to execute more jobs simultaneously as jobs with a very low schedule delay were often allocated to far fewer cores, likely due to significantly lower energy consumption compared to a relatively minor drop in value.

At the low arrival rate, the value earned was slightly higher for CBA than HBA while the energy consumed was slightly higher for the latter approach. However, this difference did not seem to be very significant at <1% change in both cases. We believe that this change would be more significant for experiments over a longer period of time for a number of reasons: the overhead of initial exploration by CBA would be a smaller fraction of the total time and the algorithm would explore less frequently as jobs were executed more times. This effect occurred at all arrival rates but was more prevalent at the low arrival rate as the initial exploration took much more time.

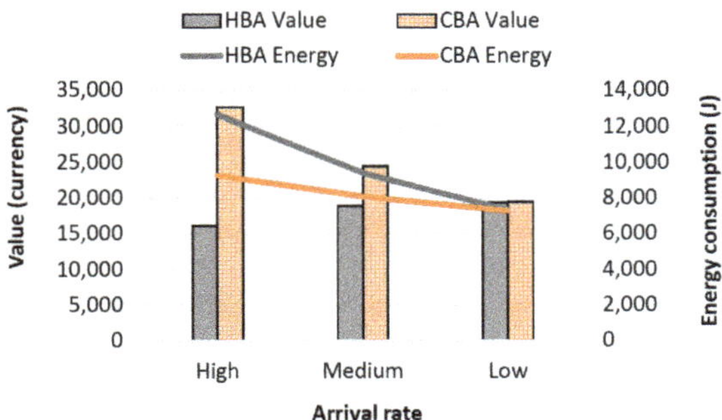

Figure 5. Value earned and energy consumed at different arrival rates.

In general, HBA resulted in higher profit and lower energy consumption when compared to CBA. On an average, HBA optimized profit and energy consumption by 40% and 20%, respectively, compared to CBA.

5.1.3. Percentage of Zero-Value Jobs

This metric is important as it relates to user satisfaction with the system: more jobs successfully serviced implies more users serviced and therefore more satisfied users. Figure 6 shows the percentage of zero-value jobs for both approaches at each of the different arrival rates. The average over each of the different arrival rates is also shown. Note that the rejections of jobs were implicit in the maximum reward check, they were not actually removed from the queue though it would be possible to add that functionality. It can be observed that, for all arrival rates, CBA had a lower percentage of zero-value jobs. This was less significant at the high and low arrival rates. For the high arrival rate, this was due to such a large number of jobs arriving that it was impossible to service a large number of them regardless of approach. For the low arrival rate, the difference was again likely due to the exploration part of the algorithm. The initial exploration caused a number of early jobs to be executed after the deadline while the algorithm learned the initial average reward for each possible allocation. The effect of exploration based on uncertainty after initial exploration was unlikely to have much effect on the number of jobs which give zero-value as this exploration favored jobs with reasonably high expected value already. On average, CBA provided 5% more user satisfaction than that of HBA.

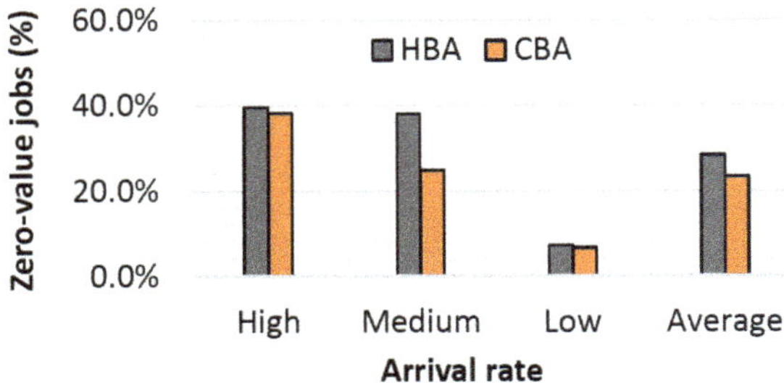

Figure 6. Percentage of zero-value jobs at various arrival rates.

5.1.4. Overhead Analysis

Computational complexity of CBA was no higher than the heuristic algorithm HBA. It did perform slightly more mathematical operations per iteration, but it was the same number of iterations required. The average time to find the allocation for a job by CBA was 0.12 milliseconds and the average energy consumption for it was 0.3 millijoules, which was quite low.

Memory overhead of CBA as compared to HBA was negligibly higher as it stored only a few extra numbers per job.

These overheads both in terms of timing and energy consumption were part of the overall profit and energy consumption results, which were better by CBA over HBA. Next, we will discuss the results related to BCC algorithm.

5.2. Experimental Results for BCC Algorithm

In this section, we discuss modeling, results and the corresponding analysis of the proposed server consolidation method. We first estimate the power reduction from proposed BCC server consolidation algorithm for both wired and wireless networks. Next, we evaluate the network-level performance with the consolidation algorithm in an HPC data center with network-level simulations. Before presenting and analyzing the results we describe the data center traffic generation procedure and simulation platform in the next subsections.

5.2.1. Traffic Generation and Simulation Platform for BCC

The BCC algorithm was evaluated with a set of traffic flows based on application demands. Real data center traffic for typical query/search type applications like map-reduce and index-search were measured in [43]. Using these measured traffic flows, a Poisson shot-noise based model to synthesize data center traffic was proposed and verified in [42]. According to [42], the new flow arrival time, the flow duration and the injection rate for each application followed a Poisson, Pareto and, Gaussian distribution respectively. The new flow arrival time was generated using a Poisson distribution with an average flow arrival rate. The average flow arrival rate was considered to be 1000 flows/s for the small-sized DCN [43]. In our evaluations, we considered a Gaussian distribution for the injection rate to have a mean of 8.0 Kbps as the base case for the simulation. Application flow duration was generated following an independent Pareto distribution having a minimum duration of 10 microseconds [43] and a mean of 1 s. We then increased the average injection rates on an incremental basis to 8 Mbps, 100 Mbps, 400 Mbps, and 650 Mbps and regenerated new traffic which represented different types of multimedia traffic and repeat the simulations. We used the Network Simulator-3 (NS-3) suite [49] to evaluate the performance of BCC for both wired fat-tree and wireless S2S-WiDCN networks. NS-3 supported the characteristics of wireless propagation as well as network-level communications. This simulation platform was used to evaluate the S2S-WiDCN with and without consolidation and compare it with the fat-tree wired DCN. For the fat-tree based wired data center architecture, we considered 1.0 Gbps links between servers to access switches and 40.0 Gbps upper-layer links. For all the cases, the migration cost for consolidation is not included in the performance analysis. We have compared the performance of the BCC consolidation algorithm for S2S-WiDCN with traditional wired fat-tree based DCN. We considered a small data center consisting of 800 servers arranged in a 20 × 8 array of racks as [25]. Each of the racks housed five servers and occupied an area of 0.6 m × 0.9 m and is 2 m high. There were 10 racks arranged in a single row and two columns of 8 rows, totaling 160 racks. In our simulations, the racks were assumed to be without any front or back door. In the traditional wired fat-tree based DCN, we considered the same number of servers arranged in same layout as S2S-WiDCN. We considered three hierarchical network layers consisting of 160 access, two aggregate, and two core layer switches, where each rack having an access layer switch.

5.2.2. Power Consumption Analysis of BCC

Here we discuss the model and parameters used in power estimation followed by the results.
Power Model for BCC:

It was not a straightforward task to estimate the actual electrical power consumed by an HPC data center. The power consumption depended on several internal factors such as utilization of computing power, the cooling mechanism, and data center networks. Data center power consumption was also affected by external parameters like the geographical location, weather, temperature, and humidity. The total IT power consumption of a data center, P_{IT} consisted of power consumption of the servers (P_{Server}) and network component ($P_{Network}$) of the HPC data center. Hence,

$$P_{IT} = P_{Servers} + P_{Network} \tag{21}$$

A major portion of P_{IT} comes from $P_{Servers}$ [50,51]. However, the power consumption of servers varies significantly with the change in CPU utilization [45]. If the utilization of i-th server is denoted by u_i, the Power consumption of that server can be given by $P_{Server}(u_i)$, where the dependence of server power on utilization is adopted from [52]. Hence, Equation (21) can be rewritten as:

$$P_{IT} = \sum_{i=1}^{N} P_{server}(u_i) + P_{network} \tag{22}$$

For the power analysis, we used the power profile of Dell Inc. PowerEdge C5220 (Intel Xeon E3-1265LV2) servers. Power consumption at different server utilization was modeled from the measurement done by the Standard Performance Evaluation Corporation's $SPEC power\_ssj2008$ database for the same server [52]. In addition to the above power model for the server, we considered an idle server to be placed in the PowerNap state [45] with minimal power consumption. The power profile of a server against different utilization is shown in Figure 7. Although compared to the server power, the power consumption of the network of an HPC data center was small, but it was not negligible [50]. One of the issues with the networking equipment was that they needed to be turned on all the time. The static power portion of the networking equipment dominated the total power consumed by the network [53]. In [53], it was shown that for a network switch, only 8% power reduced during full load to no load transition. Moreover, for the wired network, upper-level switches experienced a similar amount of traffic before and after the consolidation as the majority of the flows remained inside of the rack. For this reason, we neglected the change in networking power equipment due to the variation in injection rate or throughput. We estimated power consumption for wired DCNs using commercially available data from Cisco network switches [54] and Silicom network interface cards (NIC) [55]. The power consumption of each device used in the network is shown in the Table 2. The total network power is:

$$P_{Network} = N_{Core}P_{Core} + N_{Agg}P_{Agg} + N_{Acc}P_{Acc} + NP_{NIC}, \tag{23}$$

where N_{Core}, N_{Agg}, N_{Acc}, N are the number of cores, aggregation, access switches, and the total number of servers, respectively; P_{Core}, P_{Agg}, P_{Acc}, P_{NIC} are the power consumption of an individual core, aggregation, access switches, and network interface cards, respectively. In S2S-WiDCN, however, no core, aggregate or access layer switches were needed, but only antennas, transceivers and NICs were required for wireless communication. The power consumption of the wireless 60 GHz transceiver was modeled based upon the assessment of 60 GHz transceivers [56]. The NICs of S2S-WiDCN were equipped with transceivers for horizontal and vertical communication. In the traditional DCN, external connections were established via the two Cisco 7702 switches. To provide equivalent connectivity in S2S-WiDCN, we employed two servers to work as gateways, and their power consumption was modeled as that of the Cisco 7702 switch. The power consumption for communication per server in S2S-WiDCN was calculated as:

$$P_{Wireless} = 7P_{60GHzTran} + P_{WifiCntrl} + P_{NIC}, \tag{24}$$

where $P_{60GHzTran}$ is the power consumption of a single 60 GHz transceiver required for each of the six sectors and the horizontal link and $P_{WifiCntrl}$ is the power consumption of the IEEE802.11 2.4/5 GHz ISM adapter for the control channel. Finally, the total power consumption in S2S-WiDCN was:

$$P_{Network(WiDCN)} = N_{Core}P_{Core} + N.P_{Wireless}. \tag{25}$$

Table 2. Power consumption of different data center network (DCN) components.

Device	Model	Used in	Power (W)
Access Layer Switch	Cisco 9372	Fat-Tree	210.0
Aggregate Layer Switch	Cisco 9508	Fat-Tree	2527.0
Core Layer Switch	Cisco 7702	Fat-Tree, S2S-WiDCN	837.0
Network Interface Card	Silicom PE2G2I35	Fat-Tree, S2S-WiDCN	2.64
60 GHz Transceiver	Analog Device HMC 6300/6301	S2S-WiDCN	1.70
IEEE802.11 2.4/5 GHz Adapter	D-link DWA-171	S2S-WiDCN	0.22

Comparative Analysis of Power Consumption for BCC:

The IT power consumption of wired and S2S-WiDCN data centers with different consolidation methods including the BCC algorithm with variation in the flow injection rate are shown in Figure 8. These power consumption are computed based on the power model described in the previous sub-section for each of the simulation cases ran in the NS-3 simulator for different consolidation algorithm. In [45], it is observed that the majority of the utilization factor of a server is within the range of 20–30%. Here we adopted the utilization of the server capacity of each server without any consolidation from [45]. Figure 8a represents the power consumption of the HPC data center with no consolidation normal condition (NC) while Figure 8d represents BCC. The figure also contains the power consumption pattern if Clustered Exhaustive Search (CES) algorithm was adopted instead of BCC in Figure 8b. CES is a variant of the BCC algorithm, which finds the optimal migration within each cluster using the exhaustive search, hence being much more computationally expensive. For the sake of comparison, we also simulated a network-unaware greedy approach based consolidation (GRD) which is a variant of NICE [11] and the power consumption pattern is shown in Figure 8c. For both wired and wireless networks, at lower injection rates, all the consolidation techniques performed similarly and resulted in significant power consumption reduction compared to NC, while CES consumed the least power. BCC algorithm consumed only 2.83% more power than CES, whereas CES was significantly more computationally complex than BCC because of exhaustive search as discussed in Section 4.4. Hence during the consolidation operation, for moderate to large size of data centers, CES algorithm became impractical to implement as it was not able to perform real-time operations, whereas the BCC algorithm could. BCC algorithm for S2S-WiDCN reduced 37% power consumption compared to the NC case.

For higher injection rates, CES and GRD consumed significantly less server power compared to BCC for wired networks. This was because the CES and GRD did not consider the network bandwidth constraints while consolidating, resulting in more aggressive consolidation that in the BCC. Therefore, this difference became more apparent with increase in flow injection rates. However, this reduction of power came at the cost of the lower throughput because the CES and GRD algorithms did not consider the network traffic characteristics. This impact on performance will be discussed in detail in Section 5.2.3. Moreover, the computational complexity of CES was orders of magnitude higher than the BCC.

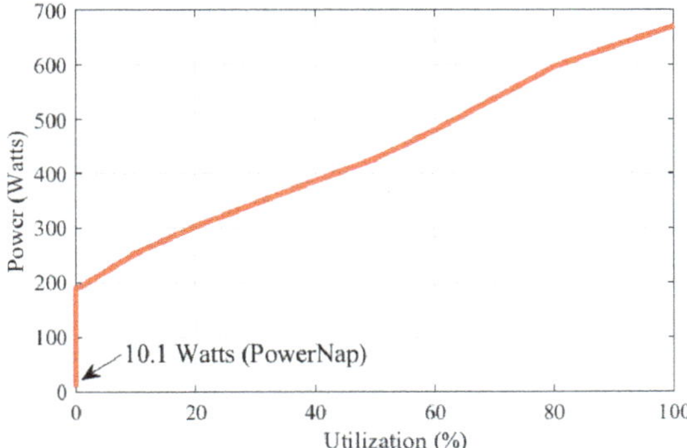

Figure 7. Power profile with varying utilization of PowerEdge C5220 server [52].

For BCC consolidation in S2S-WiDCN, similar to the wired network, with the increase in flow injection rate, the server power consumption increased, while for CES and GRD power consumption remained the same. Hence, at higher injection rate, CES and GRD consumed less power compared to BCC. However, this reduction of power came at the cost of lower throughput as CES and GRD algorithms did not consider the network traffic characteristics while consolidating. However, the increase in power consumption in BCC for S2S-WiDCN was not as drastic as in the case of wired data center. This is demonstrated by the trend arrows in Figure 8d. This was because, in the S2S-WiDCN architecture, a server had the potential to sustain a maximum of seven simultaneous links at a time with other servers in its vertical plane and horizontal line. On the contrary, in the wired architecture, there existed only one link per server albeit, of higher bandwidth. As a result, for the wired DCN with BCC, many of the VM migration attempts failed due to the violation of the inequality of (5) compared to the S2S-WiDCN. This suggested that the bandwidth constrained network-aware consolidation, BCC, was more effective on S2S-WiDCN.

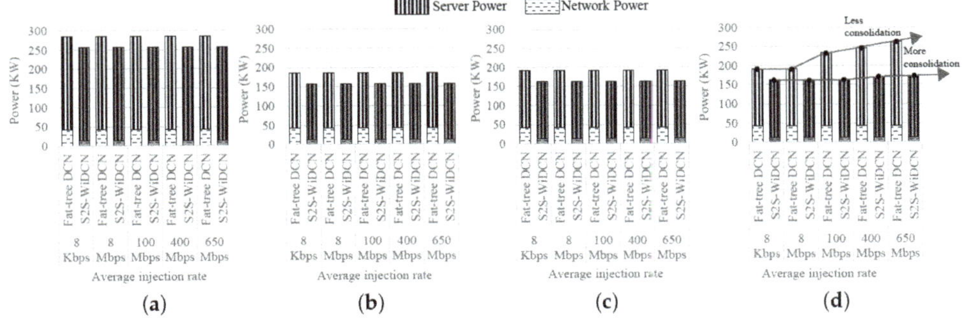

Figure 8. Power consumption comparison of different architecture for (**a**) no consolidation (NC) (**b**) clustered exhaustive search (CES) (**c**) Greedy approach base consolidation (GRD) and (**d**) bandwidth constrained consolidation (BCC). The arrows denote the power saving due to BCC.

5.2.3. Performance Analysis of BCC

Here we present the network-level performance of the S2S-WiDCN with BCC along with a comparative analysis with respect to wired fat-tree DCNs in terms of throughput.

The throughput was defined as the average rate of bit transferred per second over the DCN. The normalized throughput of both S2S-WiDCN and fat-tree architecture for different injection rates at NC, CES, GRD and BCC consolidation are shown in Figure 9. Normalized throughput was defined as the ratio of the average throughput achieved and average injection rate. For NC, although it was seen that for lower injection rate both S2S-WiDCN and fat-tree network showed similar throughput, but for both the networks, the achieved throughput started to decrease as the average injection rate went beyond 100 Mbps. However, degradation was different for wired and wireless DCNs. The throughput reduced more for the wireless DCN than the wired counterpart for higher injection rates due to the lower physical bandwidth available per channel for the wireless links of 0.563 Gbps compared to 1.0 Gbps for the wires. Further, from Figure 9, it can be seen that, for the lower injection rates, there was no significant difference in achieved throughput with BCC consolidation for both wired and wireless data centers. These throughputs were also similar compared to that NC case. However, for higher injection rates beyond 100 Mbps, for both S2S-WiDCN and fat-tree network, achieved throughput increased compared to the NC. The main contributing factor was that, due to the VM migrations, in many cases, both source and destination of flows ended up in the same physical server. Therefore, these flows are effectively eliminated from the network, which ultimately increased the average throughput of the entire network compared to NC. On the other hand, instead of BCC, if CES or GRD consolidation was implemented, at lower injection rates, there was no significant difference in achieved throughput for both S2S-WiDCN and fat-tree networks compared to BCC. For the higher injection rates beyond 100 Mbps, the performance of the wireless networks improved compared to NC, but not as well as BCC. However, the performance of the wired network degraded with the incorporation of CES or GRD consolidation algorithm. This contrasting behavior was mainly due to the number of channels available in different architectures. Due to all flows in the wired data center being channelized over the same link, the aggregate flow rate after CES or GED exceeded the physical link bandwidth violating (5). This caused degradation in throughput. On the contrary, in the S2S-WiDCN, due to the presence of multiple vertical sectors and the horizontal link a relatively larger number of flows did not violate (5) resulting in better performance compared to the wired data center.

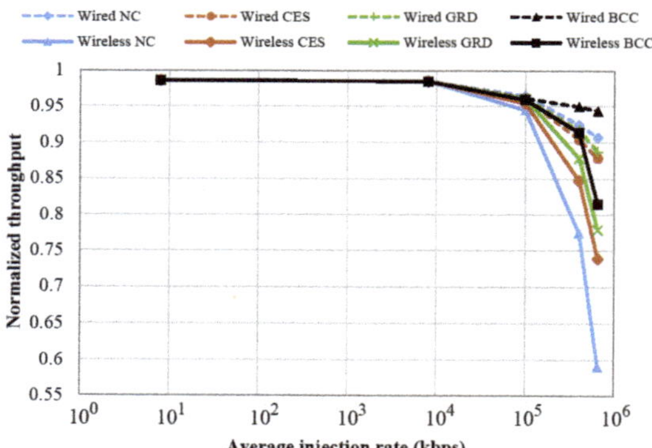

Figure 9. Average throughput for different data center architecture with NC, CES, GRD and BCC consolidation normalized with flow injection rate.

5.2.4. Accuracy of Inter-Consolidation Time Modeling

In this section, the inter-consolidation time for the BCC algorithm is analyzed and the accuracy of the mathematical estimation of inter-consolidation time is evaluated. The expected inter consolidation cost estimated from (15) is shown in Figure 10 for a data center consisting of 800 servers as discussed in Section 5.2.1.

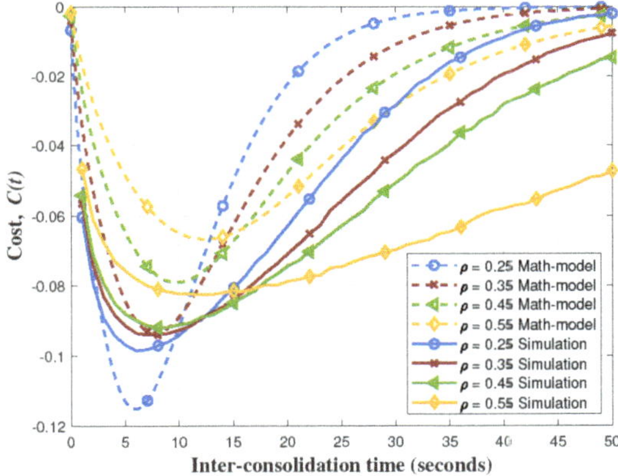

Figure 10. Consolidation cost estimated from mathematical analysis (dashed line) and from Monte Carlo simulation (solid line) with respect to inter-consolidation time.

To verify the mathematical model for cost of consolidation (15), we ran a Monte Carlo simulation for each ρ 300 times and calculated the value of the cost function. For these cases, the values for η_1, v, κ considered here were 200, 1000, and 1 respectively. The average size of migration v was considered as 1000 Megabytes which represents a practical value. For η_1, we ran the server consolidation simulation for different random initial conditions for thousands of times and used the average number of idle servers in PowerNap mode after the first cycle of server migration. We considered $\kappa = 1$ to put equal emphasis on power saving and network performance on the cost. Nevertheless, these values needed not necessarily be exactly the same across all the data centers. Depending on the capacity, performance requirements and physical limitations, these values could vary for different HPC data centers.

The average of the simulated values from the different run for each ρ are shown in Figure 10. The cost estimates in this method relied on many repetitive simulations and were highly computationally expensive as each of simulation at a particular ρ and t was repeated at least 1000 times to find the expected cost using Monte-Carlo method. On the contrary, using (15) the cost and optimal inter-consolidation time could be approximated much faster. The optimal inter-consolidation time for different ρ identified from both methods is shown in Figure 11. It was observed that for lower values of ρ ($\rho \leq 0.55$) the optimal inter-consolidation time estimated from the mathematical equation closely approximated the measured value from the Monte Carlo simulation. On the contrary, for higher values of ρ, the optimal inter-consolidation time estimated from the mathematical analysis deviated from the value measured through simulations. However, at higher ρ, the absolute value of the cost was less sensitive to the inter-consolidation time. This shows that although at higher ρ the inter-consolidation time suggested by the model may deviate from the actual optimal interval, the actual cost incurred at this non-optimal interval was not much different compared to that at the optimal interval. Hence the optimal inter-consolidation time could be estimated reasonably accurately from the mathematical form.

We used MATLAB R2018b on a system having Intel Core i7 with 16 GB memory to calculate the inter consolidation time with both Monte-Carlo simulation and mathematical model. On average, computation time required for the Monte-Carlo simulation took 1039.4 s to complete the calculation while the mathematical model took only about 0.798 s on average. Hence there was more than 1000× speedup in calculation time using the mathematical model.

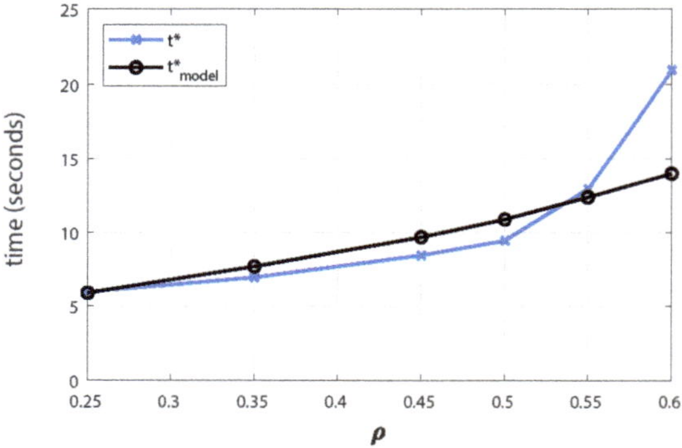

Figure 11. Comparison of optimal inter-consolidation time obtained from the mathematical model and Monte-Carlo simulations at different ρ.

5.3. Overall Power Saving with a Combination of BCC and CBA

In this subsection we discuss the combined effect of CBA and BCC together for power saving. We evaluated the power consumption of a data center consisting of 800 servers with a higher injection rate traffic having an average flow rate of 400 Mbps. For this comparison, the power consumption of each server was modeled based on Odroid-XU3 board. The power consumption of the Odroid-XU3 at PowerNap was conservatively assumed to be 1 watt and the full load power consumption was 20 Watts. We also assumed that the power saving ratio per device followed the energy saving ratio due to CBA algorithm. In Figure 12 we showed the power consumption of the data center in different cases including, normal condition (NC), utilizing BCC only, utilizing CBA only and finally, utilizing both BCC and CBA together.

At higher datarate, BCC consolidation could reduce the power consumption of the overall data center more for S2S-WiDCN compared to fat-tree network. The CBA algorithm working standalone outperformed BCC for fat-tree wired network while in S2S-WiDCN, BCC could reduce more power compared to CBA. Nevertheless, for both wired and wireless networks, combining both BCC and CBA could achieve maximum power reduction.

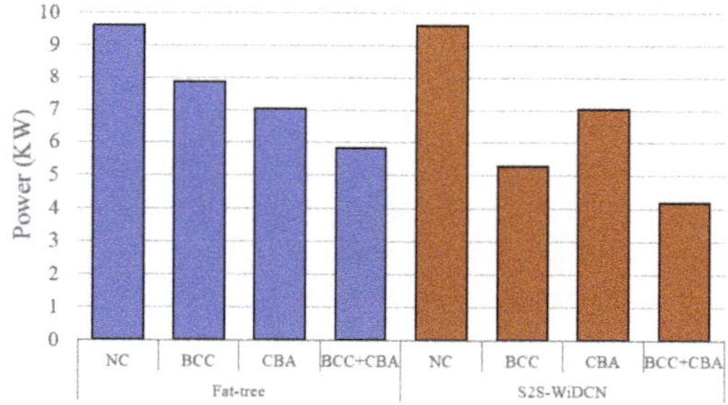

Figure 12. Power consumption comparison between normal condition, BCC, CBA and combination of BCC and CBA.

6. Conclusions

In this paper we investigated an algorithm for jointly optimizing value and energy when considering resource allocation in HPC data centers. The algorithm was created under the assumption that no prior information is available for accurate predictions of job value and energy, instead it used a technique inspired by reinforcement learning to explore the value and energy of jobs before exploiting in future allocations. It has also been shown that the percentage of zero-value jobs is lower for all of the different arrival rates for CBA. We have also incorporated a server consolidation algorithm BCC for both wired and wireless data center networks. We have shown that both the approaches significantly reduce energy and power consumption of the entire data center. It has been observed that, if BCC and CBA are adopted simultaneously for the wireless HPC data center architecture, maximum power saving can be achieved. We also derived a mathematical model for BCC for determining optimal inter-consolidation interval to enable the data center resource management unit to schedule consolidations at optimal intervals without relying on computationally expensive simulations to estimate the optimal interval.

Author Contributions: Conceptualization, A.K.S., A.G. (Amlan Ganguly), C.V.M. and B.M.A.-H.; Data curation, S.A.M. and A.G. (Alexander Gilday); Formal analysis, S.A.M. and A.G. (Alexander Gilday); Investigation, S.A.M., A.G. (Alexander Gilday), A.K.S., A.G. (Amlan Ganguly) and X.W.; Methodology, X.W.; Supervision, A.K.S., A.G. (Amlan Ganguly), G.V.M. and B.M.A.-H.; Writing—original draft, S.A.M. and A.G. (Alexander Gilday); Writing—review & editing, S.A.M. and A.G. (Alexander Gilday). All authors have read and agreed to the published version of the manuscript.

Funding: This work was supported in part by the Engineering and Physical Sciences Research Council under EPSRC Grant EP/L000563/1, and EP/K034448/1 the PRiME Programme Grant (www.prime-project.org), and US National Science Foundation(NSF) CAREER grant CNS-1553264.

Conflicts of Interest: The authors declare no conflict of interest.

References

1. Rodero, I.; Jaramillo, J.; Quiroz, A.; Parashar, M.; Guim, F.; Poole, S. Energy-efficient application-aware online provisioning for virtualized clouds and data centers. In Proceedings of the International Green Computing Conference (IGCC), Chicago, IL, USA, 15–18 August 2010; pp. 31–45. [CrossRef]

2. Benini, L.; Micheli, G.d. System-level power optimization: Techniques and tools. *ACM Trans. Des. Autom. Electron. Syst. (Todaes)* **2000**, *5*, 115–192. [CrossRef]

3. Aksanli, B.; Rosing, T. Providing regulation services and managing data center peak power budgets. In Proceedings of the 2014 Design, Automation & Test in Europe Conference & Exhibition (DATE), Dresden, Germany, 24–28 March 2014; pp. 143–147.

4. Bogdan, P.; Garg, S.; Ogras, U.Y. Energy-efficient computing from systems-on-chip to micro-server and data centers. In Proceedings of the 2015 Sixth International Green Computing Conference and Sustainable Computing Conference (IGSC), Las Vegas, NV, USA, 14–16 December 2015; pp. 1–6.

5. Koomey, J. Growth in data center electricity use 2005 to 2010. In *A Report by Analytical Press, Completed at the Request of The New York Times*; Analytics Press: Burlingame, CA, USA, 2011.

6. Verma, A.; Ahuja, P.; Neogi, A. pMapper: Power and migration cost aware application placement in virtualized systems. In Proceedings of the 9th ACM/IFIP/USENIX International Conference on Middleware, Leuven, Belgium, 1–5 December 2008; pp. 243–264.

7. Jung, G.; Hiltunen, M.A.; Joshi, K.R.; Schlichting, R.D.; Pu, C. Mistral: Dynamically managing power, performance, and adaptation cost in cloud infrastructures. In Proceedings of the 2010 IEEE 30th International Conference on Distributed Computing Systems, Genova, Italy, 21–25 June 2010; pp. 62–73.

8. Zu, Y.; Huang, T.; Zhu, Y. An efficient power-aware resource scheduling strategy in virtualized datacenters. In Proceedings of the IEEE 2013 International Conference on Parallel and Distributed Systems, Seoul, Korea, 15–18 December 2013; pp. 110–117.

9. Mann, V.; Kumar, A.; Dutta, P.; Kalyanaraman, S. VMFlow: Leveraging VM mobility to reduce network power costs in data centers. In *International Conference on Research in Networking*; Springer: New York, NY, USA, 2011; pp. 198–211.

10. Huang, D.; Yang, D.; Zhang, H.; Wu, L. Energy-aware virtual machine placement in data centers. In Proceedings of the 2012 IEEE Global Communications Conference (GLOBECOM), Anaheim, CA, USA, 3–7 December 2012; pp. 3243–3249.

11. Cao, B.; Gao, X.; Chen, G.; Jin, Y. NICE: Network-aware VM consolidation scheme for energy conservation in data centers. In Proceedings of the 2014 20th IEEE International Conference on Parallel and Distributed Systems (ICPADS), Hsinchu, Taiwan, 16–19 December 2014; pp. 166–173.

12. Ferreto, T.C.; Netto, M.A.; Calheiros, R.N.; De Rose, C.A. Server consolidation with migration control for virtualized data centers. *Future Gener. Comput. Syst.* **2011**, *27*, 1027–1034. [CrossRef]

13. Sun, G.; Liao, D.; Zhao, D.; Xu, Z.; Yu, H. Live migration for multiple correlated virtual machines in cloud-based data centers. *IEEE Trans. Serv. Comput.* **2015**, *11*, 279–291. [CrossRef]

14. Kliazovich, D.; Bouvry, P.; Khan, S.U. DENS: Data center energy-efficient network-aware scheduling. *Clust. Comput.* **2013**, *16*, 65–75. [CrossRef]

15. Khemka, B.; Friese, R.; Pasricha, S.; Maciejewski, A.A.; Siegel, H.J.; Koenig, G.A.; Powers, S.; Hilton, M.; Rambharos, R.; Poole, S. Utility maximizing dynamic resource management in an oversubscribed energy-constrained heterogeneous computing system. *Sustain. Comput. Inform. Syst.* **2015**, *5*, 14–30. [CrossRef]

16. Kim, S.; Kim, Y. Application-specific cloud provisioning model using job profiles analysis. In Proceedings of the 2012 IEEE 14th International Conference on High Performance Computing and Communication & 2012 IEEE 9th International Conference on Embedded Software and Systems (HPCC-ICESS), Liverpool, UK, 25–27 June 2012; pp. 360–366.

17. Singh, A.K.; Dziurzanski, P.; Indrusiak, L.S. Value and energy optimizing dynamic resource allocation in many-core HPC systems. In Proceedings of the 2015 IEEE 7th International Conference on Cloud Computing Technology and Science (CloudCom), Vancouver, BC, Canada, 30 November–3 December 2015; pp. 180–185.

18. Baccour, E.; Foufou, S.; Hamila, R.; Hamdi, M. A survey of wireless data center networks. In Proceedings of the 2015 49th Annual Conference on Information Sciences and Systems (CISS), Vancouver, BC, Canada, 30 November–3 December 2015; pp. 1–6.

19. Vardhan, H.; Ryu, S.R.; Banerjee, B.; Prakash, R. 60 GHz wireless links in data center networks. *Comput. Netw.* **2014**, *58*, 192–205. [CrossRef]

20. Halperin, D.; Kandula, S.; Padhye, J.; Bahl, P.; Wetherall, D. Augmenting data center networks with multi-gigabit wireless links. *ACM Sigcomm Comput. Commun. Rev.* **2011**, *41*, 38–49. [CrossRef]

21. Zaaimia, M.; Touhami, R.; Fono, V.; Talbi, L.; Nedil, M. 60 GHz wireless data center channel measurements: Initial results. In Proceedings of the 2014 IEEE International Conference on Ultra-WideBand (ICUWB), Paris, France, 1–3 September 2014; pp. 57–61.

22. Mamun, S.A.; Umamaheswaran, S.G.; Chandrasekaran, S.S.; Shamim, M.S.; Ganguly, A.; Kwon, M. An Energy-Efficient, Wireless Top-of-Rack to Top-of-Rack Datacenter Network Using 60 GHz Links. In Proceedings of the 2017 IEEE Green Computing and Communications (GreenCom), Exeter, UK, 21–23 June 2017; pp. 458–465.

23. Cheng, C.L.; Zajić, A. Characterization of 300 GHz Wireless Channels for Rack-to-Rack Communications in Data Centers. In Proceedings of the 2018 IEEE 29th Annual International Symposium on Personal, Indoor and Mobile Radio Communications (PIMRC), Bologna, Italy, 9–12 September 2018; pp. 194–198.

24. Shin, J.Y.; Kirovski, D.; Harper, D.T., III. Data Center Using Wireless Communication. U.S. Patent 9,391,716, 12 July 2016.

25. Mamun, S.A.; Umamaheswaran, S.G.; Ganguly, A.; Kwon, M.; Kwasinski, A. Performance Evaluation of a Power-Efficient and Robust 60 GHz Wireless Server-to-Server Datacenter Network. *IEEE Trans. Green Commun. Netw.* **2018**, *2*, 1174–1185. [CrossRef]

26. Auer, P. Using confidence bounds for exploitation-exploration trade-offs. *J. Mach. Learn. Res.* **2002**, *3*, 397–422.

27. Yeo, C.S.; Buyya, R. A Taxonomy of Market-based Resource Management Systems for Utility-driven Cluster Computing. *Softw. Pract. Exp.* **2006**, *36*, 1381–1419. [CrossRef]

28. Theocharides, T.; Michael, M.K.; Polycarpou, M.; Dingankar, A. Hardware-enabled Dynamic Resource Allocation for Manycore Systems Using Bidding-based System Feedback. *EURASIP J. Embed. Syst.* **2010**, *2010*, 261434.

29. Locke, C.D. Best-Effort Decision-Making for Real-Time Scheduling. Ph.D. Thesis, Carnegie Mellon University, Pittsburgh, PA, USA, 1986.

30. Aldarmi, S.; Burns, A. Dynamic value-density for scheduling real-time systems. In Proceedings of the Euromicro Conference on Real-Time Systems, York, UK, 9–11 June 1999; pp. 270–277.

31. Bansal, N.; Pruhs, K.R. Server Scheduling to Balance Priorities, Fairness, and Average Quality of Service. *SIAM J. Comput.* **2010**, *39*, 3311–3335. [CrossRef]

32. Hussin, M.; Lee, Y.C.; Zomaya, A.Y. Efficient energy management using adaptive reinforcement learning-based scheduling in large-scale distributed systems. In Proceedings of the 2011 International Conference on Parallel Processing, Taipei, Taiwan, 13–16 September 2011; pp. 385–393.

33. Lin, X.; Wang, Y.; Pedram, M. A reinforcement learning-based power management framework for green computing data centers. In Proceedings of the 2016 IEEE International Conference on Cloud Engineering (IC2E), Berlin, Germany, 4–8 April 2016; pp. 135–138.

34. Singh, A.K.; Dziurzanski, P.; Indrusiak, L.S. Value and energy aware adaptive resource allocation of soft real-time jobs on many-core HPC data centers. In Proceedings of the 2016 IEEE 19th International Symposium on Real-Time Distributed Computing (ISORC), York, UK, 17–20 May 2016; pp. 190–197.

35. Sun, X.; Ansari, N.; Wang, R. Optimizing resource utilization of a data center. *IEEE Commun. Surv. Tutor.* **2016**, *18*, 2822–2846. [CrossRef]

36. Ray, M.; Sondur, S.; Biswas, J.; Pal, A.; Kant, K. Opportunistic power savings with coordinated control in data center networks. In Proceedings of the 19th International Conference on Distributed Computing and Networking, Varanasi, India, 4–7 January 2018; p. 48.

37. Farrington, N.; Porter, G.; Radhakrishnan, S.; Bazzaz, H.H.; Subramanya, V.; Fainman, Y.; Papen, G.; Vahdat, A. Helios: A hybrid electrical/optical switch architecture for modular data centers. *ACM SIGCOMM Comput. Commun. Rev.* **2011**, *41*, 339–350.

38. Calheiros, R.N.; Buyya, R. Energy-Efficient Scheduling of Urgent Bag-of-Tasks Applications in Clouds Through DVFS. In Proceedings of the IEEE International Conference on Cloud Computing Technology and Science (CLOUDCOM), Singapore, 15–18 December 2014; pp. 342–349.

39. Irwin, D.E.; Grit, L.E.; Chase, J.S. Balancing Risk and Reward in a Market-Based Task Service. In Proceedings of the IEEE International Symposium on High Performance Distributed Computing (HPDC), Honolulu, HI, USA, 4–6 June 2004; pp. 160–169.

40. Bubeck, S.; Cesa-Bianchi, N. Regret analysis of stochastic and nonstochastic multi-armed bandit problems. *Found. Trends Mach. Learn.* **2012**, *5*, 1–122. [CrossRef]

41. Carpentier, A.; Lazaric, A.; Ghavamzadeh, M.; Munos, R.; Auer, P. Upper-confidence-bound algorithms for active learning in multi-armed bandits. In *International Conference on Algorithmic Learning Theory*; Springer: New York, NY, USA, 2011; pp. 189–203.

42. Han, Y.; Yoo, J.H.; Hong, J.W.K. Poisson shot-noise process based flow-level traffic matrix generation for data center networks. In Proceedings of the 2015 IFIP/IEEE International Symposium on Integrated Network Management (IM), Ottawa, ON, Canada, 11–15 May 2015; pp. 450–457.

43. Benson, T.; Akella, A.; Maltz, D.A. Network traffic characteristics of data centers in the wild. In Proceedings of the 10th ACM SIGCOMM Conference on Internet Measurement, Melbourne, Australia, 1–3 November 2010; pp. 267–280.

44. Dutt, S. New faster kernighan-lin-type graph-partitioning algorithms. In Proceedings of the 1993 International Conference on Computer Aided Design (ICCAD), Santa Clara, CA, USA, 7–11 November 1993; pp. 370–377.

45. Meisner, D.; Gold, B.T.; Wenisch, T.F. PowerNap: Eliminating server idle power. *ACM SIGARCH Comput. Archit. News* **2009**, *44*, 205–216. [CrossRef]

46. Zhang, J.; Yu, F.R.; Wang, S.; Huang, T.; Liu, Z.; Liu, Y. Load Balancing in Data Center Networks: A Survey. *IEEE Commun. Surv. Tutor.* **2018**, *20*, 2324–2352. [CrossRef]

47. Davidson, A.; Tarjan, D.; Garland, M.; Owens, J.D. Efficient parallel merge sort for fixed and variable length keys. In Proceedings of the 2012 Innovative Parallel Computing (InPar), San Jose, CA, USA, 13–14 May 2012; pp. 1–9.

48. Kleinrock, L. *Queueing Systems, Volume 2: Computer Applications*; Wiley: New York, NY, USA, 1976.

49. ns-3 Network Simulator. Available online: https://www.nsnam.org/ (accessed on 4 July 2019).

50. Shehabi, A.; Smith, S.J.; Sartor, D.A.; Brown, R.E.; Herrlin, M.; Koomey, J.G.; Masanet, E.R.; Horner, N.; Azevedo, I.L.; Lintner, W. *United States Data Center Energy Usage Report*; Technical Report LBNL-1005775; LBNL: Berkeley, CA, USA, 2016.

51. Pelley, S.; Meisner, D.; Wenisch, T.F.; VanGilder, J.W. Understanding and abstracting total data center power. In Proceedings of the 2009 Workshop on Energy Efficient Design (WEED), Ann Arbor, MI, USA, 20 June 2009.

52. SPECpower_ssj2008. Results for Dell Inc. PowerEdge C5220. Available online: https://www.spec.org/power_ssj2008/results/res2013q2/power_ssj2008-20130402-00601.html (accessed on 4 July 2019).

53. Heller, B.; Seetharaman, S.; Mahadevan, P.; Yiakoumis, Y.; Sharma, P.; Banerjee, S.; McKeown, N. Elastictree: Saving Energy in Data Center Networks. In Proceedings of the 7th USENIX Conference on Networked Systems Design and Implementation (NSDI), San Jose, CA, USA, 28–30 April 2010; pp. 249–264.

54. Cisco. Cisco Data Center Switches. Available online: https://www.cisco.com/c/en/us/products/switches/data-center-switches/index.html (accessed on 4 July 2019).

55. Silicom. Silicom PE2G2I35 Datasheet. Available online: http://www.silicom-usa.com/wp-content/uploads/2016/08/PE2G2I35-1G-Server-Adapter.pdf (accessed on 4 July 2019).

56. Analog Devices. Analog Devices HMC6300 and HMC6301 60 GHz Millimeter Wave Transmitter and Receiver Datasheet. Available online: http://www.analog.com/media/en/technical-documentation/data-sheets/HMC6300.pdf (accessed on 4 July 2019).

Publisher's Note: MDPI stays neutral with regard to jurisdictional claims in published maps and institutional affiliations.

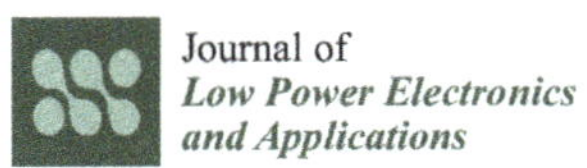

Journal of
*Low Power Electronics
and Applications*

Article

Low-Complexity Run-time Management of Concurrent Workloads for Energy-Efficient Multi-Core Systems [†]

Ali Aalsaud [1,2], Fei Xia [2], Ashur Rafiev [2], Rishad Shafik [2,*], Alexander Romanovsky [3] and Alex Yakovlev [2]

[1] School of Engineering, Al-Mustansiriyah University, Baghdad 10052, Iraq; ali.m.m.aalsaud@gmail.com
[2] School of Engineering, Newcastle University, Newcastle NE1 7RU, UK; Fei.Xia@newcastle.ac.uk (F.X.); Ashur.Rafiev@newcastle.ac.uk (A.R.); Alex.Yakovlev@newcastle.ac.uk (A.Y.)
[3] School of Computing, Newcastle University, Newcastle NE1 7RU, UK; alexander.romanovsky@newcastle.ac.uk
* Correspondence: Rishad.Shafik@ncl.ac.uk
† This paper is an extended version of our paper published in PATMOS, A. Aalsaud, A. Rafiev, F. Xia, R. Shafik and A. Yakovlev, "Model-Free Run-time Management of Concurrent Workloads for Energy-Efficient multi-core Heterogeneous Systems." In Proceedings of the 2018 28th International Symposium on Power and Timing Modeling, Optimization and Simulation (PATMOS), Platja d'Aro, 2018; pp. 206–213, doi:10.1109/PATMOS.2018.8464142.

Received: 30 June 2020 ; Accepted: 10 August 2020; Published: 25 August 2020

Abstract: Contemporary embedded systems may execute multiple applications, potentially concurrently on heterogeneous platforms, with different system workloads (CPU- or memory-intensive or both) leading to different power signatures. This makes finding the most energy-efficient system configuration for each type of workload scenario extremely challenging. This paper proposes a novel run-time optimization approach aiming for maximum power normalized performance under such circumstances. Based on experimenting with PARSEC applications on an Odroid XU-3 and Intel Core i7 platforms, we model power normalized performance (in terms of instruction per second (IPS)/Watt) through multivariate linear regression (MLR). We derive run-time control methods to exploit the models in different ways, trading off optimization results with control overheads. We demonstrate low-cost and low-complexity run-time algorithms that continuously adapt system configuration to improve the IPS/Watt by up to 139% compared to existing approaches.

Keywords: energy-efficient computing; run-time management; machine learning; concurrent workloads; multi-core systems

---

1. Introduction

Modern computing continues to evolve with increasing complexity in both hardware and software. More applications of different types are concurrently executed on platforms featuring an increasing number and type of parallel computing resources (cores) [1,2]. The advantages are clear, as parallel computing can help delay the potential saturation of Moore's Law and better use the performance and energy efficiency opportunities provided by technology scaling [3,4]. However, managing resources in this complex space for energy efficiency is proving highly challenging, especially when different application scenarios (single or concurrent) need to be taken into account [5,6].

Contemporary processors, such as those from Arm and Intel, feature dynamic voltage frequency scaling (DVFS) as a means of handling the energy and performance trade-off [7,8]. Power governors enable DVFS at the system software level. For instance, Linux includes different power governors that can be activated based on the system requirements. These include *powersave* for low-power, low-performance mode, *ondemand* for performance-sensitive DVFS, *performance* for higher performance, and *userspace* for user-specified DVFS. These governors attempt to suitably tune the voltage/frequency pairs according to performance and energy requirements and workload variations. The voltage/frequency can be tuned to just satisfy the performance requirements according to the workload, but not more, in order to reduce energy consumption.

Performance requirements continue to increase, making DVFS alone less effective [9]. As a result, DVFS is often coupled with task mapping (TM), which distributes workloads among multiple cores [10]. When satisfying the same performance requirement, using more cores means that each core has a lighter load and aggressive DVFS can be applied to reduce the overall energy consumption. On the other hand, in order to achieve such energy efficiency, it is crucial to understand the synergy between hardware and software [11].

Core allocations to threads (TM) are usually handled by a scheduler, instead of the governor which takes care of DVFS [12]. A typical Linux scheduler does load balancing, i.e., it distributes the overall workload at any time across all available cores to achieve maximum utilization. Although this objective is rational the implementation tends to be crude. For instance, there is usually no discrimination about the type of task or thread being scheduled, such as CPU- or memory-intensive [12]. Given particular performance requirements, different types of threads should be treated differently for performance and energy optimization. Indiscriminate treatment may lead to sub-optimal energy efficiency [13].

A number of approaches have been reported on the research of using DVFS and TM synergistically for energy-efficient multi-core computing [13]. These approaches broadly fit into two types: offline (OL) and run-time (RT). In OL approaches, the system is extensively reasoned to derive energy and performance models [13,14], which lead to run-time decisions based on these models which stay constant. In RT approaches, the models are typically learned using monitored information [15,16]. Since RT modeling is costly in terms of resources, often a combination of OL and RT are used [13].

Section 2 provides a brief review of these approaches. A recurring scheme in these approaches is that the focus is primarily on single-application workloads in isolation. However, the same application can exhibit different energy/performance trade-offs depending on whether it is running alone or concurrently with other different workloads. This is because: (a) a workload application may switch between memory- and CPU-intensive routines, and (b) architectural sharing between applications affect the energy/performance trade-offs (see Section 7.1.2). Table 1 summarizes the features and limitations of existing approaches.

Table 1. Features and limitations of existing methods when compared with the proposed approach.

Approach	Platforms	WLC	Validation	Apps	Controls	Size
[17,18]	homo.	No	simulation	single	TM+DVFS	P
[19]	hetero.	No	simulation	single	RT, TM+DVFS	P
[15]	homo.	No	practical	single	RT, DVFS	L
[14]	hetero.	No	simulation	single	OL, TM+DVFS	P
[13]	hetero.	OL	practical	conc.	RT, DVFS	NP
[16]	not CPUs.	RT	practical	conc.	RT, DVFS	NP
This work	hetero.	RT	practical	conc.	RT, TM+DVFS	L

Tackling energy efficiency in concurrent applications considering the workload behavior changes highlighted above is non-trivial. When mapping onto heterogeneous multi-core systems, this becomes more challenging because the state space is large and each application requires different optimization. The hardware state space of a multi-core heterogeneous system includes all possible core allocations and DVFS combinations. Here, we discuss this using the scenario of multiple parallel applications running on one of the example experimental platforms used in this paper, the Odroid XU3 (detailed in Section 4.2), which has two types of CPU cores, A7 and A15, organized into two DFVS domains, as motivational examples. Here, N_{apps} is the total number of concurrent applications running on the system; N_{A7} is the number of A7 cores used and N_{A15} is the number of A15 cores used. Table 2 shows the number of possible core allocations for a total N_{apps} number of applications running on the Odroid with $N_{A7} = 3$ and $N_{A15} = 4$. The "brute force" value represents $(N_{A7} + 1)^{N_{apps}} \cdot (N_{A15} + 1)^{N_{apps}}$ combinations, not all of which are actually allowed considering the following rules: (1) each application must have at least one thread; and (2) no more than one thread per core is allowed. However, there is no simple algorithm to iterate through only valid core allocations and an explosion of the search state space is inevitable. The number of possible core allocations is then multiplied by the number of DVFS combinations, which is calculated as $M_{A7} \cdot M_{A15}$, where M_{A7} is the number of DVFS points in the A7 domain, and M_{A15} is the number of DVFS points in the A15 domain.

Table 2. Number of possible core allocations for different multi-application scenarios.

N-apps	Brute Force	Valid
1	20	19
2	400	111
3	8000	309
4	$1.6 \cdot 10^5$	471
5	$3.2 \cdot 10^6$	405
6	$6.4 \cdot 10^7$	185
7	$1.28 \cdot 10^9$	35

In this work, we addressed these limitations with an adaptive approach, which monitors application scenarios at RT. The aim was to determine the optimal system configuration such that the power normalized performance can be maximized at all times. The approach is based on profiling single and concurrent applications through power and performance measurements. For the first time, our study reveals the impact of parallelism in different types of heterogeneous cores on performance, power consumption, and power efficiency in terms of instruction per second (IPS) per unit power (i.e., IPS/Watt) [20]. In our proposed approach, we make the following specific contributions:

1. using empirical observations and CPU performance counters, derive RT workload classification thresholds;
2. based on the workload classification and multivariate linear regression (MLR) to model power and performance trade-offs expressed as instructions per second (IPS) per Watt (IPS/Watt), propose a low-complexity approach for synergistic controls of DVFS and TM;
3. using synthetic and real-world benchmark applications with different concurrent combinations, investigate the approach's energy efficiency, measured by power-normalized performance in IPS/Watt, and implement the low-complexity approach as a Linux power governor and validate through extensive experimentation with significant IPS/Watt improvements.

To the best of our knowledge, this is the first RT optimization approach for concurrent applications based on workload classification, refined further with MLR-based modeling, practically implemented and demonstrated on both heterogeneous and homogeneous multi-core systems. The rest of the paper is organized as follows. Section 2 reviews the existing approaches. The proposed system approach is described in Section 3. Section 4 shows the configuration of systems used in the experiments and the applications. Workload classification techniques are described in Section 5, where Section 5.2 details the run-time implementations. Section 6 deals with combining workload classification with multivariant linear regression, with the decision space of the latter significantly reduced by the former. Section 7 discusses the experimental results, and, finally, Section 8 concludes the paper.

2. Related Work

Energy efficiency of multi-core systems has been studied extensively over the years. A power control approach for multi-core processors executing single application has been proposed in Reference [21]. This approach has three layers of design features also shown by other researchers: firstly, adjusting the clock frequency of the chip depending on the power budget; secondly, dynamically group cores to run the same applications (as also shown in Reference [22,23]), and finally, modifying the frequency of each core group (as also shown in Reference [11,24]). Among others, Goraczko et al. [17] and Luo et al. [18] proposed DVFS approaches with software task partitioning and mapping of single applications using a linear programming (LP) based optimization during RT to minimize the power consumption. Goh et al. [25] proposed a similar approach of task mapping and scheduling for single applications described by synthetic task graphs.

Several other works have also shown power minimization approaches using practical implementation of their approaches on heterogeneous platforms. For example, Sheng et al. [19] presented an adaptive power minimization approach using RT linear regression-based modeling of the power and performance trade-offs. Using the model, the task mapping and DVFS are suitably chosen to meet the specified performance requirements. Nabina and Nunez-Yanez [15] presented another DVFS approach for field-programmable gate array (FPGA)-based video motion compensation engines using RT measurements of the underlying hardware.

A number of studies have also shown analytical studies using simulation tools, like gem5, together with McPAT [14,26], for single applications. These works have used DVFS, task mapping, and offline optimization approaches to minimize the power consumption for varying workloads.

Over the years substantial research has been carried out addressing RT energy minimization and/or performance improvement approaches. These approaches have considered a single-metric based optimization: primarily performance-constrained power minimization, or performance improvement within a power budget [27]. For example, Shafik et al. proposed an RT DVFS control approach for power minimization of multiprocessor embedded systems [28]. Their approach uses performance and user experience constraints to derive the lowest possible operating voltage/frequency points through reinforcement learning and transfer principles. Das et al. presented another power minimization approach that models RT workload characterization to continually update the DVFS and core allocations through multinomial logic regression based predictive controls [29].

An RT classification of workloads and corresponding DVFS controls based on similar principles is proposed by Wang and Pedram for performance-constrained power minimization [16]. As far as performance optimization within a power budget is concerned, Chen and Marculescu proposed a distributed reinforcement learning algorithm to model power and performance trade-offs during RT [30]. Using this model, the DVFS and core allocations are adapted dynamically using feedback from the performance counters. Another power-limited performance optimization approach is presented by Cochran et al. showing programming model based power budget annotations and corresponding controls [31]. Based on application requirements, Nabina and Nunez-Yanez [15] presented a DVFS approach for FPGA-based video motion compensation engines. Santanue et al. [32] and Tiago et al. [33] suggested a smart load balancing technique to improve energy efficiency for single applications running on heterogeneous systems. This technique depends on the sense-predict-balance process through the variation of workload and performance/power trade-offs.

Gem5 with McPAT have been used to demonstrate four different core types, where each core operated in a fixed frequency. Petrucci et al. [14] proposed a thread scheduling algorithm called (lucky), which is based on lottary scheduling. This algorithm is implemented by using Linux 2.6.34 kernel with performance monitor to optimize the thread-to-core affinity. Matthew et al. [34] proposed a DVFS approach with different core allocated for controlling concurrent applications exercised on homogeneous systems at RT.

Numerous studies have focused on using classification-based techniques in dynamic power management with DVFS together at run-time [9,31,35–39]. For instance, Gupta et al. [9] proposed a new run-time approach based on workload classification. To build this classifier extensive offline experiments are made on heterogeneous many core platforms and MATLAB is used to determine the classifier parameters offline. Pareto function is used to determine the optimal configuration. However, this classification is heavily based on offline analysis results and assigns an application a fixed type, regardless of its operating context. It also requires the annotation of applications by application programmers through using a special API.

Dey et al. [40] suggested a new management technique for a power and thermal efficiency agent for mobile MPSoC platforms based on reinforcement learning. Fundamental to this approach is the use of software agent to explore the DVFS in mobile CPU and GPU based on user's interaction behavior. This approach has been validated on Galaxy Note 9 smartphone utilizing Exynos 9810. The experimental results show that this new management technique can increase performance while reducing temperature and power consumption.

A model-free RT workload classification (WLC) approach with corresponding DVFS controls is proposed by Wang and Pedram [16]. This approach employs reinforcement learning, with the action space size a big concern for the authors, even though for only homogeneous systems at much higher granularity than CPU cores. WLC has also been used OL, but this produces a fixed class for each application [13] and cannot deal with workload behavior changes during execution.

For a comprehensive survey on the wider related field of energy management in energy-critical systems, see Reference [41]. This paper is based on our previous work published in PATMOS 2018 [42], with substantial extensions.

3. Proposed Methodology

Our method studies concurrent application workloads being executed on various hardware platforms with parallel processing facilities, and we attempt RT management decision optimization from the results of this analysis.

The RT management (RTM) decisions consist of TM and DVFS, which influence system performance and power dissipation [11]. The RTM takes as input information derived from system monitors, including hardware performance counters and power monitors, available from modern multi-core hardware platforms. Based on this information, the RTM algorithms attempt to increase power normalized performance by tuning the TM and DVFS outputs. The general architecture of this system view is shown in Figure 1. We develop a simple RTM algorithm based on workload classification (WLC), which classifies each workload by its usage of CPU and memory (Section 5). This minimal-cost algorithm may be used on its own or be optionally augmented and refined with an MLR procedure to further optimize the power normalized performance of the execution at an additional cost (Section 6). The WLC procedure significantly reduces the decision space of any additional MLR step making the total overhead much lower than implementing the entire optimization process purely with MLR [6] (Section 7).

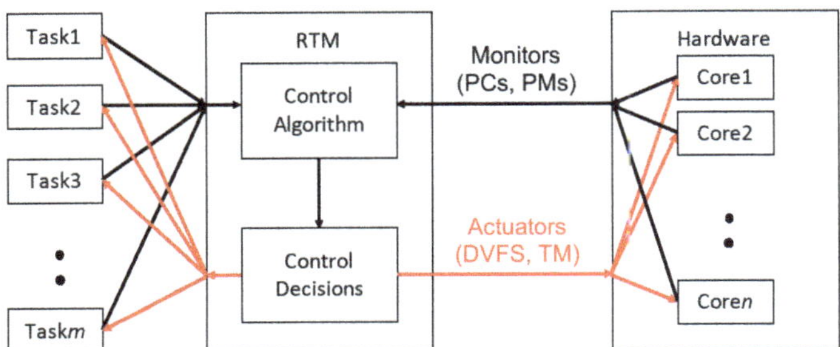

Figure 1. Run-time management (RTM) architecture showing two-way interactions between concurrent applications and hardware cores.

4. System Fundamentals

In this section, we describe the platforms, workload applications, and performance counters used in this investigation. We study a homogeneous and a heterogeneous parallel processing platform, which both provide all the performance counters and power monitors we need for the methodology. We chose standard benchmark application workloads, which provide a variety of degrees of concurrency and memory access and CPU usage scenarios. The two hardware platforms, PARSEC workload applications, and performance counters are further detailed below.

4.1. Homogeneous System

The homogeneous experimental platform is a PC based on an Intel Core i7 Sandybridge CPU which contains no on-chip GPU facility. This CPU is chosen because it has a reasonable number of hard (4) and soft (8) cores, has no on-chip GPU to complicate the power consumption and communications, and has a relatively large number of possible operating frequencies and voltages. The operating system is Ubuntu Linux.

Run-time power monitoring is developed for the experimental platform for validation purposes. This is done by inserting a precision shunt resister into the earth side of the power connection to the CPU. As high-precision current meters tend to have a 1A upper limit, which many CPU operations will exceed, the shunt resister allows the inference of current via measuring voltage.

The performance and power utility ***Likwid*** [43] is used to obtain the majority of the experimental data. ***Likwid*** makes use of on-chip performance counters (sensors) in Intel CPUs to collect performance and power data. For instance, the Running Average Power Limit (RAPL [44]) counters are accessed to infer power dissipation. The form factor of the platform allows the actual measurement of CPU power by way of an inserted shunt resister into the CPU power supply circuit, and readings from these measurements were used in initial experiments to build confidence on the RAPL readings.

Before the main experiments, ***Likwid*** was first confirmed to be accurate for the experimental platform through cross-validation with physical power measurements using the shunt resister, described above. The use of performance counters rather than external power measurement in most of the experiments is motivated by the desire of developing an RTM, which, for practicality and wide applicability, can only rely on built-in sensors and not shunt resisters.

4.2. Heterogeneous System

The popularity of heterogeneous architectures, containing two or more types of different CPU cores, continues to grow [45]. These systems offer better performance and power trade-off flexibility; however, it may be more complicated to ensure optimal energy consumption. The Odroid-XU3 board supports techniques, such as DVFS, affinity, and core disabling, commonly used to optimize system operation in terms of performance and energy consumption [46,47].

The Odroid-XU3 board is a small eight-core computing device implemented on energy-efficient hardware. The board can run Ubuntu 14.04 or Android 4.4 operating systems. The main component of Odroid-XU3 is the 28nm System-on-Chip (Soc) Exynos 5422. This SoC is based on the ARM big.LITTLE heterogeneous architecture and consists of a high performance Cortex-A15 quad core processor block, a low power Cortex-A7 quad core block, Mali-T628 MP6 GPU cluster, and 2GB DRAM LPDDR3. The board contains four real time current sensors that give the possibility of power measurement on the four separate power domains: big (A15) CPUs, LITTLE (A7) CPUs, GPU cluster, and DRAM. In addition, there are also four temperature sensors for each of the A15 CPUs and one sensor for the GPU cluster. This work only concerns the CPU blocks, and the other parts of the SoC may be investigated in future work.

On the Odroid-XU3, for each CPU power domain, the supply voltage (Vdd) and clock frequency can be tuned through a number of pre-set pairs of values. The performance-oriented Cortex-A15 block has a range of frequencies between 200 MHz and 2000 MHz with a 100 MHz step, whilst the low-power Cortex-A7 quad core block can scale its frequencies between 200 MHz and 1400 MHz with a 100 MHz step.

4.3. Workload Applications

The PARSEC [48] benchmark suite attempts to represent both current and emerging workloads for multiprocessing hardware. It is a commonly used benchmark suite for evaluating concurrency and parallel processing. We therefore use PARSEC on the Odroid-XU3 platform, in which heterogeneity can be representative of different design choices that can greatly affect workloads. PARSEC applications exhibit different memory behaviors, different data sharing patterns, and different workload partitions from most other benchmark suites in common use. The characteristics of applications, according to Reference [48], which are used in this paper can be seen in Table 3.

Table 3. Qualitative summary of the inherent key characteristics of PARSEC benchmarks [48].

Program	Application Domain	Type	Parallelization Model	Granularity	Working Set	Data Usage	Sharing Exchange
bodytrack	Computer Vision	CPU+mem	data-parallel	medium	medium	high	medium
ferret	Similarity Search	CPU	pipeline	medium	unbounded	high	high
fluidanimate	Animation	mem	data-parallel	fine	large	low	medium
canneal	Engineering	CPU	unstructured	medium	unbounded	high	high
freqmine	Data Mining	CPU	data-parallel	fine	unbounded	high	medium
streamcluster	Data Mining	mem	data-parallel	medium	medium	low	medium

Whilst we experimented with all PARSEC applications at various stages of work, six applications from the suite are selected for presentation in the paper to represent CPU-intensive, memory-intensive, and a combination of both. Such a classification reduces the effort of model characterization for combinations of concurrently running applications (Section 5). We found no surprises worth reporting in the accumulated experimental data with regard to the other PARSEC applications.

4.4. Performance Counters

In this work, we use performance counters to monitor system performance events (e.g., cache misses, cycles, instruction retired) and, at the same time, capture the voltage, current, power, and temperature directly from the sensors of Odroid-XU3. For the Intel Core i7, real power measurements with a shunt resister were used to establish confidence in the RAPL power counters initially, whilst the majority of experiments are based on performance counter readings once the confidence has been achieved. The performance counter consists of two modules: kernel module and a user space module.

For the Odroid, the hardware performance counter readings are obtained using the method presented by Walker et al. [49], with similar facilities used through *Likwid* for the Core i7.

Here, we describe the Odroid case in more detail. In the user space module, the event specification is the means to provide details of how each hardware performance counter should be set up. Table 4 lists notable performance events, some of which are explained as follows:

1. INST_RETIRED is the retired instruction executed and is part of the highly reported instruction per cycles (IPC) metric.
2. Cycles is the number of core clock cycles.
3. MEM_ACCESS is Memory Read or Write operation that causes a cache access to at least the level of data.
4. L1I_CACHE is level 1 instruction cache access.

Table 4. Performance counter events.

perf_eventt_name	Description
INST_RETIRED	Instruction architecturally executed.
BUS_CYCLE	Bus cycle
MEM_ACCESS	Data memory access.
L1I_CACHE	Instruction Cache access.
L1D_CACHE_WB	Data cache eviction.
L2D_CACHE	Level 2 data cache access
L2D_CACHE_WB	Level 2 data cache refill
L2D_CACHE_REFILL	Level 2 data cache write-back.

5. Workload Classification RTM

This section makes use of both heterogeneous and homogeneous systems in its investigations but mainly concentrates on the heterogeneous Odroid XU3 in its discourse, unless otherwise noted. Different types of cores are especially useful for demonstrating the advantages of the approach.

5.1. Workload Classification Taxonomy

The taxonomy of workload classes chosen for this work reflects differentiation between CPU-intensive and memory-intensive workloads, with high- or low-activity. Specifically, workloads are classified into the following four classes:

- Class 0: low-activity workloads;
- Class 1: CPU-intensive workloads;
- Class 2: CPU- and memory-intensive workloads; and
- Class 3: memory-intensive workloads.

Extensive exploratory experiments are run in this work to investigate the validity of these general concepts.

The experiments are based on our synthetic benchmark, called *mthreads* [50], which attempts to controllably re-create the effect of memory bottleneck on parallel execution. The tool accomplishes this by repeatedly mixing CPU-intensive and memory-intensive operations, the ratio of each type is controlled by the parameter M. The CPU-intensive operation is a simple integer calculation. The memory-intensive operation is implemented by randomly writing to a 64 MB pre-allocated array. The randomization helps reduce the effect of caching. Parameter $M = 0$ gives CPU-intensive execution, $M = 1$ leads to memory-intensive execution; the values in between provide a linear relation to the number of memory accesses per instruction. The execution is split into N identical parallel threads, each pinned to a specific core. Figure 2 presents the flowchart of the tool.

Figure 3 shows the energy efficiency of *mthreads* running on 2 to 4 A7 cores (one of the A7 cores may have a heavy operating system presence—if C0 is turned off, the operating system stops; hence, this data does not include the single core case, which would be skewed by this system behavior) with M values ranging from 0 to 1. It can be seen that it is better to use fewer cores for memory-intensive tasks (larger M), but it is better to run more cores in parallel for CPU-intensive tasks (smaller M). Characterization results sweeping through the frequency ranges and core combinations with *mthreads* confirm the validity of the classification taxonomy and establish a TM and DVFS strategy based on relative CPU and memory use rates. The full set of *mthreads* data, supported by experiments with applications other than *mthreads* including the entire PARSEC suite, is used to generate our run-time management (RTM) presented in subsequent sections.

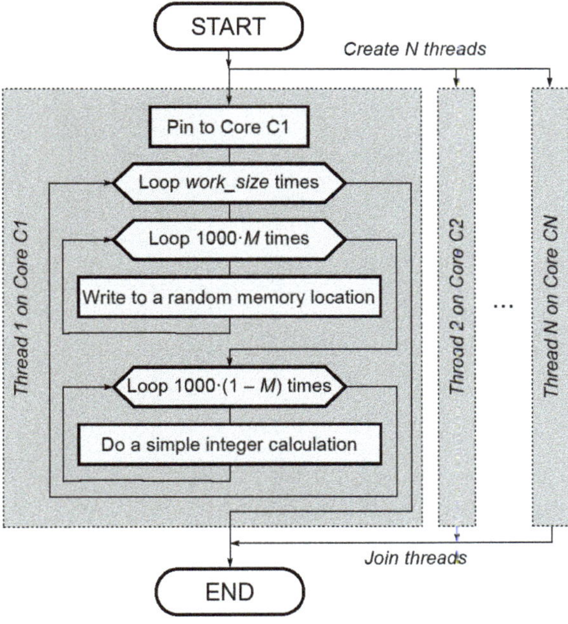

Figure 2. Flowchart of *mthreads* synthetic benchmark. M and N are controlled parameters.

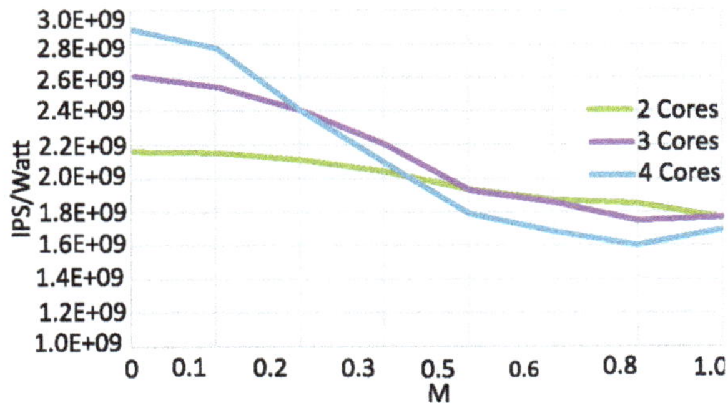

Figure 3. instruction per second (IPS)/Watt for different memory use rates ($0 \leq M \leq 1$).

5.2. Run-time Management Based on Workload Classification

Figure 1 presents the general architecture of RTM inside a system. In this section, we explain the central RTM functions—classification and control actions based on performance monitors and actuators (e.g., TM and DVFS). The general approach does not specify the exact form of the taxonomy into which workloads are classified, the monitors and actuators the system need to have, nor the design figure of merit. Our examples classify based on differentiating CPU and memory usages and the execution intensiveness, try to maximize IPS/Watt through core-allocation and DVFS, and get information from system performance counters [42].

The governor implementation is described in Figure 4, which refines Figure 1. At time t_i, task i is added to the execution via the system function *execvp*(). The RTM makes TM and DVFS decisions based on metric classification results, which depends on hardware performance counters and power monitors to directly and indirectly collect all the information needed. This helps avoid instrumenting applications and/or special API's (unlike, e.g., Reference [51]), providing wider support for existing applications. The TM actuation is carried out indirectly via system functions. For instance, core pinning is done using *sched_affinity(pid)*, where *pid* is the process ID of a task. DVFS is actuated through the *userspace* governor as part of *cpufreq* utilities.

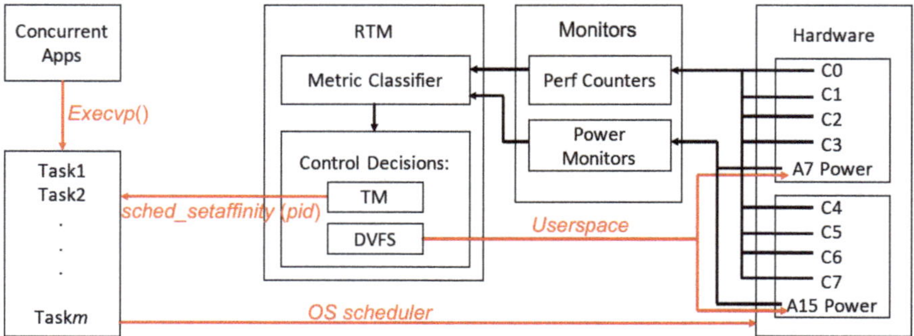

Figure 4. Governor implementation based on RTM.

5.3. Workload Classification

Real applications do not have precisely tuneable memory usage rates, unlike *mthreads*. They may also have phases during which they may appear to be one class or another during their execution; therefore, attempts at classifying each application as a whole offline (as seen in Reference [13]) may be of limited value (see Section 7.1.1 for detailed discussions). In this work, information from performance counters is used to derive the classes of all applications running on the system for each control decision cycle. The assumption is that, during a control decision cycle, the class of an application is unlikely to change. This assumption requires that the length of control cycles is sufficiently short relative to the rate of class change of the applications (according to the Nyquist/Shannon sampling principle). The choice of control cycle length therefore depends on expected application scenarios and what happens when/if Nyquist/Shannon is violated should be carefully considered by the designer. This point will be discussed in detail in Section 7.1.2, with the help of system design case studies.

The classification using performance counter readings is based on calculating a number of metrics from performance counter values recorded at set time intervals and then deriving the classes based on whether these metrics have crossed certain thresholds. Example metrics and how they are calculated are given in Table 5.

Table 5. Metrics used to derive classification.

Metrics	Definitions
nipc	$(InstRet/Cycles)(1/IPC_{max})$
iprc	$InstRet/ClockRef$
nnmipc	$(InstRet/Cycles - Mem/Cycles)(1/IPC_{max})$
cmr	$(InstRet - Mem)/InstRet$
uur	$Cycles/ClockRef$

Normalized instructions per clock (nipc) measures how intensive the computation is. It is the instructions per unhalted cycle (IPC) of a core, normalized by the maximum IPC (IPC_{max}). IPC_{max} can be obtained from manufacturer literature. *Cycles* is the unhalted cycles counted. Normalization allows nipc to be used independent of core types and architectures.

Instructions per reference clock (iprc) contributes to determining how active the computation is. *ClockRef* is the total number of clock cycles given by $ClockRef = Freq/Time$ with Freq and Time from the system software.

Normalized non-memory IPC (nnmipc) discounts memory accesses from nipc, indicating CPU activity. From experiments with our synthetic benchmark, this shows an inverse correlation to the memory use rate.

CPU to memory ratio (cmr) relatively compares CPU to memory activities.

Unhalted clock to reference clock ratio (urr) determines how active an application is.

The general relationship between these metrics and the application (workload) classes are clear, e.g., the higher *nnmipc* is, the more CPU-intensive a workload will be. A workload can be classified by comparing the values of metrics to thresholds. Decision-making may not require all metrics. The choice of metrics and thresholds can be made by analyzing characterization experiment results for each platform. From studying the relationship between M and the list of metrics from *mthreads* experiments on the Odroid XU3, we find that *nnmpic* and *cmr* show the best spread of values with regard to corresponding to different values of M (see Figure 5). Whichever one of these to use depends on designer preferences on the range of threshold values between different application classes to use. We choose *nnmipc* to differentiate CPU and memory usage rates and *urr* to differentiate low and high activity. The thresholds used are determined based on our *mthreads* characterization database and given in Table 6. We tested this approach by running PARSEC programs and obtaining values of the chosen metrics, with the results shown in Table 7. These confirm that *nnmipc* can be used to differentiate CPU- and memory-intensive applications. For instance, *ferret* is regarded as CPU-intensive [52] and its per-core *nnmipc* value is above 0.35. The other metrics may work better on other platforms and are included here as examples of potential candidates depending on how a *mthreads*-like characterization program behaves on a platform with regard to the relationships between M values and the metrics.

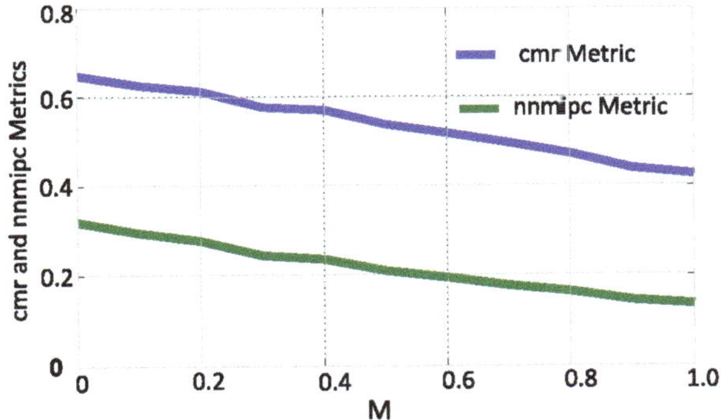

Figure 5. CPU to memory ratio (cmr) and Normalized non-memory IPC (nnmipc) metrics for different memory use rates ($0 \leq M \leq 1$).

Table 6. Classification details for Odroid XU3.

Metric Ranges	Class
urr of all cores [0, 0.11]	0: low-activity
nnmipc per-core [0.35, 1]	1: CPU-intensive
nnmipc per-core [0.25, 0.35)	2: CPU+memory
nnmipc per-core [0, 0.25)	3: memory-intensive

Table 7. PARSEC applications and their performance counter metrics on XU3.

Applications	nnmipc	nipc	iprc	cmr	urr
Bodytrack	0.306	0.417	0.503	0.754	0.603
Ferret	0.384	0.560	0.978	0.739	1.01
Fluidanimate	0.206	0.317	0.690	0.723	1.08
Streamcluster	0.166	0.286	0.570	0.465	0.995

To confirm our approach, another set of experiments were carried out on the Intel Core i7 platform, as can be seen in Table 8. These results agree with those found from the Odroid XU3. Based on these experiments, we also choose **nnmipc** to differentiate CPU and memory usage rates and **urr** for differentiating low and high activity. Threshold values are established from Core i7 characterization experiments and are different from those for Odroid XU3.

Table 8. PARSEC applications and their performance counter metrics on Intel i7 Processor.

Applications	iprc	nnmipc	cmr
Bodytrack	0.727449908	0.573472873	0.788333
Caneal	0.71442	0.58642	0.750138
Fluidanimate	0.6949938	0.50526088	0.727001
Freqmine	0.867086053	0.629553377	0.726056
Streamcluster	0.370102144	0.248135025	0.67045

In principle, for each hardware platform, based on the available performance counters, the choice of metrics and the classification threshold values should both be based on classification results obtained from that platform.

5.4. Control Decision Making

This section presents an RTM control algorithm that uses application classes to derive its decisions. The behavior is specified in the form of two tables: a threshold table (Table 6), used for determining application classes; and a decision table (Table 5), providing a preferred action model for each application class.

The introduction of new concurrent applications or any other change in the system may cause an application to change its behavior during its execution. It is therefore important to classify and re-classify regularly. The RTM works in a dedicated thread, which performs classification and decision-making action every given time frame. The list of actions performed every RTM cycle is shown in Algorithm 1.

Algorithm 1 Inside the RTM cycle.

1: Collect monitor data
2: **for** each application **do**
3:     Compute classification metrics   ▷ Section 5.3
4:     Use metric and threshold table to determine application class   ▷ Table 5
5:     Use decision table to find core allocation and frequency preferences   ▷ Table 6
6:     Distribute the resources between the applications according to the preferences
7:     Wait for $T_{control}$ ▷ Section 5.4
8: **end for**
9: **return**

In Algorithm 1, $T_{control}$ is the time between two RTM control cycles. The RTM determines the TM and DVFS of power domains once each control cycle, and these decisions keep constant before the next control cycle. The data from the system monitors (performance counters and power meters) is collected asynchronously. Every core has a dedicated monitor thread, which spends most of its time in a sleep state and wakes every $T_{control}$ to read the performance counter registers. The readings are saved in the RTM memory. This means that the RTM always has the latest data, which is at most $T_{control}$ old. This is mainly done because ARM performance counter registers can be accessed only from code on the same CPU core. In this case, asynchronous monitoring has been empirically shown to be more efficient. In our experiments, we chose $T_{control}$ = 500 ms, which has shown a good balance between RT overhead and energy minimization. The time the RTM takes (i.e., RT overhead) is negligible compared to 500 ms for the size of our system. This interval can be easily reduced with slightly higher overheads or increased with less energy efficiency trade-offs. The flowchart of the entire RTM cycle is shown in Figure 6.

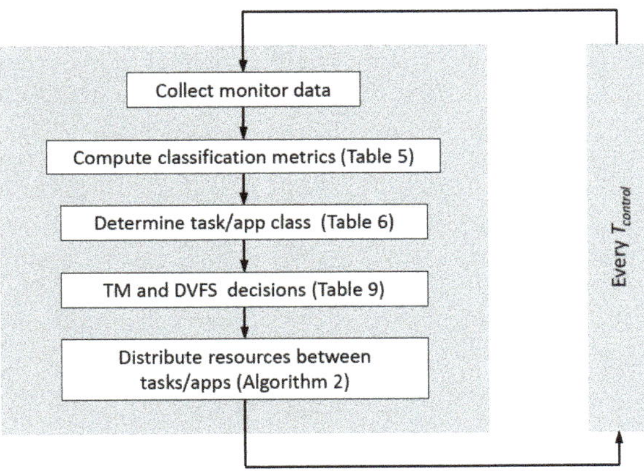

Figure 6. Flowchart of the RTM cycle.

The RTM uses monitor data to calculate the classification metrics discussed in Section 5.2. These metrics form a profile for each application, which is compared against the thresholds (Table 6). Each row of the table represents a class of applications and contains a pre-defined value range for each classification metric. Value ranges may be unbounded. A metric x can be constrained to the range $[c, +\infty)$, equivalent to $x \geq c$. An application is considered to belong to a class, if its profile satisfies every range in a row. If an application does not satisfy any class, it is marked as "unclassified" and gets a special action from the decision table. An application is also unclassified when it first joins the execution. In that case, it goes to an A15 core for classification.

The decision table (Table 9) contains the following preferences for each application class, related to system actuators (DVFS and core allocation decisions): number of A7 cores, number of A15 cores, and clock frequencies. Number of cores can take one of the following values: none, single, or maximum. Frequency preference can be minimum or maximum. The CPU-intensive application class (Class 1) runs on the maximum number of available A15 cores at the maximum frequency as this has shown to give the best energy efficiency (in terms of power normalized performance) in our previous observations [7].

Table 9. RTM control decisions.

Class	Frequency	A7	A15
0	min	single	none
1	max	none	max
2	min	max	max
3	max	max	none
unclassified	min	single	none

Tables 6 and 9 are constructed OL in this work based on large amounts of experimental data, with those involving PARSEC playing only a supporting role. For instance, although *ferret* is regarded as CPU-intensive, it is so only on average and has non CPU-intensive phases (see Section 7.1.1). Therefore, Table 9 is obtained mainly from analyzing experimental results from our synthetic benchmark *mthreads* (which has no phases), with PARSEC only used for checking if there are gross disagreements (none was found). Because of the empirical nature of the process, true optimally is not claimed.

In this work, we assume that the RTM does not have to deal with more threads than the number of cores in the system—if there are more threads than cores, some will not get scheduled by the system scheduler, which is outside the domain of the RTM. Our experiments therefore do not feature more concurrent applications than the number of cores in the system. The RTM attempts to satisfy the preferences of all running applications. In the case of conflicts between frequency preferences, the priority is given to the maximum frequency. When multiple applications request cores of the same type, the RTM distributes all available cores of that type as fairly as possible. When these conflicting applications are of different classes, each application is guaranteed at least a single core. Core allocation (TM) is done through the following algorithm.

Algorithm 2 shows the procedure APPLYDECISION for mapping the RTM decisions to the core affinity masks. RTM provides a decision for each app and for each core type $d_{j,i} \in \{NONE, MIN, MAX\}$, where $j \in \{A7, A15\}$ is the core type, and $1 \leq i \leq m$ is the app index, given the total number of apps m. The decisions are arranged in arrays $D_{A7} = (d_{A7,1}, \ldots, d_{A7,m})$ and $D_{A15} = (d_{A15,1}, \ldots, d_{A15,m})$. Additional constants used by the algorithm are: n_{A7}, n_{A15} are the total number of little and big cores, respectively, and the IDs of cores by type are listed in the pre-defined $C_{A7} = (c_{A7,1}, \ldots, c_{A7,n_{A7}})$, $C_{A15} = (c_{A15,1}, \ldots, c_{A15,n_{A15}})$. The complexity of the algorithm is linear to m. The result of the algorithm is the set of core IDs C_i, which can be used to call the `sched_setaffinity` function for the respective app i.

Algorithm 2 mapping the RTM decisions to the core affinities

1: **procedure** APPLYDECISION(D_{A7}, D_{A15})
2:　　$(r_{A7,1}, \ldots, r_{A7,m}) \leftarrow$ REQCORES (D_{A7}, n_{A7})　　　　　　$\triangleright$ Get per-app number of little cores
3:　　$(r_{A15,1}, \ldots, r_{A15,m}) \leftarrow$ REQCORES (D_{A15}, n_{A15})　　　　$\triangleright$ Get per-app number of big cores
4:　　**for** $1 \leq i \leq m$ **do**
5:　　　　$C_{i,A7} \leftarrow$ (next $r_{A7,i}$ elements from C_{A7})
6:　　　　$C_{i,A15} \leftarrow$ (next $r_{A15,i}$ elements from C_{A15})
7:　　　　$C_i \leftarrow C_{i,A7} \cup C_{i,A15}$　　　　　　$\triangleright$ Use C_i to set core affinity mask for the app i.
8:　　**end for**
9: **end procedure**

10: **function** REQCORES$((d_1, \ldots, d_m), n)$
11:　　$k_{MIN} \leftarrow$ count $(d_i = \text{MIN})$ for $1 \leq i \leq m$
12:　　$k_{MAX} \leftarrow$ count $(d_i = \text{MAX})$ for $1 \leq i \leq m$
13:　　**if** $k_{MAX} > 0$ **then**
14:　　　　$v \leftarrow \lfloor (n - k_{MIN}) / k_{MAX} \rfloor$　　　　　　$\triangleright$ v is the MAX number of cores
15:　　　　$w \leftarrow (n - k_{MIN}) \mod k_{MAX}$　　　　　　$\triangleright$ w is the remainder
16:　　**end if**
17:　　**for** $1 \leq i \leq m$ **do**
18:　　　　**if** $d_i = \text{MAX}$ **then**
19:　　　　　　**if** $w > 0$ **then**　　　　　　$\triangleright$ Distribute the remainder
20:　　　　　　　　$r_i \leftarrow v + 1$
21:　　　　　　　　$w \leftarrow w - 1$
22:　　　　　　**else**
23:　　　　　　　　$r_i \leftarrow v$
24:　　　　　　**end if**
25:　　　　**else if** $d_i = \text{MIN}$ **then**
26:　　　　　　$r_i \leftarrow 1$
27:　　　　**else**
28:　　　　　　$r_i \leftarrow 0$
29:　　　　**end if**
30:　　**end for**
31:　　**return** $(r_1, \ldots, r_m)$
32: **end function**

---

6. Low-Complexity Run-time with WLC and MLR

Although an RTM purely based on workload classification is low-cost, its coarse granularity may affect its optimality, and further improvement may be possible with an additional MLR step to refine the control decisions. Figure 7 shows the algorithm with which workload classification may be used to reduce the decision space of the subsequent MLR step to achieve a right balance of complexity reduction and optimization quality.

The first step is to update the application queue—during the preceding interval, new applications may have joined the queue. If so, Algorithm 1 is used to determine the application class of each new interval, as explained in Section 5.1. This may reduce the state space of the subsequent search for optimality. For example, for Class 0, the search of optimal configuration for Odroid XU-3 is reduced from $4 \times 13 \times 4 \times 19 = 4004$ different frequency and core configurations (four A7 cores with 13 DVFS points and four A15 with 19 different DVFS points) to one by using C0 (or the first available A7 core) and $F = 200$ MHz as the optimal configuration. For class 1, the search for optimal configuration is reduced by more than 75% because we used the A15 cores at high frequencies (800–2000 MHz), and the state space is reduced by more than 80% for Class 3 because we used the A7 cores at high frequencies (800–1400 MHz). After this reduction of search space, MLR is used to determine the optimal frequency and core allocations for each class type using the method described in Reference [6].

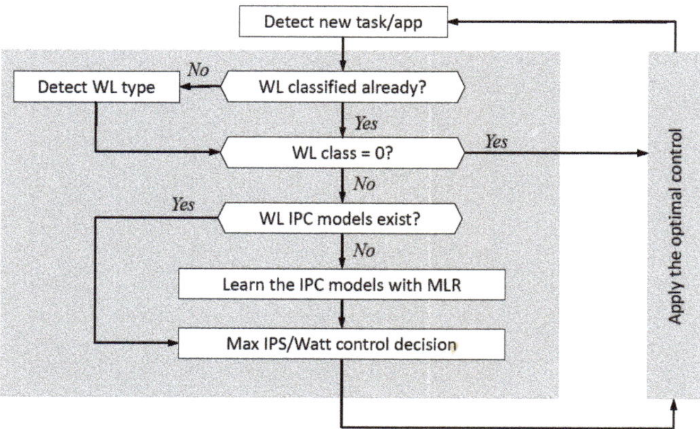

Figure 7. Flow chart for multivariate linear regression (MLR) with workload (WL) classification.

7. Experimental Results

Extensive experiments have been carried out with a large number of application scenarios running on the XU3 platform, with additional confirmatory explorations on the Intel i7 platform. These experiments include running single applications on their own and a number of concurrent applications. In the concurrent scenarios, multiple copies of the same application and different applications of the same class and different applications of different classes have all been tested.

7.1. Workload Classification-Only Results

7.1.1. A Case Study of Concurrent Applications

An example execution trace with three applications is shown in Figure 8. Parts at the beginning and end of the run contain single and dual application scenarios. The horizontal axis is time, while the vertical axis denotes TM and DVFS decisions. Cores C0–C3 are A7 cores, and C4–C7 are A15 cores. The figure shows application classes and the core(s) on which they run at any time. This is described by numbers, for instance, "2/3" on core C1 means that App 2 is classified as of Class 3 and runs on C1 for a particular time window. "1/u" means that App 1 is being classified. The lower part of the figure shows the corresponding power and IPS traces. Both parameters are clearly dominated by the A15 cores.

As can be seen, initial classifications are carried out on C4, but, when C4 is allocated to an application, C7 is reserved for this purpose. The reservation of dedicated cores for initial classification fits well for architectures where the number of cores is greater than the number of applications, as in the case of modern multi-core systems, such as Odroid XU3.

Re-classification happens for all running applications at every 500 ms control cycle, according to Algorithm 1. Each application is re-classified on the core where it is running. Figures 8 and 9 show the motivation for this. The same application can belong to different classes at different times. This proves that an OL classification method, which gives each application an invariable class, is unusable for efficient energy minimization.

Figure 9 shows example traces of the PARSEC apps ferret and fluid animate being classified whilst running as single applications. It can be seen that the same application can have different CPU/memory behaviors and get classified into different classes. This is not surprising as the same application can have CPU-intensive phases when it does not access memory and memory-intensive phases where there is a lot of memory access. In addition, it is also possible for an application to behave as belonging to different classes when mapped to different numbers of cores. The classification can also be influenced by whether an application is running alone or running in parallel with other applications, if we compare Figures 8 and reftwoapp. These are all strong motivations for RT re-classification. The result of classification affects an application's IPS and power (see Figure 8).

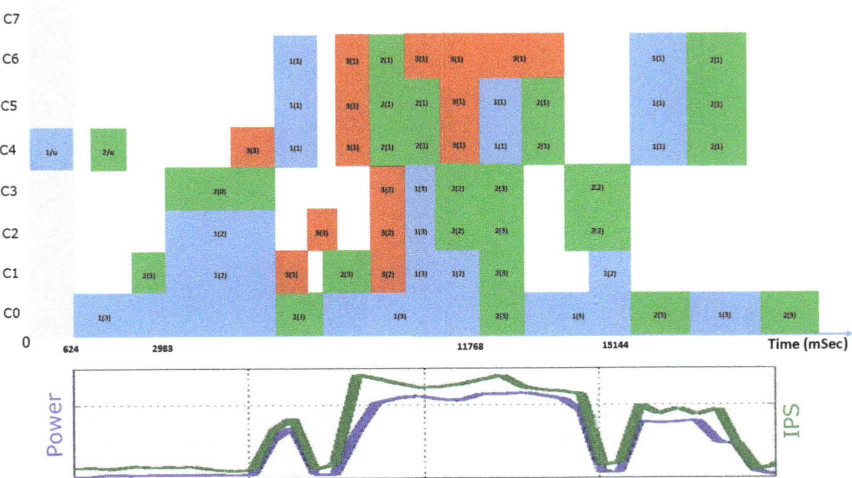

Figure 8. Execution trace with task mapping (TM) and dynamic voltage frequency scaling (DVFS) decisions and their effects on performance and power.

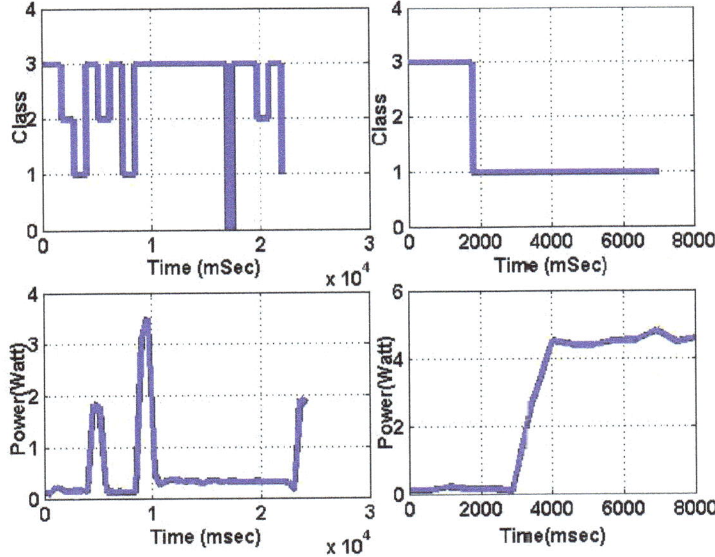

Figure 9. Fluid animate (left) and ferret (right) classification and power traces.

7.1.2. RTM Stability, Robustness and Control Decision Cycle Selection

Algorithm 1 can oscillate between two different sets of classification and control decisions in alternating cycles. This may indicate the loss of stability of the RTM approach. The reasons for such oscillations have been isolated into the following cases:

- The control cycle length coincides with an application's CPU and memory phase changes.

- An application's behavior takes it close to particular threshold values, and different instances of evaluation put it on different sides of the thresholds.
- An application is not very parallelizable. When it is classified on a single core, it behaves as CPU-intensive, but, when it is classified on multiple cores, it behaves as low-activity. This causes it to oscillate between Class 0 and Class 1 in alternating cycles.

We address these issues as follows. Case 1 rarely happens, and, when it happens, it disappears quickly because of the very low probability of an application's phase cycles holding constant and coinciding with the control cycle length. This can be addressed, in the rare case when it is necessary, by tuning the control cycle length slightly if oscillations persist. In general, if the Nyquist/Shannon sampling frequency requirement is not violated, this is not a worry.

Case 2 also happens rarely. In general, increasing the number of classes and reducing the distances between control decisions of adjacent classes reduce the RTM's sensitivity to threshold accuracy; hence, Case 2 robustness does not have to be a problem, and thresholds (Table 6) and decisions (Table 9) can be tuned both OL and during RT.

Case 3 is by far the most common. It is dealt with through adaptation. This type of oscillation is very easy to detect. We put in an extra class, "low-parallelizability", and give it a single big core. This class can only be found after two control cycles, different from the other classes, but this effectively eliminates Case 3 oscillations.

Empirically, the PARSEC applications used in this paper as examples tend to have relatively stable periods during which their classes do not change. These periods can run from hundreds of ms to multiple seconds. We chose a control decision cycle of 500 ms such that it may, on rare occasions, violate the Nyquist/Shannon sampling principle for some applications, in order to expose potential oscillatory behavior and test the effectiveness of our mitigating methods. The experimental results confirm the validity of our methods of dealing with the different cases of oscillatory behavior.

7.1.3. Comparative Evaluation of the WLC-Only RTM

Complexity: Our RTM has a complexity of $O(N_{app} * N_{class} + N_{core})$, where N_{app} is the number of applications (tasks) running, N_{class} is the number of classes in the taxonomy, and N_{core} is the number of cores. N_{class} is usually a constant of small value, which can be used to trade robustness and quality with cost. The RTM's computation complexity is therefore linear to the number of applications running and the number of cores. In addition, the basic algorithm itself is a low-cost, lookup-table approach with the table sizes linear to N_{class}.

Schemes found in existing work, with, e.g., model-based [6], machine-learning [53], linear programming [18], or regression techniques [6,19], have a decision state space size of $O((N_{A7DVFS} * N_{A15DVFS}) * (N_{A7} * N_{A15})^{N_{app}})$, where N_{A7} and N_{A15} are the numbers of A7 and A15 cores and N_{A7DVFS} and $N_{A15DVFS}$ are the numbers of DVFS points of the A7 and A15 power domains, for this type of platform. This NP complexity is sensitive to system heterogeneity, unlike our approach.

Overheads: We compared the time overheads (OH) of our method with the linear-regression (LR) method found in, e.g., Reference [6,19]. For each 500 ms control cycle, our RTM, running at 200 MHz, requires 10 ms to complete for the trace in Figure 8. Over 90% of this time is spent on monitor information gathering. In comparison, LR requires 100 ms to complete the same actions. It needs a much larger set of monitors. The computation, also much more complex, evenly divides its time in model building and decision-making. In addition, a modeling control, such as LR, requires multiple control intervals to settle and the number of control decision cycles needed is combinatorial with N_{A7}, N_{A15}, N_{A7DVFS}, and $N_{A15DVFS}$.

Scalability: Our RTM is scalable to any platform as it is (a) agnostic to the number and type of application running in concurrently and (b) independent of the number or type of cores in the platform, and their power domains. This is because the complexity of the RTM only grows linearly with increased number of concurrent applications and cores. Our experiments on the Intel i7 platform confirm this.

7.2. Comparative Results between Our Three RTM Types

In this section, we compare the IPS/Watt results from MLR-only, workload classification-only, and the combined workload classification plus MLR RTM types.

7.2.1. MLR-Only RTM Results

We previously explored an MLR-only RTM with PARSEC applications on the Odroid XU3 in comparable experimental conditions [6]. This power governor/RTM aims to improve IPS/Watt, the same as the RTM's developed in this paper. The results from Reference [6] are compared to those obtained from this work in Table 10.

Table 10. Percentage IPS/Watt imporovement of the RTM over the Linux *ondemand* governor, all with Odroid XU3.

Application Scenarios	Workload Classification (WLC)	Multivariate Linear Regression (MLR)	MLR + WLC
Fluidanimate alone	127%	127%	139%
Two different class applications	68.60%	61.74%	128.42%
Three different class applications	46.60%	29.30%	61.27%
Two Class 3 applications	24.50%	19.81%	40.33%
Three Class 3 applications	44.40%	36.40%	58.25%
Two Class 1 applications	31.00%	26.53%	41.74%

7.2.2. WLC-Only and WLC Combined with MLR RTM Results

In this work, we propose two new power governors (RTMs). The first is the light-weight WLC-only approach described in Section 5. The second is the more sophisticated approach of combining WLC with a further step of MLR-based optimization, described in Section 6.

Figure 10 shows the results obtained from running the WLC-only RTM on the Odroid XU3, comparing the IPS/Watt metric obtained with the performance of the Linux *ondemand* governor [54]. These results show IPS/Watt improvements of 24 to 127% over the benchmark *ondemand* governor in the application scenarios included in the figure.

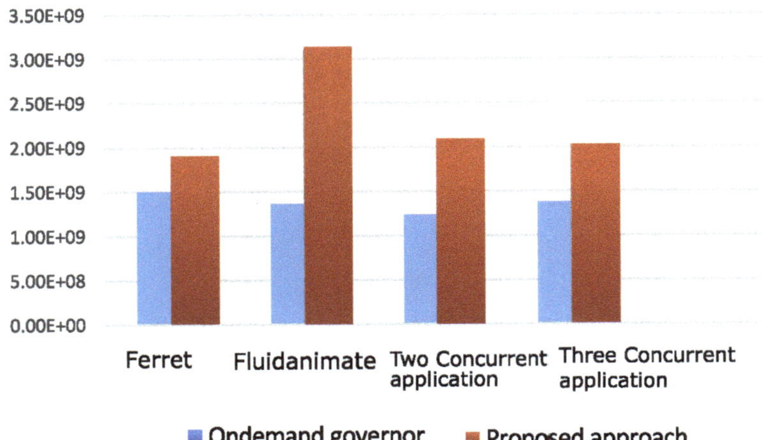

Figure 10. IPS/Watt between the proposed WLC-only power governor and the *ondemand* governor on Odroid XU3.

Experiments with the combined WLC+MLR approach demonstrate that it is possible to further improve IPS/Watt by supplementing the WLC method with additional MLR optimization. Figure 11 show the IPS/Watt comparisons between this method and the Linux *ondemand* governor on the Odroid XU3. It can be seen from these results that further improvements over Figure 10 are evident.

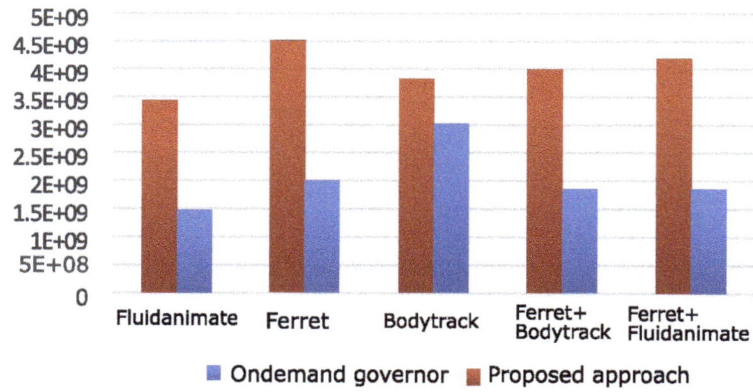

Figure 11. IPS/Watt Comparison between the proposed WLC+MLR and *ondemand* [54] governors on Odroid XU3 .

This combined method is also applied to the Intel Core i7 platform and the IPS/Watt results obtained are compared with those from running the Linux *ondemand* governor in Figure 12. The improvements on IPS/Watt range from 20% to 40%.

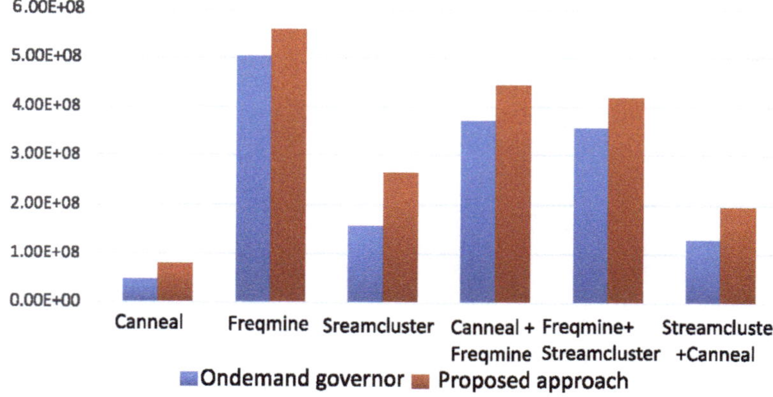

Figure 12. IPS/Watt comparison between the proposed WLC+MLR and *ondemand* [54] governors on Intel i7, with all cores allocated to the tasks/apps.

In general, it is found that the heterogeneous Odroid XU3 platform demonstrates the methods proposed in this paper better than the Intel Core i7 platform. This is mainly because the latter is not specifically designed for CPU power efficiency, and there is a limited scope for IPS/Watt improvement by tuning TM and DVFS. There is a comparatively high background power dissipation, whatever the TM and DVFS decisions are. On the other hand, the Odroid platform, based on ARM big.LITTLE architecture, has CPU energy efficiency at the core of its hardware design philosophy and provides a much wider scope of IPS/Watt improvements via TM and DVFS decisions.

As a result, we concentrate on comparing the different RTM methods based on data obtained from the Odroid XU3 experiments. Table 10 compares the results of all three RTMs against the *ondemand* governor on the Odroid XU3 platform.

From Table 10, it can be seen that the improvements in IPS/Watt obtained by the combined WLC with MLR approach is higher than the WLC-only and MLR-only methods.

The main problem with the MLR-only approach is that it does not take changes of application behavior in each control decision cycle into account. An MLR model typically takes multiple control cycles to settle, and, after it settles, it may no longer be optimal.

The WLC-only approach improves on this by re-classifying every control cycle and this improves the optimality of the control decisions and reduces the controller overhead at the same time. However, because of its coarse-grain nature the decisions tend to be sub-optimal leaving further improvements possible.

By combining WLC and MLR modeling, the WLC+MLR method makes use of the WLC technique to provide a coarse-grain pre-decision which is then potentially refined through MLR modeling for further IPS/Watt improvements. This results in quick decisions, vastly reduced MLR learning space and more up to date MLR model results that approximate true optima much better.

By comparing with the *ondemand* governor, we seek a vehicle for indirect comparisons with a relatively broad range of existing and upcoming research, as this governor is popular target for result comparisons in most related types of work. To demonstrate the efficacy of this approach, we look at the following example. Gupta [9] proposed a run-time approach consisting of a combination of offline characterization and run-time classification. The thesis describes experimental results showing an average increase of 81% in IPS/Watt compared to the *ondemand* governor for memory intensive applications running alone. Results, such as this, can be compared with our results listed in Table 10. Although the experimental scenarios may not be entirely like-for-like, much can be inferred as to the effectiveness of different methods from this kind of indirect comparison. In this specific case, the Gupta improvement figure of 81% is most appropriately compared with the fluid animate alone figure in Table 10, where our approaches obtain over 120% of improvements.

Data collected from our large number of validation runs shows the RTM out-performing the Linux *ondemand* governor by considerable margins on IPS/Watt, as shown in Table 10. The method can be generalized to other optimization targets, such as throughput, energy-delay product, and any energy and throughput trade-off metric. It is also possible to switch targets at RT. This will require constructing multiple decision tables and switching between them during RT. This is a subject for future work.

8. Conclusions

An optimization scheme targeting power-normalized performance was developed for controlling concurrent application executions on platforms with multiple cores.

In the first instance, models are obtained off-line from experimental data. Explorations with model simplification are shown to be successful as by and large optimal results are obtained from using these models in RT control algorithms compared with existing Linux governors. In many cases, the improvements obtained are quite significant.

A run-time workload classification management approach is proposed for multiple concurrent applications of diverse workloads running on heterogeneous multi-core platforms. The approach is demonstrated by a governor aimed at improving system energy efficiency (IPS/Watt). This governor classifies workloads according to their CPU and memory signatures and makes decisions on core allocation and DVFS. Due to model-free approach, it leads to low RTM complexity (linear with the number of applications and cores) and cost (lookup tables of limited size). The governor implementation does not require application instrumentation, allowing for easy integration in existing systems. Experiments show the governor provides significant energy efficiency advantage compared to existing approaches. Detection of low-parallelizability improves the stability of the governor. A synthetic benchmark with tunable memory use supports the characterization process.

This method is further improved with tuning the results of workload classification by a learning-based optimization using multivariant linear regression. With the workload classification having drastically reduced the modeling space, the regression-based learning has been shown to work effectively. This RTM is demonstrated on both heterogeneous and homogeneous platforms.

For experimental purposes of homogeneous and heterogeneous systems, we demonstrated a novel RT approach, capable of workload classification and power-aware performance adaptation under sequential and concurrent application scenarios in heterogeneous multi-core systems. The approach is based on power and performance models that can be obtained during RT by multivariate linear regression based on low-complexity hypotheses of power and performance for a given operating frequency. The approach is extensively evaluated using PARSEC-3.0 benchmark suite running on the Odroid-XU3 heterogeneous platform.

A selection of experimental results was presented to illustrate the kinds of trade-offs in a variety of concurrent application scenarios, core allocations, and DVFS points, highlighting an improvement of power normalized performance which produced IPS/Watt improvements between 26% and 139% for a range of applications. It is expected that modern embedded and high-performance system designers will benefit from the proposed approach in terms of a systematic power-aware performance optimization under variable workload and application scenarios. Our future work will include investigating the scalability of the approach to more complex platforms and higher levels of concurrency.

Author Contributions: Conceptualization, all authors; methodology, all authors; software, A.R. (Ashur Rafiev) and A.A.; validation, A.A., A.R. (Ashur Rafiev) and F.X.; formal analysis, F.X., A.R. (Ashur Rafiev) and A.A.; investigation, A.A., A.R. (Ashur Rafiev) and F.X.; resources, A.Y. and A.R. (Alexander Romanovsky); data curation, A.A.; writing–original draft preparation, A.A., A.R. (Ashur Rafiev) and F.X.; writing–review and editing, F.X. and R.S.; visualization, A.A., F.X. and A.R. (Ashur Rafiev); supervision, A.Y., R.S., A.R. (Alexander Romanovsky) and F.X.; project administration, A.Y., A.R. (Alexander Romanovsky), R.S. and F.X.; funding acquisition, A.Y., A.R. (Alexander Romanovsky) and A.A. All authors have read and agreed to the published version of the manuscript.

Funding: This work is funded by the EPSRC (project PRiME, grant EP/K034448/1 and Project STRATA, grant EP/N023641/1). Aalsaud is also supported by studentship funding from the Ministry of Iraqi Higher Education and Scientific Research.

Conflicts of Interest: The authors declare no conflict of interest.

References

1. Prakash, A.; Wang, S.; Irimiea, A.E.; Mitra, T. Energy-efficient execution of data-parallel applications on heterogeneous mobile platforms. In Proceedings of the 2015 33rd IEEE International Conference on Computer Design (ICCD), New York, NY, USA, 18–21 October 2015; pp. 208–215.
2. Plyaskin, R.; Masrur, A.; Geier, M.; Chakraborty, S.; Herkersdorf, A. High-level timing analysis of concurrent applications on MPSoC platforms using memory-aware trace-driven simulations. In Proceedings of the 2010 18th IEEE/IFIP International Conference on VLSI and System-on-Chip, Madrid, Spain, 27–29 September 2010; pp. 229–234.
3. Shafik, R.; Yakovlev, A.; Das, S. Real-power computing. *IEEE Trans. Comput.* **2018**, *67*, 1445–1461. [CrossRef]
4. Borkar, S. Design challenges of technology scaling. *IEEE Micro* **1999**, *19*, 23–29. [CrossRef]
5. Orgerie, A.C.; Assuncao, M.D.D.; Lefevre, L. A survey on techniques for improving the energy efficiency of large-scale distributed systems. *ACM Comput. Surv. (CSUR)* **2014**, *46*, 47. [CrossRef]
6. Aalsaud, A.; Shafik, R.; Rafiev, A.; Xia, F.; Yang, S.; Yakovlev, A. Power–aware performance adaptation of concurrent applications in heterogeneous many-core systems. In Proceedings of the 2016 International Symposium on Low Power Electronics and Design, San Francisco, CA, USA, 8–10 August 2016; pp. 368–373.
7. Mittal, S. A survey of techniques for improving energy efficiency in embedded computing systems. *Int. J. Comput. Aided Eng. Technol.* **2014**, *6*, 440–459. [CrossRef]
8. Intel Corporation. *Timeline of Processors*; Intel Corporation: Santa Clara, CA, USA, 2012.

9. Gupta, U. Power-Performance Modeling and Adaptive Management of Heterogeneous Mobile Platforms. Ph.D. Thesis, Arizona State University, Tempe, AZ, USA, 2018.
10. Rafiev, A.; Al-Hayanni, M.; Xia, F.; Shafik, R.; Romanovsky, A.; Yakovlev, A. Speedup and Power Scaling Models for Heterogeneous Many-Core Systems. *IEEE Trans. Multi-Scale Comput. Syst.* **2018**, *4*, 436–449. [CrossRef]
11. Shafik, R.A.; Al-Hashimi, B.M.; Kundu, S.; Ejlali, A. Soft Error-Aware Voltage Scaling Technique for Power Minimization in Application-Specific Multiprocessor System-on-Chip JOLPE **2009**, *5*, 145–156. [CrossRef]
12. Torrey, A.; Cleman, J.; Miller, P. Comparing interactive scheduling in Linux. *Softw. Pract. Exp.* **2007**, *34*, 347–364. [CrossRef]
13. Reddy, B.K.; Singh, A.K.; Biswas, D.; Merrett, G.V.; Al-Hashimi, B.M. Inter-cluster Thread-to-core Mapping and DVFS on Heterogeneous Multi-cores. *IEEE Trans. Multi-Scale Comput. Syst.* **2018**, *4*, 369–382. [CrossRef]
14. Petrucci, V.; Loques, O.; Mossé, D. Lucky scheduling for energy-efficient heterogeneous multi-core systems. In Proceedings of the 2012 USENIX conference on Power-Aware Computing and Systems, Hollywood, CA, USA, 7 October 2012; p. 7.
15. Nabina, A.; Nunez-Yanez, J.L. Adaptive voltage scaling in a dynamically reconfigurable FPGA-based platform. *ACM Trans. Reconfigurable Technol. Syst. (TRETS)* **2012**, *5*, 20. [CrossRef]
16. Wang, Y.; Pedram, M. Model-Free Reinforcement Learning and Bayesian Classification in System-Level Power Management. *IEEE Trans. Comput.* **2016**, *65*, 3713–3726. [CrossRef]
17. Goraczko, M.; Liu, J.; Lymberopoulos, D.; Matic, S.; Priyantha, B.; Zhao, F. Energy-optimal software partitioning in heterogeneous multiprocessor embedded systems. In Proceedings of the 45th Annual Design Automation Conference, Anaheim, CA, USA, 8–13 June 2008; pp. 191–196.
18. Luo, J.; Jha, N.K. Power-efficient scheduling for heterogeneous distributed real-time embedded systems. *IEEE Trans. Comput. Aided Des. Integr. Circuits Syst.* **2007**, *26*, 1161–1170. [CrossRef]
19. Yang, S.; Shafik, R.A.; Merrett, G.V.; Stott, E.; Levine, J.M.; Davis, J.; Al-Hashimi, B.M. Adaptive energy minimization of embedded heterogeneous systems using regression-based learning. In Proceedings of the 2015 25th International Workshop on Power and Timing Modeling, Optimization and Simulation (PATMOS), Salvador, Brazil, 1–4 September 2015; pp. 103–110.
20. Wang, A.; Chandrakasan, A. A 180-mV subthreshold FFT processor using a minimum energy design methodology. *IEEE J. Solid-State Circuits* **2005**, *40*, 310–319. [CrossRef]
21. Ma, K.; Li, X.; Chen, M.; Wang, X. Scalable power control for many-core architectures running multi-threaded applications. *ACM SIGARCH Comput. Archit. News* **2011**, *39*, 449–460. [CrossRef]
22. Xu, Z.; Tu, Y.C.; Wang, X. Exploring power-performance trade-offs in database systems. In Proceedings of the 2010 IEEE 26th International Conference on Data Engineering (ICDE 2010), Long Beach, CA, USA, 1–6 March 2010; pp. 485–496.
23. Rafiev, A.; Iliasov, A.; Romanovsky, A.; Mokhov, A.; Xia, F.; Yakovlev, A. Studying the Interplay of Concurrency, Performance, Energy and Reliability with ArchOn—An Architecture-Open Resource-Driven Cross-Layer Modelling Framework. In Proceedings of the 2014 14th International Conference on Application of Concurrency to System Design, Tunis La Marsa, Tunisia, 23–27 June 2014; pp. 122–131.
24. Wong, H.; Aamodt, T.M. The Performance Potential for Single Application Heterogeneous Systems. In Proceedings of the 8th Workshop on Duplicating, Deconstructing, and Debunking, Austin, TX, USA, 21 June 2009.
25. Goh, L.K.; Veeravalli, B.; Viswanathan, S. Design of fast and efficient energy-aware gradient-based scheduling algorithms heterogeneous embedded multiprocessor systems. *IEEE Trans. Parallel Distrib. Syst.* **2009**, *20*, 1–12.
26. Ben Atitallah, R.; Senn, E.; Chillet, D.; Lanoe, M.; Blouin, D. An efficient framework for power-aware design of heterogeneous MPSoC. *IEEE Trans. Ind. Inform.* **2013**, *9*, 487–501. [CrossRef]
27. Hankendi, C.; Coskun, A.K. Adaptive power and resource management techniques for multi-threaded workloads. In Proceedings of the 2013 IEEE International Symposium on Parallel & Distributed Processing, Workshops and Phd Forum, Cambridge, MA, USA, 20–24 May 2013; pp. 2302–2305.
28. Shafik, R.A.; Yang, S.; Das, A.; Maeda-Nunez, L.A.; Merrett, G.V.; Al-Hashimi, B.M. Learning transfer-based adaptive energy minimization in embedded systems. *IEEE Trans. Comput. Aided Des. Integr. Circuits Syst.* **2016**, *35*, 877–890. [CrossRef]

29. Das, A.; Kumar, A.; Veeravalli, B.; Shafik, R.; Merrett, G.; Al-Hashimi, B. Workload uncertainty characterization and adaptive frequency scaling for energy minimization of embedded systems. In Proceedings of the Design, Automation & Test in Europe Conference & Exhibition (DATE), Grenoble, France, 9–13 March 2015; pp. 43–48.

30. Chen, X.; Zhang, G.; Wang, H.; Wu, R.; Wu, P.; Zhang, L. MRP: Mix real cores and pseudo cores for FPGA-based chip-multiprocessor simulation. In Proceedings of the 2015 Design, Automation & Test in Europe Conference & Exhibition, Grenoble, France, 9–13 March 2015; pp. 211–216.

31. Cochran, R.; Hankendi, C.; Coskun, A.K.; Reda, S. Pack & Cap: Adaptive DVFS and thread packing under power caps. In Proceedings of the 44th Annual IEEE/ACM International Symposium on Microarchitecture, Porto Alegre, Brazil, 3–7 December 2011; pp. 175–185.

32. Sarma, S.; Muck, T.; Bathen, L.A.; Dutt, N.; Nicolau, A. SmartBalance: A sensing-driven linux load balancer for energy efficiency of heterogeneous MPSoCs. In Proceedings of the 2015 52nd ACM/EDAC/IEEE Design Automation Conference (DAC), San Francisco, CA, USA, 8–12 June 2015; pp. 1–6.

33. Mück, T.; Sarma, S.; Dutt, N. Run-DMC: Run-time dynamic heterogeneous multicore performance and power estimation for energy efficiency. In Proceedings of the 10th International Conference on Hardware/Software Codesign and System Synthesis, Amsterdam, The Netherlands, 4–9 October 2015; pp. 173–182.

34. Travers, M.; Shafik, R.; Xia, F. Power-Normalized Performance Optimization of Concurrent Many-Core Applications. In Proceedings of the 2016 16th International Conference on Application of Concurrency to System Design (ACSD), Torun, Poland, 19–24 June 2016; pp. 94–103.

35. Kyrkou, C.; Bouganis, C.S.; Theocharides, T.; Polycarpou, M.M. Embedded hardware-efficient real-time classification with cascade support vector machines. *IEEE Trans. Neural Netw. Learn. Syst.* **2015**, *27*, 99–112. [CrossRef]

36. Reddy, B.K.; Merrett, G.V.; Al-Hashimi, B.M.; Singh, A.K. Online concurrent workload classification for multi-core energy management. In Proceedings of the 2018 Design, Automation & Test in Europe Conference & Exhibition (DATE), Dresden, Germany, 19–23 March 2018; pp. 621–624.

37. Bitirgen, R.; Ipek, E.; Martinez, J.F. Coordinated management of multiple interacting resources in chip multiprocessors: A machine learning approach. In Proceedings of the 2008 41st IEEE/ACM International Symposium on Microarchitecture, Lake Como, Italy, 8–12 November 2008; pp. 318–329.

38. Van Craeynest, K.; Jaleel, A.; Eeckhout, L.; Narvaez, P.; Emer, J. Scheduling heterogeneous multi-cores through performance impact estimation (PIE). In Proceedings of the 2012 39th Annual International Symposium on Computer Architecture (ISCA), Portland, OR, USA, 9–13 June 2012; pp. 213–224.

39. Wen, Y.; Wang, Z.; O'boyle, M.F. Smart multi-task scheduling for OpenCL programs on CPU/GPU heterogeneous platforms. In Proceedings of the 2014 21st International Conference on High Performance Computing (HiPC), Dona Paula, India, 17–20 December 2014; pp. 1–10.

40. Dey, S.; Singh, A.; Wang, X.; McDonald-Maier, K. User Interaction Aware Reinforcement Learning for Power and Thermal Efficiency of CPU-GPU Mobile MPSoCs. In Proceedings of the 2020 Design, Automation & Test in Europe Conference & Exhibition (DATE), Grenoble, France, 9–13 March 2020.

41. Pasricha, S.; Ayoub, R.; Kishinevsky, M.; Mandal, S.K.; Ogras, U.Y. A Survey on Energy Management for Mobile and IoT Devices. *IEEE Des. Test* **2020**. [CrossRef]

42. Aalsaud, A.; Rafiev, A.; Xia, F.; Shafik, R.; Yakovlev, A. Model-free run-time management of concurrent workloads for energy-efficient many-core heterogeneous systems. In Proceedings of the 2018 28th International Symposium on Power and Timing Modeling, Optimization and Simulation (PATMOS), Platja d'Aro, Spain, 2–4 July 2018; pp. 206–213.

43. Likwid—Light Weight Performance Tools. Available online: http://github.com/RRZE-HPC/likwid/wiki (accessed on 20 June 2020).

44. Hähnel, M.; Döbel, B.; Völp, M.; Härtig, H. Measuring energy consumption for short code paths using RAPL. *ACM Sigmetrics Perform. Eval. Rev.* **2012**, *40*, 13–17. [CrossRef]

45. Kumar, S.; Djie, M.; van Leuken, R. Low Overhead Message Passing for High Performance Many-Core Processors. In Proceedings of the 2013 First International Symposium on Computing and Networking, Matsuyama, Japan, 4–6 December 2013; pp. 345–351. [CrossRef]

46. Odroid XU3. Available online: https://www.hardkernel.com/shop/odroid-xu3/ (accessed on 11 August 2020).

47. Skalicky, S.; Lopez, S.; Lukowiak, M.; Schmidt, A.G. A Parallelizing MATLAB Compiler Framework and Run time for Heterogeneous Systems. In Proceedings of the 2015 IEEE 17th International Conference on High Performance Computing and Communications, 2015 IEEE 7th International Symposium on Cyberspace Safety and Security, and 2015 IEEE 12th International Conference on Embedded Software and Systems, New York, NY, USA, 24–26 August 2015; pp. 232–237.

48. Bienia, C.; Li, K. PARSEC 2.0: A New Benchmark Suite for Chip-Multiprocessors. Available online: http://www-mount.ece.umn.edu/~jjyi/MoBS/2009/program/02E-Bienia.pdf (accessed on 20 June 2020).

49. Walker, M.J.; Diestelhorst, S.; Hansson, A.; Das, A.K.; Yang, S.; Al-Hashimi, B.M.; Merrett, G.V. Accurate and stable run-time power modeling for mobile and embedded cpus. *IEEE Trans. Comput. Aided Des. Integr. Circuits Syst.* **2017**, *36*, 106–119. [CrossRef]

50. Mthreads Benchmark. Available online: https://github.com/ashurrafiev/PThreads (accessed on 20 June 2020).

51. Gupta, U.; Patil, C.A.; Bhat, G.; Mishra, P.; Ogras, U.Y. Dypo: Dynamic pareto-optimal configuration selection for heterogeneous mpsocs. *ACM Trans. Embed. Comput. Syst. (TECS)* **2017**, *16*, 1–20. [CrossRef]

52. PARSEC Benchmark Suite. Available online: https://parsec.cs.princeton.edu/ (accessed on 20 June 2020).

53. Singh, A.K.; Leech, C.; Reddy, B.K.; Al-Hashimi, B.M.; Merrett, G.V. Learning-based run-time power and energy management of multi/many-core systems: Current and future trends. *J. Low Power Electron.* **2017**, *13*, 310–325. [CrossRef]

54. Pallipadi, V.; Starikovskiy, A. The ondemand governor. In Proceedings of the Linux Symposium, Ottawa, ON, Canada, 19–22 July 2006; pp. 215–230.

MDPI

St. Alban-Anlage 66

4052 Basel

Switzerland

Tel. +41 61 683 77 34

Fax +41 61 302 89 18

www.mdpi.com

Journal of Low Power Electronics and Applications Editorial Office

E-mail: jlpea@mdpi.com

www.mdpi.com/journal/jlpea